Magnetostatic Waves and Their Application

Magnetostatic Waves and Their Application

P. KABOŠ

Slovak Technical University Bratislava, Slovakia

V. S. STALMACHOV

Saratov State University, Russia

CHAPMAN & HALL

London · Glasgow · New York · Tokyo · Melbourne · Madras

Published by Chapman & Hall, 2–6 Boundary Row, London SE1 8HN, UK

Chapman & Hall, 2–6 Boundary Row, London SE1 8HN, UK

Blackie Academic & Professional, Wester Cleddens Road, Bishopbriggs, Glasgow G64 2NZ, UK

Chapman & Hall Inc., One Penn Plaza, 41st Floor, New York NY 10119, USA

Chapman & Hall Japan, Thomson Publishing Japan, Hirakawacho Nemoto Building, 6F, 1-7-11 Hirakawa-cho, Chiyoda-ku, Tokyo 102, Japan

Chapman & Hall Australia, Thomas Nelson Australia, 102 Dodds Street, South Melbourne, Victoria 3205, Australia

Chapman & Hall India, R. Seshadri, 32 Second Main Road, CIT East, Madras 600 035, India

Published in co-edition with Ister Science Press Limited
Ister Science Press Limited, Staromestská 6, Bratislava, Slovak Republic

First edition 1994

Typeset by Forma, Pribinova 25, Bratislava, Slovak Republic
Printed in England by Clays Ltd, St Ives plc

ISBN 0 412 54710 4

A catalogue record for this book is available from the British Library

Printed on acid-free text paper, manufactured in accordance with ANSI/NISO Z39.48-1992 (Permanence of Paper).

CONTENTS

PREFACE

The text presented deals with the basic problems of high-frequency magnetoelectronics, which include the problems of the excitation, propagation and transformation of magnetostatic spin waves in magnetically ordered matter as well as their application in the various fields of high-frequency technology. Recently, in addition to acoustic waves in solids, magnetic spin waves have taken an active part in the high-frequency electronics field. There are a lot of monographs in the former field, but monographic literature is almost non-existent in the latter — in the field of magnetoelectronics and especially on the wave processes in magnetically ordered matter. Only in 1990 was the monograph of A.V. Vashkovsky, V.S. Stalmachov and J.V. Sharaevsky, *Magnetostatic waves in high-frequency technology* published in Russia, which is, in the main, a first textbook for university students whose specialization is radioelectronics.

Our first idea was to translate the book by Vashkovsky et al. However, we became convinced that translation would not be appropriate at least for the two reasons. First, the monograph mentioned above, with respect to the choice of material and the method of interpretation is a university textbook which contains lots of formulae, calculation and graphic dependences. Second, the scope of that monograph is too wide for a potential reader who is interested in the new branches of high-frequency electronics. For these reasons we decided to write a new book in such a way that it would be acceptable to a wide readership, would give basic information about the new field mentioned above and would be of reasonable extent. Special attention has been paid to the choice of equations and graphic material. Since the monograph presented and the monograph by Vashkovsky et al. include roughly the same complex of problems, we have used the results from this book also for the preparation of this manuscript. In addition, the book includes new chapters on spin wave spectroscopy, the problems of the technology of thin ferromagnetic layer preparation and the construction of high-frequency elements and devices.

The manuscript was prepared taking into account the contemporary level of knowledge in magnetoelectronics and electrodynamics. Especially recently, magnetoelectronics has begun to develop rapidly, to which new magnetoelectric materials with the necessary high-frequency properties have contributed. In connection with this development the problem of education of specialists in the field of solid state electronics and especially in the field of high-frequency magnetoelectronics has come to the fore. We took this into account in preparing this monograph. On the other hand, a lot of research workers — physicists and radiotechnical engineers — who do research or work in technical praxis also need information about the new modern fields of high-frequency radioelectronics. This information is also useful for workers with boundary specialization. The book is devoted to these workers.

This book was written by two authors. Chapters 1–5 were written by V.S. Stalmachov and Chapters 6–8 by P. Kabos. The editing was carried out by both authors. We are very grateful to J.P. Sharaevsky, K.V. Gretchushkin, V.N. Prokurin, A.V. Stalmachov, P. Hyben and others who participated in the preparation of this book.

We have mentioned already that the development of this field is still in progress. For this reason there were a lot of problems with the choice of material for this book; it is unlikely to be perfect and of course the subjective approach of the authors can be seen. It is understandable that some shortcomings may also emerge because we had to process a wide range of material. That is why the authors will be grateful for any comments, requirements and suggestions.

All equations are expressed in the S.I. system of units. This may cause some problems when the results are compared with those in the professional literature, in which the CGS system of units is traditionally used.

List of signs and principal symbols

a	lattice constant
A	insertion attenuation
a_k	weighting coefficient
$\vec{B}$	magnetic field induction vector
$\vec{b}$	high-frequency component of the vector of magnetic field induction
C_j	contrast of Fabry–Perot interferometer
C	capacity
b_k	amplitude of k^{tanh} spin wave
c	velocity of light
$\vec{D}$	electric induction vector
D	exchange parameter
d	the size of the dielectric gap between two magnetic layers
d_h	the distance between microstrip and layer
$\vec{E}$	electric field intensity vector
f	frequency
FMR	ferromagnetic resonance
FPI	Fabry–Perot interferometer
G	gain coefficient
$\vec{H}$	magnetic field intensity vector
ΔH	the resonance line width
$\vec{h}$	high-frequency component of the magnetic field intensity vector
h_c	critical high-frequency magnetic field intensity
GGG	gallium gadolinium garnet
$\overline{\overline{I}}$	identity tensor
$\vec{J}$	mechanical moment of atom, surface current density
J_V	current density
J_S	saturation polarization
H(f)	transmission function
K_i	anisotropy constants
$\vec{k}$	wave vector
k	wave number
L_0	self-inductance per unit length
L	the length of a beam

l_i	the thicknesses of dielectric layers
l	the length of the microstrip line
$\vec{M}$	magnetization vector
M_0	saturation magnetization
$\vec{M}_i$	components of the magnetization vector
MSW	magnetostatic wave
$\vec{m}$	magnetic moment of atom, high-frequency component of magnetization vector
m_q	density of fictitious magnetic charge
$\overline{\overline{N}}$	demagnetizing tensor
n	index of refraction
n_i	components of the complex index of refraction
$\vec{P}$	vector of polarization
P	power
p	the order of diffraction
R_i	radiation resistance
Q	quality factor
S_{ij}	coefficients of dispersion matrix
T	transmission coefficient of FPI
T_k	relaxation time of the spin wave
t	time, the thickness of a ferrite layer
X	reactance
v_p	phase velocity
v_g	group velocity
V	volume
w	the width of microstrip line
x,y,z	co-ordinates of the rectangular co-ordinate system
Y	admittance
Z	impedance
YIG	yttrium iron garnet
α	Landau–Lifshitz loss parameter
α_k	loss parameter of the kth spin wave
γ	gyromagnetic ratio
γ_p	propagation constant
δ	skin depth
$\overline{\overline{\varepsilon}}$	permittivity tensor
ε_{ij}	components of permittivity tensor
θ, θ_0	the angle of gradient of the mirror, the angle of incidence
θ_k	the angle between the magnetic field induction vector

	and the wave vector (the co-ordinate of the wave vector in the spherical co-ordinate system)
ϱ_V	volume charge density
ϱ	frequency dependence of the density of spin wave states
λ	wavelength
λ_s	magnetostriction coefficients
$\overline{\overline{\mu}}$	tensor of permeability
μ, μ_a	components of the permeability tensor
μ_S	surface permeability
$\overline{\overline{\chi}}$	susceptibility tensor
χ, χ_a	components of susceptibility tensor
η	attenuation factor
σ	conductivity
σ_i	internal strain(tension)
$\varkappa$	attenuation constant
τ	delay
φ	MSW deviation angle from the normal to the direct external magnetic field
ϕ_k	co-ordinate of the wave vector in spherical co-ordinate system
ψ	scalar magnetic potential, the deviation angle of group velocity
ω	angular frequency
Ω	normalized frequency
Ω_i	normalized quantities e.g. frequency, magnetization
e	electron charge 1.60×10^{-19} C
m	electron mass 9.11×10^{-31} kg

INTRODUCTION

Electronics of very high frequencies is a field of science and technology which includes questions concerning generation, amplification, transmission, reception, and transformation of high-frequency signals. During its existence (since the early 1940s), high-frequency electronics has undergone a number of evolutionary changes connected with the enrichment of scientific knowledge and with the development of technical and technological possibilities. During the first stage, research and development were concentrated exclusively on vacuum high-frequency electronics connected with the study of the interaction of electromagnetic fields with waves excited by flows of free charge carriers. Electromagnetic fields were generated with the use of waveguide structures (cavity resonator, waveguide, strip line, periodic delay structures, etc.). The flow of charge carriers (electrons) was generated by means of an emitter and an arbitrary corresponding vacuum system. Using this scheme, high-frequency vacuum devices (klystron, vacuum tube with forward and backward waves of type O and M, magnetron, etc.) were constructed. The basic principle of these devices is as follows. In the flow of free charge carriers so-called electron waves are excited. These waves interact with the waves in waveguide structures and cause their mutual transformation. In the flows of free charge carriers waves of various types may exist, for example synchronous and cyclotron waves, waves of space charge, etc. For instance, the interaction of synchronous waves or space charge waves with electromagnetic waves is used in powerful amplifiers or generators of O or M type. This is one of the oldest fields of high-frequency electronics, which is also under development at the present time.

The interaction of high-frequency fields with cyclotron waves was used in vacuum parametric amplifiers, phase shifters, delay lines, etc. In the development of vacuum electronics in the 1950s–1960s important results were achieved in various countries.

With the gradual development of high-frequency electronics it became clear that in many applications, and mainly in the field of low-power engineering, vacuum high-frequency electronics is not very suitable. This was mainly the consequence of the very complex construction of vacuum devices — with a cathode which required a relatively high input power and the necessary presence of extensive waveguide structures which can be constructed only with great difficulties in the short-wave frequency band (e.g. the millimetre band), etc. These facts led to an intensive search for new ways of realizing basic high-frequency electronics using new technological possibilities. Since the mid-1950s successful development of solid state semiconductor electronics has begun, in which, as an active element, wave phenomena in the flows of charge carriers in solids are used. This enabled the abandonment of the demanding vacuum devices. But, as before, electromagnetic structures remained in the form of cavity resonators, waveguides, periodic delay structures, etc. In this way generators and amplifiers with Gunn diodes, avalanche diodes, various modifications of high-frequency parametric devices, etc. were realized. The field of very high-frequency electronics connected with wave phenomena in solid state plasmas is currently under development. By means of it the construction of devices has moved into the field of microwave and even higher frequencies.

The next step in the development of high-frequency radioelectronics was the transition from extensive artificial structures for the transmission of electromagnetic waves (resonators, waveguides, etc.) to natural structures — periodic crystal lattices. In this way a new era of high-frequency electronics development, connected with a new branch — solid state wave electronics — started. The idea that natural periodic structures created by a periodic crystal lattice can be an interesting field for the study of the use of high-frequency wave processes in them was formulated long ago. It is based on the knowledge that in a solid state medium (in a crystal), depending on its characteristics, external conditions and exciting frequency, waves of various types can propagate, for example electromagnetic (fast), acoustic (slow) and magnetostatic spin waves (very slow). The first of these represent ordinary electromagnetic waves which can propagate in the given medium, the second group are deformation waves connected with the motion of atoms in crystal lattices, and the third group represent propagating excitations (the change of precession angle) of magnetic moments in the nodes of a crystal lattice of magnetically ordered structures. These waves can interact with each other and also with waves due to the flow of free charge carriers in the solid state plasma, which creates possibilities for the construction of elements for the control of amplitude, phase, pass band, delay time etc., i.e. parts for high-frequency signal processing. The use of acoustic waves has extended into a new branch of high-frequency engineering — acoustic electronics of very high frequencies. This field has proved to be very promising in the band up to gigahertz frequencies ($1\,\text{GHz} = 10^9\,\text{Hz}$). With the use of surface and volume acoustic wave excitation, passive high-frequency elements, for example delay lines, pass and stop band filters, phase shifters, etc. were constructed.

The second group of slow waves in a solid state medium — magnetostatic spin waves, which can in principle be divided into waves with and without exchange interaction — can exist in a solid state medium in a very wide frequency range, practically overlapping all microwave bands (from hundreds of MHz to hundreds of GHz). For long wavelengths (small wave numbers) of propagating magnetostatic waves, the exchange interaction, effective over short distances, does not have a substantial influence on the shape of the spectrum and therefore these waves can be called magnetostatic waves without exchange interaction. Exactly these kinds of spin wave were called magnetostatic waves (MSW) in the literature. For small wavelengths of spin waves (big wave numbers) the influence of exchange interaction is substantial and in magnetic material spin waves with exchange interaction are excited, which in most cases are simply called spin waves (SW). It should be noted that MSW were known long ago, but they penetrated into the field of high-frequency technology much later, mainly in connection with the discovery of new magnetic materials and especially of yttrium-iron garnet.

In contrast to acoustic waves, magnetostatic waves are very disperse and they can be controlled very easily by means of an external magnetic field. In principle, three basic types are distinguished: (a) forward volume MSW, which are excited in ferrites magnetized perpendicularly to the plane of a magnetic layer, (b) backward volume waves, excited in ferrites which are magnetized in the layer plane in the direction of MSW propagation, and (c) surface MSW excited in ferrites magnetized in a layer plane perpendicular to the MSW propagation direction. All the given types of MSW are used in high-frequency elements as delay lines, various filters, phase shifters, magnetostatic resonators, elements for noise suppression, phase and amplitude correctors etc. Magnetostatic spin wave elements have a lot of advantages in comparison with analogous acoustic elements. Among the advantages are the possibility of their use in higher frequency bands, simplicity of their excitation and detection (insertion attenuation is lower), control by an external bias magnetic field, the possibility of dispersion characteristics shaping by external boundary conditions, the possibility to construct elements with inverse dynamic nonlinearities, etc. This has caused growing interest in research and development during the last ten years.

Now that an intensive search for new magnetic materials is going on, these materials are supposed to have more advantageous properties (concerning mainly the pass band width) and they should be promosing enough to allow the transition into higher frequency bands. The authors hope that this monograph will be useful not only for students, but also for teachers, researchers and engineers working in the field of spin wave electronics and that it will serve as a complementary stimulus for the study of the new field of high-frequency electronics — solid state wave electronics.

Chapter 1

OSCILLATIONS AND WAVES IN MAGNETICALLY ORDERED MATTER

1.1 Magnetically ordered matter and its properties

Magnetically ordered matter is characterized by high spontaneous magnetization if the external magnetic field is not present. It includes ferromagnetic, antiferromagnetic and ferrimagnetic materials.

Ferromagnetic matter is characterized by the spontaneous parallel orientation of the elementary magnetic moments, which results from the so-called exchange forces. Four metals — iron, cobalt, nickel, gadolinium — their alloys and a lot of oxides are included in it.

The essence of the spontaneous magnetization in antiferromagnetic materials is the same as in ferromagnetic materials, but antiferromagnetic materials are characterized by the "chessboard" arrangement of sublattices, i.e. each magnetic moment is surrounded by oppositely oriented magnetic moments. The difference between two antiparallel moments is small and for this reason the resultant magnetic moment (the magnetization per unit volume) will be small, too. Some metals (chromium, magnesium, germanium) and their alloys are included in antiferromagnetic materials. There are a number of antiferromagnetic dielectrics and semiconductors.

Ferrimagnetic materials (ferrites) are characterized by the spontaneous "chessboard" structure, but as a result of different magnitudes of individual magnetic moments in sublattices, which are oriented in various directions, the resultant magnetic moment can be relatively big. Non-metallic ferrites have found wide application in high-frequency technology. Thus, for example, in the high-frequency devices which take advantage of magnetostatic waves, non-metallic ferrite (magnetodielectric material) — yttrium ferrite with the garnet structure — is used almost exclusively. Its composition is described by the formula $Y_3Fe_5O_{12}$ and in the literature it is called yttrium-iron garnet (YIG). There are also other non-metallic magnetic materials with somewhat different properties applicable for high-frequency technology — barium ferrites, magnesium spinels etc. Those who are interested in them are referred to the book by Kvasnica [3].

We shall now give some basic information about the theory of magnetism in order to understand better the following text. In accordance with quantum mechanical theory, the electron in an atomic electron shell is characterized by the intrinsic (spin) and orbital angular momenta. The first one has a pure quantum mechanical character and the eigenvalues of the projections of the spin moment S in the z-axis are quantized as follows

$$S_z = \hbar S, \hbar(S-1), \ldots, (-\hbar S), \tag{1.1}$$

where $\hbar = h/2\pi$, h is the Planck's constant, S is the spin quantum number (spin) of the particle. For an electron the value S_z expressed in $\hbar$ units can be

$$S_z = \pm\frac{1}{2} \tag{1.2}$$

The orbital angular momentum $\vec{l}$, although it is a classical quantity, is also quantized with values in the range

$$l, (l-1), \ldots, (-l) \tag{1.3}$$

where l is the so-called orbital momentum quantum number.

The magnetic moments of a particle are associated with its angular momenta. The spin magnetic moment associated with the spin moment is given by

$$\vec{m}_s = -\gamma_s \cdot \hbar \cdot \vec{S}, \tag{1.4}$$

where γ_s, the gyromagnetic ratio is

$$\gamma_s = \frac{g \cdot e}{2 \cdot m_e} \tag{1.5}$$

where e and m_e are the charge magnitude and mass of the electron, respectively, and $g = 2.0023$ is the so-called gyromagnetic constant.

The orbital magnetic moment is associated with the orbital angular momentum

$$\vec{m}_l = -\gamma_l \cdot \hbar \cdot \vec{l}, \tag{1.6}$$

where γ_l in analogy with (1.5) is

$$\gamma_l = \frac{g_l \cdot e}{2 \cdot m_e}, \tag{1.7}$$

and $g_l \approx g/2 \approx 1$.

The total mechanical moment of an electron $\vec{j}$ is

$$\vec{j} = \vec{S} + \vec{l}, \tag{1.8}$$

and the total magnetic moment $\vec{m}$ will be

$$\vec{m} = \vec{m}_s + \vec{m}_l, \tag{1.9}$$

The magnetic moment of an atom consists of the orbital angular and spin momenta of all electrons in the atomic electron shell and the spin magnetic moment of the atomic nucleus. Magnitudes of the spin moments of an electron and a nucleus are of the same order. However, the magnetic spin moment of the nucleus is small because the gyromagnetic ratio for heavy particles (protons) is smaller (approximately 1800 times smaller than for electrons) depending on the ratio m_p/m_e. For this reason we can consider the influence of the magnetic moments of nuclei on the magnetic properties of atoms to be small.

The mechanical moment $\vec{J}$ and magnetic moment $\vec{m}$ of an atom are the vector sums of the moments (1.8) and (1.9) of all electrons in the atom, i.e.

$$\vec{S} = \sum_i \vec{S}_i, \vec{L} = \sum_i \vec{l}_i, \vec{J} = \vec{S} + \vec{L} \tag{1.10}$$

$$\vec{m}_s = \sum_i \vec{m}_{si}, \vec{M}_l = \sum_i \vec{m}_{li}, \vec{m} = \vec{m}_s + \vec{m}_l \tag{1.11}$$

which represents the model of Russel–Saunders's bond [210].

In solids the atoms (ions, molecules) are in the nodes of the crystal lattice, and they bind each other and interact. How does this interaction arise?

In accordance with classical physics an electric and magnetic force (dipole–dipole) interaction exists between atoms (ions). The electric force is stronger: its order is $e^2/a \cong 10^{-18}$ J (a is the lattice constant). The magnetic force is much weaker and its order is $m_s^2/a^3 \cong 10^{-23}$ J, i.e. weaker by five orders. Both of these interactions obey the inverse square law and they are long-range interactions.

From the quantum mechanical point of view, besides the interactions mentioned above, there are also strong so-called exchange forces (between the atoms in a crystal). We would like just to mention that the exchange interaction is a pure quantum mechanical phenomenon which has no classical analogy. The exchange energy is minimal if all spins in the matter have parallel (ferromagnetism) or antiparallel orientation (antiferromagnetism). The cause of magnetic order in magnetic crystals is the fact that the system is trying to achieve the state with minimal energy. Thus, in the equilibrium state the spins are trying to achieve parallel or antiparallel alignment. The exchange energy is of a much higher order than the energy of magnetic interaction, and for this reason the role of the latter in the magnetic alignment is relatively small. On the other hand the exchange forces are short-range interactions and that is why they are effective at much shorter distances than magnetic dipolar forces. An estimation of the energy of exchange interaction can be made on the basis of the following considerations. Since at the Curie temperature T_c ferromagnetic order is broken, it is obvious that the energy of exchange interaction for one atom is of the same order as kT_c, where k is Boltzmann's constant. For ferromagnetic materials it is of order 10^{-20} J, i.e. it is three orders higher than the energy of magnetic interaction.

On the basis of all the facts we have mentioned we can say that in magnetically ordered matter spins can interact in two ways — by means of the common magnetic dipolar interaction and exchange interaction. Two kinds of forces — magnetic and exchange forces — are associated with these interactions in magnetic dielectrics and these forces are also responsible for the wave processes in magnetic crystals. The magnetic, long-range forces describe very well the long wave excitations of the crystal lattice i.e. in the case that the wavelength of excitation $\lambda \gg a$. In this case a lot of crystal lattice nodes take part in the interaction, and because of this it is called the collective interaction. There are also short-range forces (between adjacent atoms) but they are small for long wave excitations because the deflection between the spins of adjacent atoms is negligibly small.

On the other hand the short-range exchange forces describe very well the short wave excitations of the lattice, i.e. if $\lambda \cong a$. In this case the deflection of adjacent spins in nodes of the crystal lattice is big and the exchange forces are predominant. Thus the long wave excitations are the consequence of the magnetic forces (dipole–dipole interaction) and short wave excitations are the consequence of the exchange forces (exchange viscosity). Since there are two types of interactions in magnetic crystals, there are also two types of waves — the spin waves due to magnetic interaction and the spin exchange waves which result from exchange interaction. The former are often called magnetostatic waves (MSW). We will deal almost exclusively with magnetostatic waves. It is obvious that the classification of spin waves into two groups is an approximation and strictly speaking it is always necessary to take into account both types of interactions.

1.2 Oscillation processes in ferromagnetic materials The Landau–Lifshitz equation Eigenoscillations of the magnetization vector

Since there are elastic forces in magnetic matter, an oscillation process characterized by eigenfrequency and losses can occur. Two approaches are possible in the investigation of these processes. The first one stems from quantum mechanical considerations and its basis is as follows.

Let us consider a particle with the spin magnetic moment $\vec{m}_s$. According to (1.4) we can write

$$\vec{m}_s = -\gamma_s \cdot \hbar \cdot \vec{S}$$

Let us apply an external static magnetic field $\vec{H}$ to this moment. The energy of interaction of magnetic moment $\vec{m}_s$ with the field $\vec{H}$ will be

$$W = -\mu_o \vec{m}_s \cdot \vec{H} = \gamma_s \hbar \mu_0 S H \cos\alpha \tag{1.12}$$

where μ_0 is the permeability of vacuum ($4\pi.10^{-7}$ H/m) and α is the angle between $\vec{m}_s$ and $\vec{H}$.

Let the external magnetic field be oriented in the z-axis direction, i.e. $\vec{H} = H\vec{z}_0$ and $(\hbar \cdot \vec{S})_z = S_z$. Then from (1.12) it follows that

$$W = +\gamma_s \cdot S_z \cdot \mu_0 \cdot H \tag{1.13}$$

In accordance with quantum mechanics the change of total energy in the system can be only discrete. Thus in the case we are investigating the S_z values can be only $\pm\hbar/2$ and also the energy can have only two allowed values (Further, we omit the subscript s in γ_s. The consequence of this will be discussed later.):

$$W_1 = \frac{1}{2}\gamma \cdot \hbar \cdot \mu_0 \cdot H \quad \text{or} \quad W_2 = -\frac{1}{2}\gamma \cdot \hbar \cdot \mu_0 \cdot H \tag{1.14}$$

and the particle in an external magnetic field can be in a state with energy W_1 or energy W_2 and for an atom in a magnetic field it will result in spectral line splitting — the Zeeman effect.

The transition from one energy state to another causes the energy quantum $\hbar \cdot \omega_0$ to be emitted (absorbed). If we compare the energy difference $(W_1 - W_2)$ with the energy of the energy quantum $\hbar \cdot \omega_0$ we obtain

$$\omega_0 = \gamma \cdot \mu_0 \cdot H = \gamma \cdot B \tag{1.15}$$

Let us assume that we apply an external magnetic field with a frequency equal to the transition frequency (1.15) of the system. In this case we can observe in the system the phenomenon called ferromagnetic resonance — matter is absorbing electromagnetic energy and the spin system is excited. The spins are transferring from state 2 to state 1. The transition frequency is given by (1.15) and at ferromagnetic resonance it is equal to the frequency of the external magnetic field. This is a microscopic (quantum mechanical) description of ferromagnetic resonance phenomenon.

The processes above can also be described in a classical way. In this case the quantum mechanical character of the spin is not taken into account and the spontaneous macroscopic magnetization is defined as

$$\vec{M} = \frac{\Delta\vec{m}}{\Delta V}, \tag{1.16}$$

where $\Delta\vec{m}$ is the magnetic moment of a small but still macroscopic volume ΔV. Then $\vec{M}$ is called the magnetization vector or the magnetic moment of the unit volume. If $\vec{m}$ is the magnetic moment of an atom (see (1.10), (1.11)) and n is the concentration of atoms, then

$$\vec{M} = \sum_i n \cdot \vec{m}_i \tag{1.17a}$$

and the corresponding mechanical moment $\vec{I}$ for the volume unit will be

$$\vec{I} = \sum_i n \cdot \vec{J}_i \tag{1.17b}$$

In general the magnetization $\vec{M}(\vec{r}, t)$ can be a function of co-ordinates and time.

In the case that magnetic matter is magnetized to saturation then in the steady state the vector $\vec{M}$ is parallel to the vector $\vec{H}$. If the vector $\vec{M}$ is deflected from the equilibrium state, then an elastic force arises whose moment is proportional to $(\vec{M} \times \mu_0\vec{H})$ and it tries to turn the magnetic moment into the equilibrium state again. However, since at the same time the mechanical moment of the system $\vec{I}$ is present, a gyroscopic force arises and it causes the precession of $\vec{I}$ and also $\vec{M}$ around the magnetic field direction (see Fig. 1.1a). The precession frequency will be

$$\omega_0 = \gamma \cdot \mu_0 \cdot H = \gamma \cdot B,$$

so it is equal to (1.16). Now, if an external magnetic field with the frequency $\omega = \omega_0$ is applied to the system, then the effect of ferromagnetic resonance occurs and electromagnetic energy is absorbed (emitted) in the process. In this process the angle of the gradient of the precessing vector $\vec{M}$ changes but its length remains constant, i.e. $|\vec{M}| = \text{const}$.

Now we present a simple "derivation" of the equation of motion for the magnetization vector $\vec{M}(\vec{r}, t)$ in an external magnetic field $\vec{H}(\vec{r}, t)$. The basis for our consideration will be the mechanics of rotational motion. Let us consider magnetodielectrics as a system of magnetic gyroscopes whose elementary angular momenta $\vec{J}$ are associated with their magnetic moments $\vec{m}$ in the form

$$\vec{J} = -\frac{\vec{m}}{\gamma}, \tag{1.18}$$

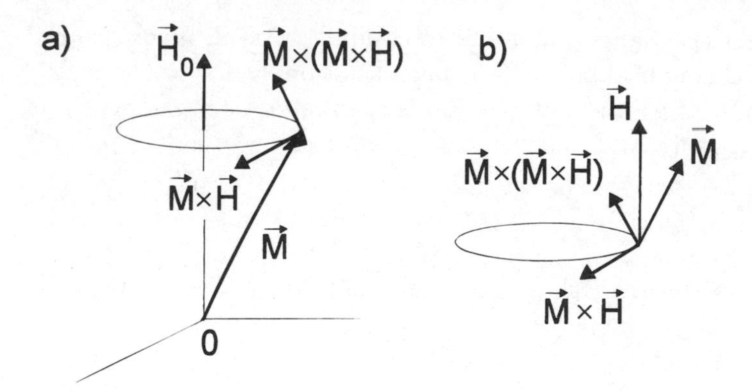

Fig. 1.1. a) Precession of the magnetization vector in an external magnetic field, b) Graphical representation of vectors $\vec{M}$, $\vec{H}$, $(\vec{M} \times \vec{H})$, and $\vec{M} \times (\vec{M} \times \vec{H})$.

In accordance with mechanics the equation of rotational motion is

$$\frac{\partial \vec{J}}{\partial t} = \vec{N}, \tag{1.19}$$

where $\vec{N}$ is the moment of external forces. In our case the external moment $\vec{N}$ is the result of the external magnetic field's effect on the magnetic moment $\vec{m}$. In the S.I. system of units the following equation [5] holds for the moment of external forces

$$\vec{N} = \vec{m} \times \vec{B} \tag{1.20}$$

where $\vec{B} = \mu_0(\vec{H} + \vec{m})$. If we substitute (1.18) and (1.20) for $\vec{J}$ and $\vec{N}$, respectively in (1.19), then we obtain

$$\frac{\partial}{\partial t}\left(-\frac{\vec{m}}{\gamma}\right) = \vec{m} \times \vec{B},$$

and if we multiply this equation by the concentration n and use (1.17), we obtain

$$\frac{\partial \vec{M}}{\partial t} = -\gamma \cdot (\vec{M} \times \vec{B})$$

Since $\vec{M} \times \vec{B} = \vec{M} \times \mu_0(\vec{M} + \vec{H})$, the equation of motion for the magnetization vector is given by

$$\frac{\partial \vec{M}}{\partial t} = -\gamma \cdot \left(\vec{M} \times \mu_0 \vec{H}\right) \tag{1.21}$$

which is called the Landau–Lifshitz equation. A strict microscopic derivation of this equation can be found in [3].

In equation (1.21) $\vec{M}$ is the magnetic moment (magnetization) of the volume unit. However, there is a question over what $\vec{H}$ is. It is not easy to answer this question, especially if we use a classical approach. In $\vec{H}$ we have to include not only external (constant and alternating) magnetic fields, magnetic fields connected with the finite size and shape of the sample (so-called demagnetizing fields) and magnetic fields of crystallographical anisotropy, but also effective magnetic fields associated with the subsidiary exchange energy caused by the deflection of the magnetic moment $\vec{M}$ from its equilibrium position. This is why the strict computation of the resulting effective value of the magnetic field intensity $\vec{H}$ is a complicated and difficult task. In the case which we examine it can be simplified. The exchange effects have to be taken into account if very small changes of $\vec{M}$ occur in space ($< 10^{-7}$ m). For magnetostatic waves which have a bigger space period (10^{-4}–10^{-5} m) we can neglect them. For this reason in our case we will include in $\vec{H}$ only external and internal (demagnetizing, anisotropy etc.) magnetic fields.

There is a certain arbitrariness in the choice of the quantity γ in (1.21). Formally, this quantity is defined as the gyromagnetic ratio (1.18). In spite of that this quantity is written for a single atom (ion) and in fact equation (1.21) describes the collective motion of the magnetic moments which interact with each other. Omitting details we can say that the values γ can be calculated using equations (1.5) or (1.7). It is necessary only to keep in view that the values of the quantity g which is present in these equations are in the interval $1 \leq g \leq 2$; the less the contribution of the angular momentum $\vec{L}$ to $\vec{J}$, the less g differs from 2.

Equation (1.21) is valid for an ideal situation in magnetically ordered matter in which the losses whose physical essence is complicated and influenced by many factors are not taken into account. In particular these factors can be the dispersion of magnetic excitation caused by the thermal lattice excitations, the presence of various chemical and geometrical inhomogeneities and the transformation of regular signals into different types of oscillations, e.g. the parametrical excitation of thermal spin waves by high-frequency signals etc. [21]. The final result of all these effects is that the forced oscillations of the magnetic system will be attenuated not only in space but also in time. Formally we can draw the following conclusion. It is known that in mechanics the dissipation of energy is taken into account by means of braking forces which are proportional to the velocities involved in the equation of motion. We will use the same idea in our case considering some effective magnetic fields of the "braking forces" proportional in their magnitudes and opposite in their signs to the velocity of the change of $\vec{M}$. These supplementary fields will create subsidiary moments which have to be included into the equation of motion. Then

$$\frac{\partial \vec{M}}{\partial t} = -\gamma \cdot \mu_0 \cdot (\vec{M} \times \vec{H}) + \frac{\alpha}{M} \cdot \left(\vec{M} \times \frac{\partial \vec{M}}{\partial t} \right), \tag{1.22}$$

where α is the dimensionless loss parameter. In the case where losses are small, i.e. $\alpha \ll 1$, then $\partial \vec{M}/\partial t$ in the second term in (1.22) can be substituted for (1.21) and then the equation of motion for the magnetization vector (taking into account the losses) will be:

$$\frac{\partial \vec{M}}{\partial t} = -\gamma \cdot \mu_0 \cdot (\vec{M} \times \vec{H}) - \frac{\alpha \cdot \gamma}{M} \cdot \mu_0 \cdot (\vec{M} \times (\vec{M} \times \vec{H}) \tag{1.23}$$

In Fig. 1.1b the vectors $\vec{M}$, $\vec{H}$, $(\vec{M} \times \vec{H})$, and $(\vec{M} \times (\vec{M} \times \vec{H})$ are depicted. Since $\vec{M} \perp (\vec{M} \times (\vec{M} \times \vec{H})$ the dissipation of energy does not change the length of the vector $\vec{M}$ but it changes (decreases) the angle of deflection of the precession. Equation (1.23) can be rewritten in the form

$$\frac{\partial \vec{M}}{\partial t} = -\gamma \cdot \mu_0 \cdot (\vec{M} \times \vec{H}) + \omega_\alpha \cdot \mu_0 \cdot \left[\vec{H} - \frac{\vec{M}(\vec{M} \cdot \vec{H})}{M^2} \right], \tag{1.24}$$

where $\omega_\alpha = \alpha.\gamma.M$.

This equation was presented for the first time in 1935 by L. D. Landau and E. M. Lifshitz as the equation of motion for the magnetization vector [2].

Now, we will solve the equation of motion for isotropic ferromagnetic materials with the equilibrium magnetization $\vec{M}_0$, magnetized to saturation by a static magnetic field. We will examine the uniform precession of magnetization, i.e. we will assume that the constant and time-variable components of the magnetization will not depend on

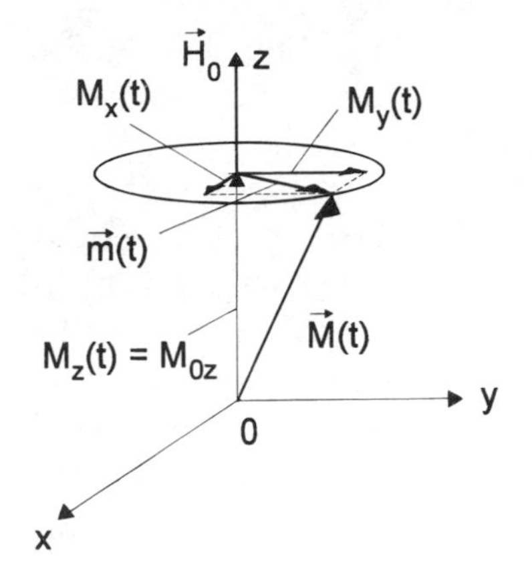

Fig. 1.2. Co-ordinate system used for solution of the equation of motion.

the co-ordinates. Then we will present the linear theory of magnetostatic oscillations, i.e. we will assume that variable quantities are much smaller than the corresponding constant quantities. In deriving these forms we will take into consideration losses in the given material.

Let a static magnetic field $\vec{H}_0$ be applied in the z-axis direction (see Fig. 1.2). In the equilibrium state $\vec{M} = \vec{M}_0$ and $\partial\vec{M}/\partial t = 0$, so from (1.21)

$$\vec{M}_0 \times \vec{H}_0 = 0 \text{ or } \vec{M}_0 \| \vec{H}_0$$

i.e. the equilibrium magnetization $\vec{M}_0$ is always oriented in the direction of the applied magnetic field. The magnitude of the $\vec{M}$ vector does not depend on $\vec{H}_0$ and is considered to be known. Further, scalar multiplication of (1.21) by $\vec{M}$ gives $\vec{M} \cdot (\partial\vec{M}/\partial t) = -\gamma \cdot \mu_0 \cdot (\vec{M} \times \vec{H}) \cdot \vec{M} \equiv 0$ or $(\partial/\partial t)(M^2) = 0$, so the length of the vector $\vec{M}$ during its arbitrary changes (oscillations) remains constant, i.e.

$$|\vec{M}| = |\vec{M}_0| = \text{const.} \tag{1.25}$$

The constant length of the vector $\vec{M}$ gives evidence that its motion represents the precession of a rigid magnetization vector around its equilibrium position $\vec{M}_0$. Let us tranform equation (1.21) into the axis projection, taking into account that $\vec{H}_0$ has only one component in the z-axis direction. Then

$$\frac{\partial M_x}{\partial t} = -\gamma.\mu_0.H_0.M_y,$$
$$\frac{\partial M_y}{\partial t} = \gamma.\mu_0.H_0.M_x, \tag{1.26}$$
$$\frac{\partial M_z}{\partial t} = 0$$

From the last equation, M_z is a constant quantity which it is necessary to compute from the initial conditions. Further, we will divide the first equation by the second and after integration we obtain

$$M_x^2 + M_y^2 = R^2 = \text{const.} \tag{1.27}$$

Equation (1.27) is the equation of a circle. At excitation the end point of the vector $\vec{M}$ begins to move on a circle in the plane perpendicular to $\vec{H}_0$ and the radius of this circle is R. If we substitute M_y in the first equation of (1.26) for the M_y value computed from (1.27) and if we denote $\omega_H = \gamma.\mu_0.H_0$, then we obtain the equation

$$\frac{\partial M_x}{\partial t} = -\gamma.\mu_0.H_0.(R^2 - M_x^2)^{1/2}$$

or

$$\frac{\partial M_x}{(R^2 - M_x^2)^{1/2}} = -\omega_H \partial t$$

Its solution is

$$\arcsin \frac{M_x}{R} = \omega_H.t + \varphi_{01}, \tag{1.28}$$

where φ_{01} is the integration constant, in this case an initial angle. In an analogous way M_y is given by

$$\arcsin \frac{M_y}{R} = -\omega_H.t + \varphi_{02}. \tag{1.29}$$

Let the position of vector $\vec{M}$ at the initial time $t = 0$ be determined as follows: $M_{x0} = 0$, $M_{y0} = R$, $M_{z0} = M_{0z}$, i.e. the vector $\vec{M}$ is deflected from the equilibrium position in the xyz plane (see

Fig. 1.2). Then using (1.28), (1.29) we find $\varphi_{01} = 0$, $\varphi_{02} = \pi/2$ and for M_x and M_y the following equations hold:

$$M_x = -R \cdot \sin(\omega_H t); \quad M_y = R \cdot \cos(\omega_H t);$$

i.e. the rotation of the vector $\vec{M}$ is clockwise (if observed in the $\vec{H}_0$ direction). The rotation frequency is determined by the formula $\omega_H = \gamma.\mu_0.H_0$. In fact this motion is a precession of the $\vec{M}$ vector around the equilibrium position $\vec{M}_0$. The longitudinal component M_{0z} of the $\vec{M}$ vector (i.e. in the magnetic field direction) is constant and M_x and M_y change in a harmonic way.

On the basis of the results mentioned above this task can be solved in a simpler way if complex functions are used. In accordance with Fig. 1.2 we can write the following equations for the $\vec{M}(x)$ vector:

$$\vec{M}(t) = M_{0z}\vec{z}_0 + \vec{m}.\mathrm{e}^{j\omega_H t}; \quad m_x^2 + m_y^2 \ll M_{0z}^2 \tag{1.30}$$

i.e. from the very beginning we assume that the component $M_{0z} =$ const. and the $\vec{m}.\mathrm{e}^{j\omega_H t}$ vector lying in the xy plane, i.e. perpendicular to the magnetic field, is changing in a harmonic way with the frequency ω_H. The $\vec{m}$ vector is the complex amplitude of the oscillations whose phase is determined by the initial conditions. If we substitute (1.30) in (1.21) we obtain the complex algebraic equation

$$j\omega_H \vec{m} = -\gamma.\mu_0(\vec{m} \times \vec{H}_0)$$

Separating this into components, we obtain the system of equations

$$\begin{aligned} j\omega_H m_x + \gamma\mu_0 H_0 m_y &= 0\ , \\ -\gamma\mu_0 H_0 m_x + j\omega_H m_y &= 0\ , \\ j\omega_H m_z &= 0. \end{aligned} \tag{1.31}$$

From these we see that the end point of the $\vec{M}$ vector moves in the plane perpendicular to the z-axis, i.e. the $\vec{m}$ vector lies in the xy plane (see Fig. 1.2) perpendicular to $\vec{H}_0$. Let the determinant of the system (1.31) be equal to zero; then for the eigenfrequency ω_H we can obtain the equation $\omega_H = \gamma\mu_0 H_0$, which is equivalent to the results obtained above. If this is substituted in (1.31) for ω_H, then we obtain

$$m_y + jm_x = 0,$$

i.e. the $\vec{m}$ vector is rotating clockwise around the magnetic field direction. So, we can conclude that the proper uniform motion of magnetization in isotropic ferromagnetic materials is a circular clockwise precession (as can be observed in the $\vec{H}_0$ direction) of the magnetization vector around the static magnetic field direction $\vec{H}_0$ with the frequency

$$\omega_H = \gamma\mu_0 H_0.$$

1.3 Forced oscillations of the magnetization vector High-frequency magnetic susceptibility Ferromagnetic resonance

Let an external magnetic field have constant and variable components

$$\vec{H}(r,t) = \vec{H}_0 + \vec{h}.\mathrm{e}^{j\omega t} \tag{1.32}$$

where $\vec{h}$ is the complex amplitude and ω is the frequency of an alternating (exciting) magnetic field. We will try to find the solution of the equation of motion in the following form:

$$\vec{M}(r,t) = M_{0z}\vec{z}_0 + \vec{m}.\mathrm{e}^{j\omega t.} \tag{1.33}$$

It is noteworthy that in (1.33) ω is not the eigenfrequency but the frequency of the exciting force. Up to now we have examined small oscillations, i.e. we consider all processes to be linear, and the following inequalities result from this:

$$|\vec{h}| \ll |\vec{H}_0|, \qquad |\vec{m}| \ll |M_{0z}|.$$

Let us substitute (1.32) and (1.33) in (1.23), neglect small terms of the second order (linear approximation) and separate the linearized vector equation into component equations taking into account that $\vec{H}_0$ and $\vec{M}_0$ are oriented in the z-axis direction. Then we obtain the system ($M_{0z} \approx M_0$)

$$\begin{aligned} j.\omega.m_x + \omega_H.m_y &= \gamma\mu_0 M_0 h_y, \\ -\omega_H.m_x + j.\omega.m_y &= -\gamma\mu_0 M_0 h_x, \\ j.\omega.m_z &= 0, \end{aligned} \tag{1.34}$$

where $\omega_H = \gamma\mu_0 H_0$.

The solution of the set of equations (1.34) referring to m_x, m_y, m_z will be

$$\begin{aligned} m_x &= \frac{\gamma\mu_0 M_0 \omega_H}{\omega_H^2 - \omega^2} h_x + j\frac{\gamma\mu_0 M_0 \omega}{\omega_H^2 - \omega^2} h_y \\ m_y &= -j\frac{\gamma\mu_0 M_0 \omega}{\omega_H^2 - \omega^2} h_x + \frac{\gamma\mu_0 M_0 \omega_H}{\omega_H^2 - \omega^2} h_y \\ m_z &= 0 \end{aligned} \tag{1.35}$$

The coefficients in the components of the magnetic field intensity are called components of the high-frequency magnetic susceptibility and they are

$$\chi = \frac{\gamma\mu_0 M_0 \omega_H}{\omega_H^2 - \omega^2}, \qquad \chi_a = \frac{\gamma\mu_0 M_0 \omega}{\omega_H^2 - \omega^2} \tag{1.36}$$

The equations (1.35) can be separated into the following equations:

$$\begin{aligned} m_x &= \chi.h_x + j.\chi_a.h_y, \\ m_y &= -j.\chi_a.h_x + \chi.h_y, \\ m_z &= 0 \end{aligned}$$

or into the tensor form

$$\vec{m} = \overline{\overline{\chi}}.\vec{h} \tag{1.37}$$

where $\overline{\overline{\chi}}$ is the tensor of high-frequency magnetic susceptibility and its form is

$$\overline{\overline{\chi}} = \begin{pmatrix} \chi & j\chi_a & 0 \\ j\chi_a & \chi & 0 \\ 0 & 0 & 0 \end{pmatrix} \tag{1.38}$$

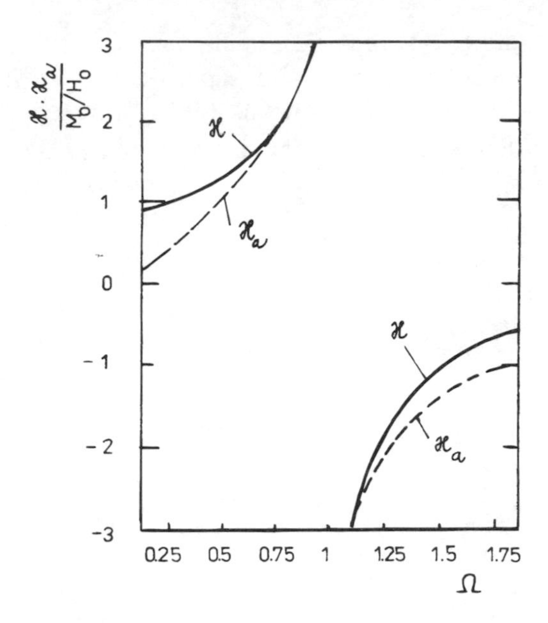

Fig. 1.3. The dependence of the susceptibility components $\varkappa$ and $\varkappa_a$ on normalized frequency $\Omega = \omega/\omega_H$.

Fig. 1.3 depicts the dependences $\chi(M, H)$ and $\chi_a(M, H)$ on the normalized applied frequency $\Omega = \omega/\omega_H$. Similar dependences can be obtained when the magnetic field H_0 is changed at the constant frequency ω. As follows from (1.36) and Fig. 1.2 that an essential peculiarity of the solution of the dependence obtained is that the magnetic susceptibility tensor elements depend on the frequency of the magnetic field. This dependence has a resonance character, i.e. in ferromagnetic materials resonance absorption of the electromagnetic field energy depending on the frequency occurs. It is the essence of the ferromagnetic resonance effect mentioned above.

Let us derive a form for the high-frequency permeability which will associate the magnetic field induction $\vec{B}$ with the magnetic field intensity $\vec{H}$. For the high-frequency components we can write

$$\vec{b} = \mu_0(\vec{m} + \vec{h}) \tag{1.39}$$

where $\vec{b}$, $\vec{m}$ and $\vec{h}$ are the complex amplitudes of the magnetic field induction, alternating magnetic field and high-frequency magnetization, respectively, and μ_0 is the permeability of vacuum. Substituting

$\vec{m}$ from (1.37) in (1.34) we obtain

$$\vec{b} = \overline{\overline{\mu}}.\vec{h} \tag{1.40}$$

where $\overline{\overline{\mu}}$ is the permeability tensor

$$\overline{\overline{\mu}} = \mu_0(\overline{\overline{1}} + \overline{\overline{\chi}}) \tag{1.41}$$

Considering (1.38), (1.40) and (1.41) the following equation for the magnetic permeability tensor holds:

$$\vec{\vec{\mu}} = \mu_0 . \begin{pmatrix} \mu & j\mu_a & 0 \\ -j\mu_a & \mu & 0 \\ 0 & 0 & 1 \end{pmatrix} \tag{1.42}$$

where the diagonal element μ of the permeability tensor is

$$\mu = 1 + \chi = \frac{\omega_H.(\omega_H + \omega_M) - \omega^2}{\omega_H^2 - \omega^2}, \tag{1.43}$$

and the off-diagonal element of the permeability tensor is

$$\mu_a = \chi_a = \frac{\omega_M.\omega}{\omega_H^2 - \omega^2} \tag{1.44}$$

In equations (1.43) and (1.44) we have denoted

$$\omega_M = \gamma\mu_0 M_0 \tag{1.45}$$

As an example, in Fig. 1.4 there are dependences of the elements of the high-frequency permeability tensor on $\Omega = \omega/\omega_H$ for various values of $\Omega_M = \omega_M/\omega_H$. The dependence of off-diagonal elements μ_a on Ω is similar. As will be mentioned later it is possible to observe the magnetostatic waves at $\Omega > 1$ for various values $\mu \lessgtr 0$. At $\mu < 0$ volume magnetostatic waves (MSW) can exist and at $\mu > 0$ surface MSW.

It is noteworthy that a relatively simple form for the high-frequency permeability tensor is connected with the orientation of the magnetic field in the z-axis direction. If there is an arbitrary orientation of the magnetic field, then the elements of this tensor are

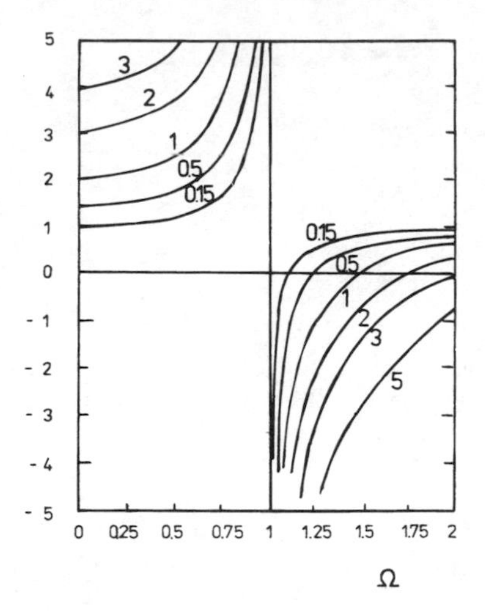

Fig. 1.4. The dependence of the diagonal component of high-frequency permeability μ on normalized frequency Ω. Parameter $\Omega_M = \omega_M/\omega_H$.

complex functions of frequency, wave number, medium parameters, magnitude and orientation of the magnetic field H_0.

Now let us examine forced oscillations considering losses. In this case it is necessary to analyze the equations of motion (1.22). If we assume that $\vec{H}(t)$ and $\vec{M}(t)$ are given by equations (1.32) and(1.33), respectively, then after substitution in (1.22) and neglecting second order terms we obtain

$$j\omega.\vec{m} + \gamma\mu_0(\vec{m} \times \vec{H}_0) + j\omega(\alpha/M_0)(\vec{M}_0 \times \vec{m}) = -\gamma\mu_0(\vec{M}_0 \times \vec{h}) \quad (1.46)$$

Separating into components we obtain

$$\begin{aligned} j\omega m_x + (\omega_H + j\omega\alpha)m_y &= \gamma\mu_0 M_0 h_y, \\ -(\omega_H + j\omega\alpha)M_x + j\omega m_y &= -\gamma\mu_0 M_0 h_x, \\ j\omega m_z &= 0 \end{aligned} \quad (1.47)$$

The only difference between the sets of equations (1.47) and (1.34) is that instead of ω_H there is

$$\omega_H \to (\omega_H + j\omega\alpha). \quad (1.48)$$

Taking advantage of this formal difference and considering losses we can write immediately, analogously to (1.36), equations for the high-frequency magnetic susceptibility tensor elements:

$$\begin{aligned}\chi &= \frac{\gamma\mu_0 M_0(\omega_H + j\omega\alpha)}{\omega_H^2 - (1+\alpha^2)\omega^2 + 2j\omega\omega_H\alpha},\\ \chi_a &= \frac{\gamma\mu_0 M_0\omega}{\omega_H^2 - (1+\alpha^2)\omega^2 + 2j\omega\omega_H\alpha}\end{aligned} \tag{1.49}$$

If $\alpha \Rightarrow 0$ (lossless case), then (1.49) will be the same as (1.36). If we separate (1.49) into real and imaginary components, assuming that the system is almost in ferromagnetic resonance ($\omega \simeq \omega_H$) and losses are very small ($\alpha \ll 1$), then

$$\begin{aligned}\chi &= \chi' + j\chi'',\\ \chi_a &= \chi_a' + j\chi_a'',\end{aligned}$$

where

$$\begin{aligned}\chi' &= \frac{\gamma\mu_0 M_0\omega_H(\omega_H^2 - \omega^2)}{(\omega_H^2 - \omega^2)^2 + 4\alpha^2\omega_H^2\omega^2},\\ \chi'' &= \frac{\gamma\mu_0 M_0\omega_H\alpha(\omega_H^2 + \omega^2)}{(\omega_H^2 - \omega^2)^2 + 4\alpha^2\omega_H^2\omega^2},\\ \chi_a' &= \frac{\gamma\mu_0 M_0\omega(\omega_H^2 - \omega^2)}{(\omega_H^2 - \omega^2)^2 + 4\alpha^2\omega_H^2\omega^2}\\ \chi_a'' &= \frac{2\gamma\mu_0 M_0\omega^2\omega_H\alpha}{(\omega_H^2 - \omega^2)^2 + 4\alpha^2\omega_H^2\omega^2}\end{aligned} \tag{1.50}$$

If $\alpha = 0$, then $\chi'' = \chi_a'' = 0$ and the expressions for χ' and χ'' are the same as (1.36). As follows from (1.50) the real parts of the high-frequency magnetic susceptibility tensor are zero for $\omega = \omega_H$ and the imaginary parts have a resonance character. The resonant frequency depends on the magnitude of the losses:

$$\omega_{\text{res}} = \omega_H\sqrt{(1 - 2\alpha^2)} \tag{1.51}$$

i.e. if losses are increasing the resonant frequency decreases approximately in a linear way.

In Fig. 1.5 the dependences $\chi''/(\gamma\mu_0 M_0/2\omega_H)$ on the quantity ω/ω_H are depicted. The maximum value of magnetic susceptibility decreases if losses are increasing, and for $\omega = \omega_H$ it is given by

$$\|\chi_{\max}''\| = \|\chi_{a\ \max}''\| = \frac{\gamma\mu_0 M_0}{2\alpha\omega_H}, \tag{1.52}$$

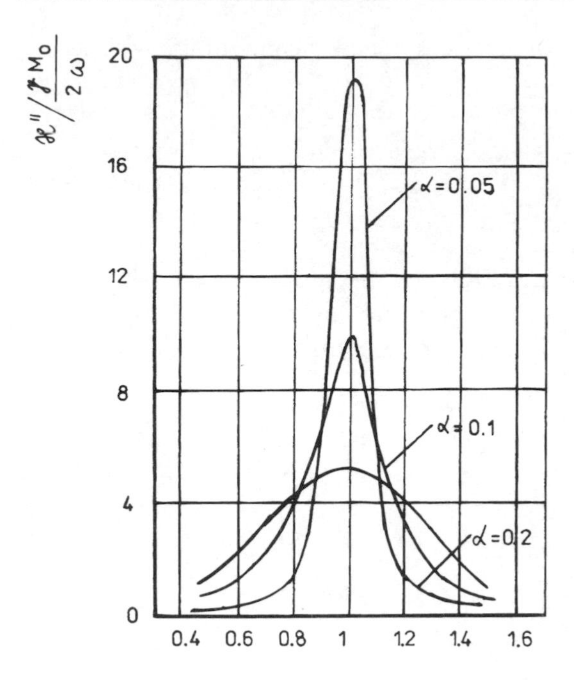

Fig. 1.5. The dependence of the imaginary part of the diagonal component of the susceptibility tensor on normalized frequency Ω for various values of the loss parameter α.

i.e. at resonance, if losses are increasing, then the absorbed energy decreases. On the other hand, the resonance line width increases and the quality factor of the ferromagnetic resonance system decreases. In descriptions of the properties of gyromagnetic media the width of the ferromagnetic resonance line is used to characterize losses. This is quite an appropriate description because the width of the resonance line can be measured. We will determine the resonance line width using (1.50). In accordance with (1.52) the half-maximum value $\chi''_{\max}$ (at level 3 dB) is given by the expression

$$\frac{1}{2}\chi''_{\max} = \frac{\gamma\mu_0 M_0}{4\alpha\omega_H}.$$

Substituting this value in equation (1.50), for $\chi''(\omega)$ we will find a mistuning $\Delta\omega_H$ at which χ'' acquires the value $\chi'' = (1/2)\chi''_{\max}$. We obtain

$$\Delta\omega_H = \alpha\omega_H, \quad or \quad \Delta H_0 = \alpha H_0. \tag{1.53}$$

For a typical magnetodielectric which is used in high-frequency technology, e.g. YIG, α has the values 10^{-3} - 10^{-4} and values of ΔH_0 in the frequency band.

Finally we will deal with the problem of how to determine the magnetic field in a magnetized sample, i.e. how to compute the field which is present in equations (1.21), and the resonant frequency of that ferromagnetic resonance which results from this field.

Let the sample have the shape of an ellipsoid whose main axes are identical with the co-ordinate axes. Let the sample be magnetized homogeneously. Before computing the internal magnetic field we denote N_x, N_y, N_z demagnetizing factors. The components of the resultant internal magnetic field $\vec{H}_i$ are connected with the external magnetic field $\vec{H}$ and the magnetization vector $\vec{M}$ by the expressions

$$H_{ix} = H_x - N_x M_x; \; H_{iy} = H_y - N_y M_y; \; H_{iz} = H_z - N_z M_z \quad (1.54)$$

or in the tensor form

$$\vec{H}_i = \vec{H} - \overline{\overline{N}}.\vec{M} \quad (1.55)$$

where $\overline{\overline{N}}$ is the so-called demagnetizing tensor. The vectors $\vec{H}$ and $\vec{M}$ can have both constant and variable components and the demagnetizing field $(-\overline{\overline{N}}.\vec{M})$ always weakens $\vec{H}$.

To elucidate the structure of (1.55) let us turn to Fig. 1.6 which explains the creation of the total field H_i in the magnetized sample. As can be seen, the demagnetizing fields arc proportional to M and in the sample they are oriented against the external field (demag-

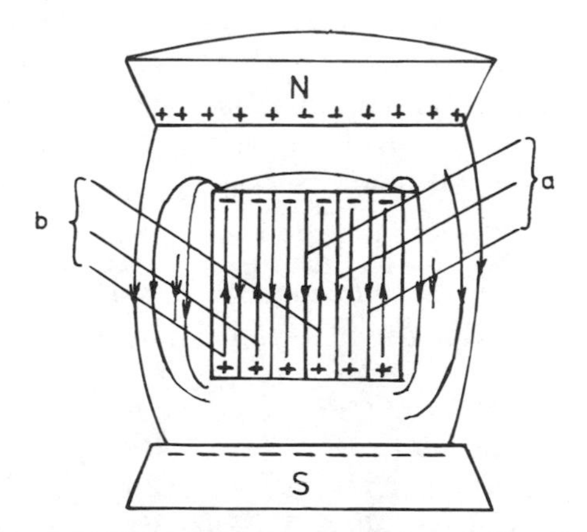

Fig. 1.6. Schematic representation of the situation in a magnetized body in an external magnetic field, a — magnetizing field, b — demagnetizing field, N — north, S — south magnetic pole.

netizing field) and outside the sample they are oriented in the field direction. For a long cylinder whose ends are far apart the internal magnetic field is equal to $-\vec{M}$, i.e. its own field is almost zero and a similar situation holds for other geometrical arrangements. Similarly, as we have already mentioned, the bodies with the shape of an ellipsoid with homogeneous magnetization in the homogeneous external magnetic field create homogeneous demagnetizing fields. In (1.55) this field is taken into consideration by the term $-\overline{\overline{N}}.\vec{M}$. An important consequence results from the facts mentioned above. The magnetic field in samples of finite size differs strongly from the external field. This means that the frequency of ferromagnetic resonance will not be given by equation (1.16), where $\vec{H}$ is the external magnetic field. In magnetized samples of finite size the frequency of ferromagnetic resonance will depend on the external magnetic field as well as the shape and size of the sample and its orientation with regard to the external magnetic field. In [4] we can find the resonant frequency ω_0 of the magnetized ellipsoid whose main axes are identical with the x, y, z co-ordinates and where the external magnetic field H_0 is oriented along the z-axis. If demagnetizing factors are determined using (1.55), then the resonant frequency will be

$$\omega_0^2 = \mu_0^2\gamma^2[H_0 + (N_y - N_z)M_0].[H_0 + (N_x - N_z)M_0] \tag{1.56}$$

and it is also called the frequency of uniform precession. For demagnetizing factors the following equation holds:

$$N_x + N_y + N_z = 1 \tag{1.57}$$

If we consider a sphere, then $N_x = N_y = N_z$ and the corresponding ω_0 is

$$\omega_0 = \gamma\mu_0 H_0 \tag{1.58}$$

i.e. as in an infinite medium. For a sample shaped as a thin plate magnetized perpendicularly to its plane the following equations hold: $N_x + N_y = 0$, $N_z = 1$ and

$$\omega_0 = \gamma\mu_0(H_0 - M_0) \tag{1.59}$$

and if it is magnetized in its plane, then $N_x - N_z = 0$, $N_y = 1$ and

$$\omega_0 = \gamma\mu_0[H_0(H_0 + M_0)]^{1/2} \tag{1.60}$$

If a cylinder is magnetized longitudinally, then

$$\omega_0 = \gamma\mu_0(H_0 + (1/2)M_0) \tag{1.61}$$

and if a cylinder is magnetized perpendicularly, then

$$\omega_0 = \gamma\mu_0[H_0(H_0 - \frac{1}{2}M_0)]^{1/2}. \tag{1.62}$$

As follows from formulae (1.58)–(1.62), the computation of the ferromagnetic resonance frequency is rather difficult in a real case. In Figs. 1.8 and 1.9 the distribution of the longitudinal components of demagnetizing field $N_z(z)$ and $N_z(x)$ for the in-plane magnetized slab of Fig. 1.7 is depicted [6]. In Fig. 1.8 there are dependences

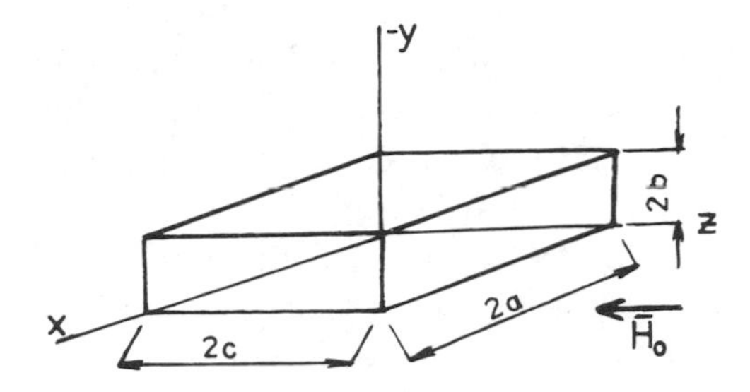

Fig. 1.7. In-plane magnetized slab.

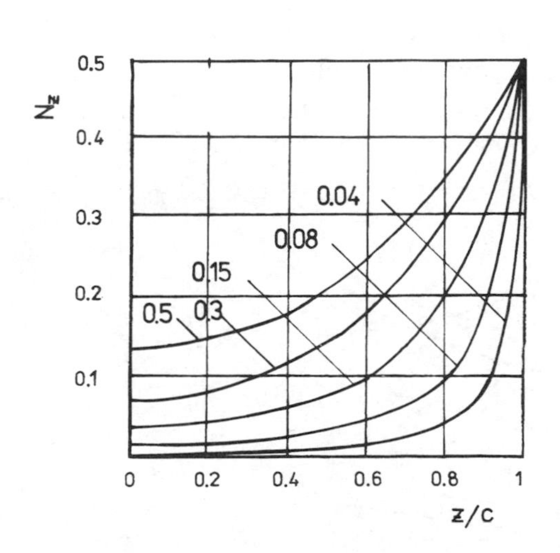

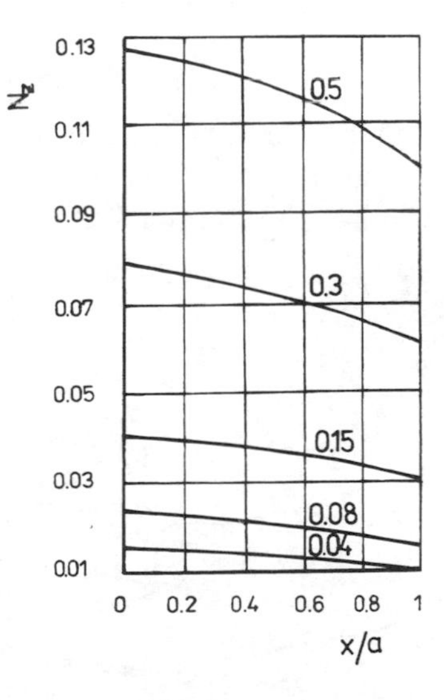

Fig. 1.8. The dependence of the demagnetizing factor N_z on z/c for $a/c = 0.5$, b/c parameter.

Fig. 1.9. The dependence of the demagnetization factor N_z on the ratio x/a for $a/c = 0.5$, b/c parameter.

$N_z(z)$ for $x = y = 0$ for various values b/c and $a/c = 0.5$. As can be seen, the demagnetizing factor N_z and also the demagnetizing field $H_{z\text{dem}}$ increase from the middle of the slab to its edges. At the same time, the thinner this slab, the more intensive this increase will be and it is located near the edge of the sheet. For very thin slabs (thin layers) $H_{z\text{dem}}$ remains almost constant along the length, and the thinner the layer, the lower the field will be. For thin YIG layers whose thickness is of the order 10^{-5} m, the quantity b/c is 10^{-2} – 10^{-3} and the N_z value (and also $H_{z\text{dem}}$) remains negligibly small. In this case the field in the sheet is approximately equal to the external field along the length of the sheet except the parts near the sheet edges. Fig. 1.9 depicts the dependence $N_z(x)$, i.e. along the slab. As can be seen, the dependence $H_{z\text{dem}}$ on the longitudinal co-ordinate differs from that on the transverse co-ordinate. For high values of b/c the demagnetizing field decreases in the direction towards the edges of the slab, i.e. the internal field is increasing. On decreasing the thickness $H_{z\text{dem}}$ is almost unchanged along the length and for very thin slab (layers) it remains almost constant along the length but with decreasing absolute magnitude. It seems that for practical application it is not necessary to consider this field in the cases mentioned above. In conclusion — for thin layers, in-plane magnetized (along the z-axis, Fig. 1.7), the longitudinal component of the internal field is equal to the external field both along the length of the layer and along the width of the layer except at the edge of the layer. Qualitative distribution of the magnetizing field in the plane x,z in the layer is depicted in Fig. 1.10. As can be seen, the field distribution has the shape of a saddle which is very shallow (for thin layers).

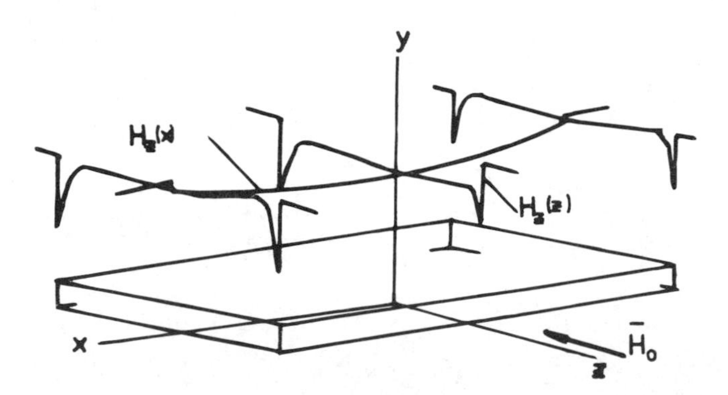

Fig. 1.10. Qualitative distribution of the magnetic field inside the layer.

The changes of $H_z(x, z)$ are more conspicuous as the thickness of the layer is increased.

1.4 Wave propagation in magnetized magnetic media

As a result of the elasticity of the spin (magnetic) system, oscillations (precession) of the magnetic moments with the frequency of the exciting force can exist and they are in resonance for the frequency equal to $\mu_0 \gamma H_i$, where H_i is the internal field in the magnetic material. If these oscillations are excited in a limited region of the ferrite sample, then due to the elasticity of this system they will propagate with a defined velocity in the sample. This propagating disturbance represents a magnetostatic wave.

The physical description of wave propagation in ferromagnetic materials is analogous in many ways to the description of wave propagation in an artificial magnetic medium — a system of magnetic dipoles. Even in 1939 L. Brillouin and M. Parodi [7] dealt with the problem of wave propagation in a longitudinally magnetized chain of magnetic dipoles. Let us imagine a chain of magnetic dipoles placed in an external magnetic field, as is depicted in Fig. 1.11a.

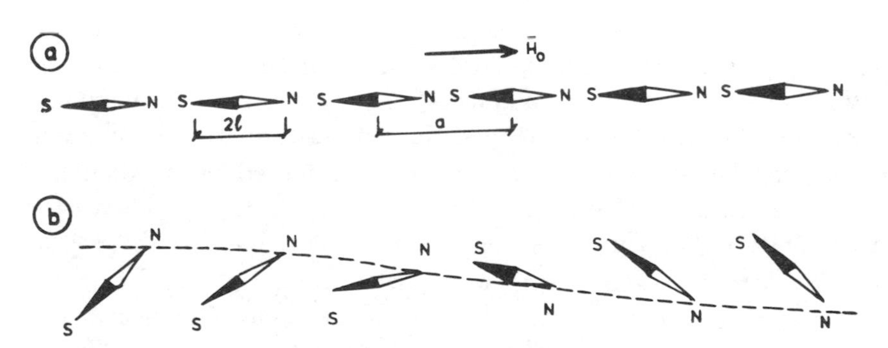

Fig. 1.11. Schematic representation of a magneto-mechanic wave excitation. N — north, S — south magnetic pole. a) steady state, b) excited wave.

Each of these dipoles can turn round its fixed centre freely in the plane of the figure. In an unexcited state all dipoles are oriented parallel to the external magnetic field $\vec{H}_0$. If the left-most dipole is deflected from the equilibrium position a mechanical moment arises which is caused both by the interaction with the external field and adjacent dipoles. This moment will try to turn the first dipole to its equilibrium position, but in so doing it will deflect the second dipole from its equilibrium position and then the next one, etc. A magnetomechanical wave will propagate along the dipole chain (Fig. 1.11b) and its velocity will be determined by the frequency of the external exciting force (i.e. the frequency with which the left-most dipole will be deflected), the magnitude of the dipole magnetic moment and the coupling both among dipoles and between dipoles and the external magnetic field. The second case, when the wave is excited perpendicularly to the magnetized chain of magnetic dipoles, is analyzed in detail in [8]. We will not deal with these solutions in detail. We would like only to mention that a so-called backward wave propagates in a longitudinally magnetized chain of magnetic dipoles and that its phase and group velocities are oriented against one another. This wave, as will be mentioned later, is analogous to a slow electromagnetic wave in the longitudinally magnetized ferrite slabs. In high-frequency magnetoelectronics terminology used in practice it will be a backward volume magnetostatic wave. A wave having normal dispersion propagates in a perpendicularly magnetized chain of magnetic dipoles. This wave is analogous to a slow electromagnetic wave in perpendicularly magnetized ferrite slabs. It is a forward volume magnetostatic wave.

Now, let us examine qualitatively wave processes in a magnetized ferrite slab. We will consider propagation of the wave impulse in simple ferromagnetic materials when in the basic state all magnetic moments of atoms $\vec{m}$ are parallel and oriented in the direction of the external magnetic field $\vec{H}_0$, which is oriented perpendicularly to the ferrite slab (see Fig. 1.12a). If we deflect the side vector $\vec{m}$ from its equilibrium position, then in accordance with (1.2) it will begin to precess around the magnetic field $\vec{H}_0$ with frequency $\omega_H = \gamma\mu_0 H_0$. Due to this a high-frequency magnetic field arises which excites the adjacent atom, whose magnetic moment will begin to precess with frequency ω_H, etc. The magnetic moments precess, describing running cones in such a way that the following one will have a defined phase shift in comparison with the previous one and the phase shift will be constant from one dipole to another along the chain (Fig. 1.12b). As a result a wave — magnetostatic spin wave — will run along the magnetic moment chain. The wave is depicted by the solid line in Fig. 1.12b. Its wavelength is also indicated in

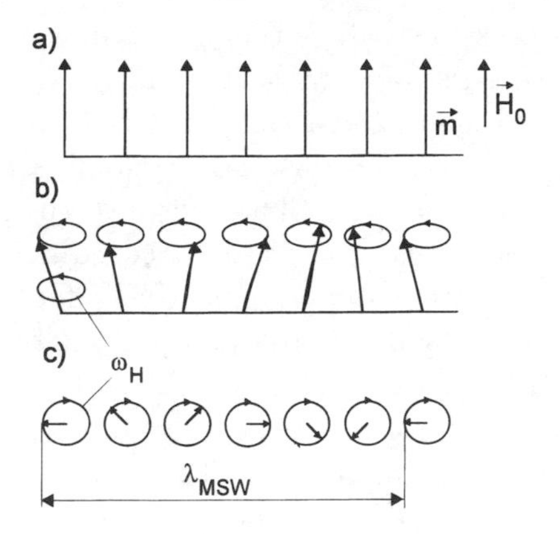

Fig. 1.12. Schematic representation of a magnetostatic spin wave. a) Uniform orientation of spins, b) side view of precessing magnetic moments, c) view as observed in the $\vec{H}_0$ direction.

Fig. 1.12c.

Fig. 1.13 helps to understand better the propagation of a magnetostatic wave. "Snapshots" of the end points of the $\vec{m}$ vectors are "taken" at constant time intervals dt during the propagation of a forward magnetostatic wave in the chain. Each $\vec{m}$ vector precesses in its phase around $\vec{H}_0$ with the frequency ω_H. The magnitude of the phase shift between the vectors is determined by the conditions of excitation of the magnetostatic spin wave. Let the end points of vectors be distributed at the initial moment as can be seen in Fig. 1.13a, which represents, as we have mentioned above, a "snapshot" of the phase magnetostatic wave. During the time interval dt each $\vec{m}$ vec-

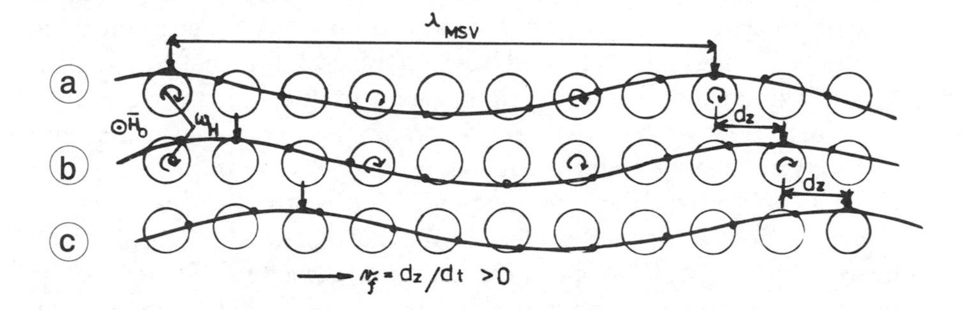

Fig. 1.13. Qualitative representation of MSW propagation. The wave front is indicated by arrows for various time moments. v_p — phase velocity.

tor turns through an angle dφ (precession) while the whole phase chain is "shifted" to the right by the distance dz (Fig. 1.13b). In Fig. 1.13c another stage of the shift of the phase wave to the right is depicted. In fact, all dipoles stay in their positions: no mechanical replacement occurs (unlike acoustic waves). The phase wave of the impulse runs to the right; its phase velocity is $v_p = \mathrm{d}z/\mathrm{d}t$. According to the theory the phase velocity of this wave depends on the difference between the frequency ω and the resonance frequency ω_0 and for $\omega \to \omega_0 v_p \to \infty$. In this case all magnetic dipoles in Fig. 1.13 will be distributed with the same phase and since they rotate synchronously the phase chain will be a straight line and the velocity of propagation and the velocity of precession phase along the z-axis go to infinity. Of course this picture is simplified. In fact, the processes in progress at the excitation of magnetostatic waves are more complicated. However, Fig. 1.13 gives a practical illustration of the propagation of magnetostatic waves in real thin-layer crystals.

In addition, in the same way we can imagine the physics of propagation of backward magnetostatic waves. To examine them it is necessary in the beginning to locate the end points of the $\vec{m}$ vectors in Fig. 1.13 in a different way. In this figure they are turned to the left with phase shift $\mathrm{d}\varphi_0 = \pi/4$ from one cell to another. If at the start we distribute them with the same phase shift but to the right, and then the rotation of each dipole begins with frequency ω_H to the right (orientation of the precession does not change), the phase wave will "run" back, to the left. It will be analogous to the backward magnetostatic wave [8]. As can be seen from Fig. 1.13 the character of wave propagation depends on the choice of the initial phase shifts $\mathrm{d}\varphi_0$ of the end points of vectors from one cell to another. Here the initial configuration of $\vec{m}$ vectors has been done artificially. In practice it is determined by the conditions of MSW excitation (orientation of magnetic field and the difference between the exciting frequency and the ferromagnetic resonant frequency).

Up to now we have examined "transverse" waves in a perpendicularly magnetized slab. Analogously we can examine a "longitudinal" spin wave when the magnetization vector and the external magnetic field are parallel to the propagating wave. The physical basis of the MSW propagation is the same as before. There is only one difference — in this case the phase chain will be represented by a space spiral, unlike the plane curve mentioned before, and this spiral is wound around the magnetic field direction and rotated as a solid body with frequency ω_H. Basically, both forward and backward waves can exist in this scheme. However, we will see later that in such a system only backward waves can be excited.

1.5 Basic equations and boundary conditions Magnetostatic approximation Walker's equation

The magnetostatic waves examined can be derived from the solution of Maxwell's equations for magnetically ordered matter:

$$\mathrm{rot}\vec{E} = -\frac{\partial \vec{B}}{\partial t}; \qquad \mathrm{div}\vec{B} = 0;$$

$$\mathrm{rot}\vec{H} = \vec{J}_V + \frac{\partial \vec{D}}{\partial t}; \qquad \mathrm{div}\vec{D} = \rho_V. \tag{1.63}$$

It is necessary to complete these equations with the material equations by means of which the $\vec{D}$ and $\vec{J}_V$ vectors and $\vec{B}$ vector are connected with the $\vec{E}$ and $\vec{H}$ vectors, respectively. These equations can be written as follows:

$$\vec{D} = \varepsilon_0 \overline{\overline{\varepsilon}}.\vec{E}; \quad \vec{B} = \mu_0 \overline{\overline{\mu}}.\vec{H}; \quad \vec{J}_V = \sigma \vec{E}. \tag{1.64}$$

In (1.63)–(1.64) $\vec{E}$ and $\vec{H}$ are the vectors of the electric and magnetic field intensities, respectively and $\vec{D}, \vec{B}$ are the vectors of the electric and magnetic field inductions, respectively. $\vec{J}_V$ is the current density, ϱ_V the volume charge density and $\overline{\overline{\varepsilon}}$ and $\overline{\overline{\mu}}$ are the tensors of permittivity and permeability, respectively. Hereinafter we will only deal with magnetic dielectrics (ferrites) in which free charge carriers do not exist and the current density is negligible, i.e. $\vec{J}_V = 0$, $\varrho_V = 0$. In addition we will assume that this matter is electrically isotropic, i.e. the tensor $\overline{\overline{\varepsilon}}$ is reduced to the scalar ε. If we divide the quantities in (1.63) into constant and time-dependent components

$$\vec{H} = \vec{H}_0 + \vec{h}(\vec{r}, t); \quad \vec{E} = \vec{E}_0 + \vec{e}(\vec{r}, t);$$
$$\vec{B} = \vec{B}_0 + \vec{b}(\vec{r}, t); \quad \vec{D} = \vec{D}_0 + \vec{d}(\vec{r}, t)$$

then we obtain

$$\begin{aligned}
&\mathrm{rot}\vec{e} = -\frac{\partial \vec{b}}{\partial t} \quad (1); &&\mathrm{div}\vec{b} = 0 \quad (3);\\
&\mathrm{rot}\vec{h} = \frac{\partial \vec{d}}{\partial t} \quad (2); &&\mathrm{div}\vec{d} = 0 \quad (4);
\end{aligned} \tag{1.65}$$

$$\vec{d} = \varepsilon_0 \varepsilon \vec{e}, \qquad \vec{b} = \mu_0 \overline{\overline{\mu}}.\vec{H}, \tag{1.66}$$

where the $\overline{\overline{\mu}}$ tensor is given by a formula similar to (1.42) or (1.46). Now, we will simplify the system (1.65). We will take advantage of the fact that hereinafter we will be interested only in slow electromagnetic waves in ferromagnetic media (i.e. $k \gg k_0.\varepsilon$). The following condition can be written for such waves:

$$\frac{2\pi}{\lambda} \gg \frac{\omega}{c} \quad or \quad \omega \ll \frac{2\pi c}{\lambda}, \tag{1.67}$$

where λ is the wavelength of the slow electromagnetic wave in the medium. The form (1.67) is valid when ω is small (quasistationary approximation) or when the wavelength of the excitation in medium λ is much smaller than the wavelength in vacuum, i.e. $\lambda \ll (2\pi c/\omega)$ (quasistatic approximation). If we take this into account, then the system (1.65) can be split into two independent systems for electric and magnetic fields:

$$\begin{aligned}
&\mathrm{rot}\vec{e} = 0, \quad (1) &&\mathrm{div}(\mu_0 \overline{\overline{\mu}}\vec{h}) = 0, \quad (2)\\
&\mathrm{rot}\vec{h} = 0, \quad (3) &&\mathrm{div}(\varepsilon_0 \varepsilon \vec{e}) = 0, \quad (4)
\end{aligned} \tag{1.68}$$

The system of equations (1.68) is analogous to the equations of magnetostatics, the only difference being that the permeability in (1.68.(2)) is frequency-dependent. For this reason slow electromagnetic waves in magnetically ordered matter are called magnetostatic waves. But now it is necessary to make a crucial comment. In [1] we can find the exact solution of the complete system of equations (1.65) for special cases and the formulae for the components of high-frequency electric and magnetic fields are found. If the condition (1.67) is applied to those solutions, i.e. quasistatic approximation, we will see that the magnetic fields obtained fulfil the magnetostatic equations ($\mathrm{rot}\vec{h} = 0$) but the electric fields do not fulfil the equation

$\mathrm{rot}\vec{e} = 0$, except in the limit in case when the high-frequency electric field itself is equal to zero. This requirement is not realized since magnetostatic waves, although they are slow, are electromagnetic waves and they have to contain both magnetic and electric components of the high-frequency field. That is why equation (1.68.(1)) for magnetostatic waves has no physical basis and has to be replaced by the equation (1.65.(1)). Then we can write Maxwell's equations for magnetostatic approximation as follows:

$$\begin{aligned} &\mathrm{rot}\vec{e} = -\mu_0\overline{\overline{\mu}} \cdot \frac{\partial \vec{h}}{\partial t}, \quad (1) \qquad \mathrm{div}(\mu_0\overline{\overline{\mu}}.\vec{h}) = 0 \quad (3) \\ &\mathrm{rot}\vec{h} = 0, \quad (2) \qquad \mathrm{div}(\varepsilon_0\varepsilon\vec{e}) = 0 \quad (4) \end{aligned} \tag{1.69}$$

Now, let us assume again that all components of $\vec{h}$ and $\vec{e}$ vectors change in time as harmonic functions:

$$\vec{h}(\vec{r},t) = \vec{h}(\vec{r}).\mathrm{e}^{j\omega t}; \qquad \vec{e}(\vec{r},t) = \vec{e}(\vec{r}).\mathrm{e}^{j\omega t}$$

Substituting these in (1.69) we obtain the system of Maxwell's equations in a magnetostatic approximation:

$$\begin{aligned} &\mathrm{rot}\vec{e} = -j\omega\mu_0\overline{\overline{\mu}}.\vec{h}, \quad (1) \qquad \mathrm{div}(\mu_0\overline{\overline{\mu}}.\vec{h}) = 0 \quad (3) \\ &\mathrm{rot}\vec{h} = 0, \quad (2) \qquad \mathrm{div}(\varepsilon_0\varepsilon\vec{e}) = 0, \quad (4) \end{aligned} \tag{1.70}$$

In this system of equations (1.70) the quasistatic high-frequency magnetic field is described by equations (2) and (3) while the only source of magnetic field is the magnetization $\vec{M}$ of the sample. The relationship between the electric and magnetic fields is given by equation (1.70.(1)), i.e. the only source of a high-frequency electric field is a time-varying magnetic field. With respect to the fact that $\mathrm{rot}\vec{h} = 0$ we can use for the description of the magnetic field the high-frequency scalar potential ψ given by the formula

$$\vec{h} = \nabla\psi = \mathrm{grad}\psi = \frac{\partial\psi}{\partial x}\vec{i} + \frac{\partial\psi}{\partial y}\vec{j} + \frac{\partial\psi}{\partial z}\vec{k} = h_x\vec{i} + h_y\vec{j} + h_z\vec{k} \tag{1.71}$$

Substituting (1.71) in (1.70.(3)) we obtain

$$\mathrm{div}(\mu_0\overline{\overline{\mu}}.\nabla\psi) = 0.$$

For a gyrotropic medium magnetized to saturation in the z-axis direction the $\overline{\overline{\mu}}$ tensor is given by (1.42) where μ and μ_a are given by the terms (1.43) and (1.44). By scalar multiplication of the $\overline{\overline{\mu}}$ tensor and $\nabla\psi$ vector we obtain a new vector with the components [8]

$$(\overline{\overline{\mu}}.\nabla\psi)_x = \mu\frac{\partial\psi}{\partial x} + j\mu_a\frac{\partial\psi}{\partial y},$$

$$(\overline{\overline{\mu}}.\nabla\psi)_y = -j\mu_a\frac{\partial\psi}{\partial x} + \mu\frac{\partial\psi}{\partial y},$$

$$(\overline{\overline{\mu}}.\nabla\psi)_z = \frac{\partial\psi}{\partial z}.$$

The divergence of this vector will be

$$\begin{aligned}\mathrm{div}(\overline{\overline{\mu}}.\nabla\psi) &= \frac{\partial}{\partial x}(\overline{\overline{\mu}}.\nabla\psi)_x + \frac{\partial}{\partial y}(\overline{\overline{\mu}}.\nabla\psi)_y + \frac{\partial}{\partial z}(\overline{\overline{\mu}}.\nabla\psi)_z \\ &= \mu\left(\frac{\partial^2\psi}{\partial x^2} + \frac{\partial^2\psi}{\partial y^2}\right) + \frac{\partial^2\psi}{\partial z^2}\end{aligned}$$

and if we substitute this term in (1.70.(3)) we obtain

$$\mu\left(\frac{\partial^2\psi}{\partial x^2} + \frac{\partial^2\psi}{\partial y^2}\right) + \frac{\partial^2\psi}{\partial z^2} = 0 \tag{1.72}$$

In the literature, equation (1.72) is called Walker's equation. It describes the distribution of the high-frequency magnetic potential (and also high-frequency magnetic field) in magnetized matter. Note that in (1.72) only the diagonal element of the $\overline{\overline{\mu}}$ tensor is present.

For free space $\mu = 1$ and (1.72) becomes the common Laplace equation

$$\frac{\partial^2\psi}{\partial x^2} + \frac{\partial^2\psi}{\partial y^2} + \frac{\partial^2\psi}{\partial z^2} = 0 \tag{1.73}$$

If the wave propagates in the plane (x,z), then the following function fulfils equations (1.72), (1.73)

$$\psi = \left(A.\mathrm{e}^{-k_y y} + B.\mathrm{e}^{k_y y}\right).\mathrm{e}^{j(k_x x + k_z z)}. \tag{1.74}$$

where k_x, k_y, k_z are the components of the $\vec{k}$ wave vector. The value of the transverse wave number k_y is real for surface waves and imaginary for volume waves.

To find the constants in (1.74) it is necessary to use the boundary conditions. If free currents are absent the boundary conditions for the magnetic field are represented by the continuity of the transverse and normal components of the magnetic field at the boundary. If we denote boundary media 1 and 2 and $\vec{n}$ is the external normal to the boundary, in our case oriented in the y-axis direction, then the boundary conditions will be

$$\begin{aligned} (\vec{n} \times \vec{h}_1) - (\vec{n} \times \vec{h}_2) &= \vec{0} \\ (\vec{n}.\vec{b}_1) - (\vec{n}.\vec{b}_2) &= 0 \end{aligned} \tag{1.75}$$

It can be easily proved that $(\vec{n} \times \vec{h}) = \vec{h}_x + \vec{h}_z$ is the tangential component of the magnetic field $\vec{h}$, and $(\vec{n}.\vec{b}) = n_y b_y = b_y$ is the perpendicular component of magnetic induction $\vec{b}$.

Chapter 2

ELECTRODYNAMICS OF MAGNETOSTATIC WAVES

2.1 The method of setting up the dispersion equations for MSW in multilayer structures

The widely used method of setting up the dispersion equations, i.e. the equations relating the wave number and frequency, for magnetostatic waves in space-bounded stratified structures is based on "sewing together" of the tangential high-frequency components of magnetic field intensity h_x and normal components of magnetic induction b_y on the boundaries. The dispersion equation can be obtained by setting up a linear equation system with regard to coefficients in the formulae for high-frequency field components, and then the determinant of this system of equations is put to zero. If the number of layers increases the rank of the determinant also grows and the resulting formulae get more complicated and less transparent. There is, however, a method which simplifies the solution of this problem. It is better, instead of "sewing together" the high-frequency fields on the boundaries of media, to "sew together" their ratios. In the case investigated it is useful to "sew together" the ratio of the normal component of magnetic induction and the tangential component of magnetic field intensity [9]. This idea is not a new one and was used by one of the authors in [10] to obtain the dispersion equations for electron waves propagating in surface electron flows. The only difference is that in that case the ratios of electric and not magnetic fields were "sewn together". In an analogy with [9], in this case "surface" magnetic permeability will be obtained by the following equation:

$$\mu_s = -j\frac{b_y}{h_x} \tag{2.1}$$

where μ_s does not have a physical meaning but since variables b_y and h_x are continuous on the interface of media, expression (2.1) is finite at the interface and changes continuously from the area under the interface to the area over the interface.

As an example, Fig. 2.1 depicts the most common ferrite structure in practical applications (thin YIG film — 1) on a dielectric substrate (thin Gallium Gadolinium Garnet (GGG) layer — 2) loaded through dielectric layer 3 by metal 4. The notation in the system is obvious from the figure. At the interface $y = 0$ the surface permeability from the side of YIG is $\mu_s(y = 0)$ and from the side of the substrate 2 is μ_{s0} (it is supposed to be known). On the interface $y = t$ the surface permeability from the side of YIG is μ_s $(y = t)$ and from the side of dielectric gap 3 is μ_{s1} (this is to compute taking into account the metallic plane 4). To obtain a dispersion equation these permeabilities have to be "sewn together" in pairs by substituting gradually one into the other.

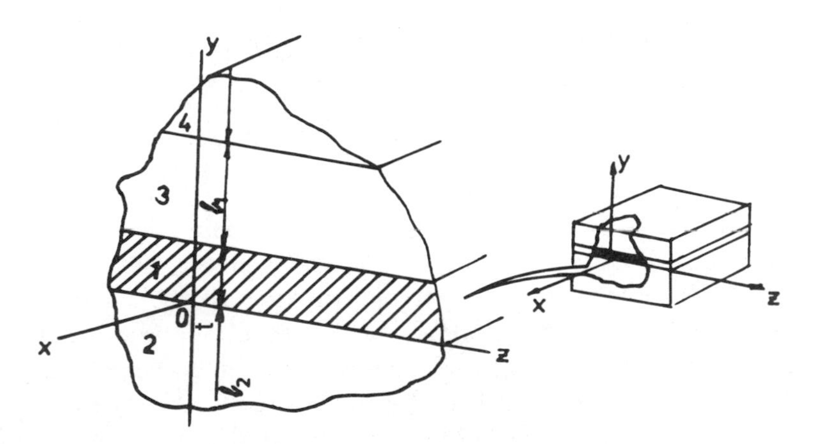

Fig. 2.1. The most frequently investigated structure. 1 — YIG layer, 2 — GGG substrate, 3 — dielectric layer, 4 — metal.

A few notes should be made in connection with the permeability tensor $\overline{\overline{\mu}}$. A magnetostatic wave will be supposed to propagate always in the x-axis direction. The magnetic field can be oriented in the direction of the x, y, or z axes. Then if $\vec{H}_0$ is oriented in the z-axis direction, surface MSW can propagate in ferrite in the ±x-axis direction. In this case, for the high-frequency components of magnetic induction, in accordance with (1.40) and (1.42):

$$\begin{aligned} b_x &= \mu_0 \mu h_x + j\mu_0 \mu_a h_y, \\ b_y &= -j\mu_0 \mu_a h_x + \mu_0 \mu h_y, \\ b_z &= \mu_0 h_z, \end{aligned} \tag{2.2}$$

where μ and μ_a are given by (1.43) and (1.44). If H_0 is oriented in the y-axis direction (perpendicularly to the plate), then forward volume waves can propagate in ferrite in the x-axis direction. It can be shown that in this case the high-frequency components of magnetic induction are

$$\begin{aligned} b_x &= \mu_0 \mu h_x + j\mu_0 \mu_a h_z \\ b_y &= \mu_0 h_y \\ b_z &= -j\mu_0 \mu_a h_x + \mu_0 \mu h_z \end{aligned} \tag{2.3}$$

If $\vec{H}_0$ is oriented in the x-axis direction, backward volume waves can exist in ferrite propagating in the x-axis direction. In this case

$$\begin{aligned} b_x &= \mu_0 h_x \\ b_y &= -j\mu_0 \mu_a h_z + \mu_0 \mu h_y \\ b_z &= \mu_0 \mu h_z + j\mu_0 \mu_a h_y \end{aligned} \tag{2.4}$$

Formulae (2.2), (2.3) and (2.4) transform to each other by a cyclic substitution. They can be written in a more compact form:

$$(b_x; b_y; b_z) = \mu_0 \cdot \begin{pmatrix} \mu_{11} & \mu_{12} & \mu_{13} \\ \mu_{21} & \mu_{22} & \mu_{23} \\ \mu_{31} & \mu_{32} & \mu_{33} \end{pmatrix} \cdot \begin{pmatrix} h_x \\ h_y \\ h_z \end{pmatrix} \tag{2.5}$$

Then, for surface waves in ferrite ($H_0 \| z$):

$$\begin{aligned} &\mu_{11} = \mu_{22} = \mu; \ \mu_{12} = j\mu_a; \ \mu_{21} = -j\mu_a; \\ &\mu_{33} = 1; \ \mu_{31} = \mu_{32} = \mu_{23} = \mu_{13} = 0. \end{aligned} \tag{2.6}$$

For forward volume waves ($H_0 \| y$):

$$\begin{aligned} &\mu_{11} = \mu_{33} = \mu; \ \mu_{13} = j\mu_a; \ \mu_{31} = -j\mu_a; \\ &\mu_{22} = 1; \ \mu_{12} = \mu_{21} = \mu_{23} = \mu_{32} = 0; \end{aligned} \tag{2.7}$$

And, at last, for backward volume MSW waves $(H_0 \| x)$:

$$\mu_{11} = 1; \; \mu_{22} = \mu_{33} = \mu; \; \mu_{23} = -j\mu_a;$$
$$\mu_{32} = j\mu_a; \; \mu_{12} = \mu_{13} = \mu_{21} = \mu_{31} = 0. \tag{2.8}$$

Coefficients (2.5) – (2.8) have a much simpler form if it is assumed that the ferrite structure is infinite in the z-axis direction. Then (2.5) – (2.8) change into

$$(b_x; b_y) = \mu_0 \begin{pmatrix} \mu_{11} & j\mu_{12} \\ -j\mu_{12} & \mu_{22} \end{pmatrix} \cdot \begin{pmatrix} h_x \\ h_y \end{pmatrix} \tag{2.9}$$

or

$$\begin{aligned} b_x &= \mu_0\mu_{11}h_x + j\mu_0\mu_{12}h_y \\ b_y &= -j\mu_0\mu_{12}h_x + \mu_0\mu_{22}h_y \end{aligned} \tag{2.10}$$

In this case formulae (2.6)–(2.8) will also be simplified and for surface waves:

$$\mu_{11} = \mu_{22} = \mu; \quad \mu_{12} = \mu_a; \tag{2.11}$$

for forward volume waves:

$$\mu_{11} = \mu; \quad \mu_{22} = 1; \quad \mu_{12} = 0; \tag{2.12}$$

and for backward volume waves:

$$\mu_{11} = 1; \; \mu_{22} = \mu; \; \mu_{12} = 0; \tag{2.13}$$

It is noteworthy that for the dielectric layer

$$\mu_{11} = \mu_{22} = 1; \; \mu_{12} = 0 \tag{2.14}$$

Walker's equation (1.72) obtained at the end of the previous chapter relates the parameters of the medium and magnetostatic potential for the case that the magnetic field is oriented in the z-axis direction. Analogous equations were obtained in [1] for arbitrary orientation of magnetic field $\vec{H}_0$ taking into account that the system is infinite in the z-axis direction. This equation is

$$\mu_{11}\frac{\partial^2\psi}{\partial x^2} + \mu_{22}\frac{\partial^2\psi}{\partial y^2} = 0$$

and its solution is

$$\psi = (A\mathrm{e}^{-\beta ky} + B\mathrm{e}^{\beta ky}) \cdot e^{jksx} \tag{2.15}$$

where $k > 0$ is wave number, $\beta = \sqrt{(\mu_{11}/\mu_{22})}$, $s = \pm 1$.

If formulae (2.15), (1.71) and (2.10) are used for the surface permeability μ_s defined by (2.1), then

$$\mu_s(y) = -s\frac{uA\mathrm{e}^{-\beta ky} + vB\mathrm{e}^{\beta ky}}{A\mathrm{e}^{-\beta ky} + B\mathrm{e}^{\beta ky}} \tag{2.16}$$

where $u = s\mu_{12} - \beta\mu_{22}$, $v = s\mu_{12} + \beta\mu_{22}$. Using the formula (2.16) values of μ_s on the ferrite layer interface from the ferrite side can be written in the following form

For $y = 0$ $\qquad \mu_s(0) = -s\dfrac{uA + vB}{A + B}$

For $y = t$ $\qquad \mu_s(t) = -s\dfrac{uA\mathrm{e}^{-\beta kt} + vB\mathrm{e}^{\beta kt}}{A\mathrm{e}^{-\beta kt} + B\mathrm{e}^{\beta kt}}$ $\qquad$ (2.17)

Surface permeabilities from the side of the layers applied are

For $y = 0$ $\qquad \mu_s = \mu_{s0}$

For $y = t$ $\qquad \mu_s = \mu_{s1}$ $\qquad$ (2.18)

In agreement with the boundary conditions (1.75) and with the formula (2.1) permeabilities (2.17) and (2.18) have to change continuously on the interface. Comparison of (2.17) and (2.18) and the elimination of coefficients A and B result in

$$\mu_{s0} = s\frac{\beta s\mu_{s1}\mu_{22} + (\mu_{11}\mu_{22} - \mu_{s1}\mu_{12} - \mu_{12}^2) \cdot \tanh(\beta kt)}{\beta\mu_{22} + s(\mu_{12} + \mu_{s1}) \cdot \tanh(\beta kt)} \tag{2.19}$$

μ_{s0} and μ_{s1} are unknown in (2.19). If they are found or given, then (2.19) is a general dispersion equation for magnetostatic waves propagating in a ferrite structure in the x-axis direction. To find μ_{s1} and μ_{s0} the formulae (2.16) will be used again. In this case, however, they will be applied to a dielectric. Taking into account (2.14) we obtain

$$\mu_s = s\frac{A\mathrm{e}^{-ky} - B\mathrm{e}^{ky}}{A\mathrm{e}^{-ky} + B\mathrm{e}^{ky}} \tag{2.20}$$

Equation (2.20) will be applied to the dielectric layer 2 (Fig. 2.1). If (2.20) is used for $y = t$ and $y = t + l_1$ and the coefficients A and B are eliminated from the given system, the relationship between μ_{s1} and μ_{s2} is given by

$$\mu_{s1} = s\frac{s\mu_{22} + \tanh(kl_1)}{1 + s\mu_{s2}\tanh(kl_1)} \tag{2.21}$$

An analogous formula can be written also for the dielectric layer under the ferrite layer where $y < 0$. That formula would relate the permeability μ_{s0} with the permeability on the other boundary of the lower layer. How do we calculate μ_{s2} and μ_{s0}? If the dielectric is loaded (for $y = t + l_1$) by an infinite ideally conductive metal then on its surface $b_y = 0$ and the surface permeability is $\mu_s = 0$

If the ferrite structure is loaded by a semi-infinite ideal dielectric (for $y < 0$ and $y > t$), then for $\mu_s(\infty)$ and $\mu_s(-\infty)$ we find

$$\mu_s(\infty) = 1/s, \qquad \mu_s(-\infty) = -1/s \tag{2.22}$$

In this case the permeability on the ferrite interface ($y = 0$, $y = t$) from the side of the load from (2.21) at the presence of a metal will be

$$\mu_{s1}(y = t) = s\tanh(kl_1), \qquad \mu_{s0}(y = 0) = -s\tanh(kl_1) \tag{2.23}$$

and for free space

$$\mu_{s1}(y = t) = s, \qquad \mu_{s0}(y = 0) = -s \tag{2.24}$$

Formulae (2.19)–(2.24) enable the simple determination of dispersion formulae for any multilayer structure with arbitrarily distributed loads, including ferrite, dielectric, and metallic layers.

In the monograph [1] general dispersion equations for structures containing two arbitrary ferrite layers with widths t_1, t_2, respectively, coupled through a dielectric gap d and loaded by metallic layers through gaps l_1 and l_2 are given. The resulting equation is correct for any orientation of magnetization; only values of μ_{11}, μ_{22}, and μ_{12} depend on this orientation (see also formulae (2.11)–(2.13)). Then the dispersion equation for the given structure is [1]

$$F_{+}(M_{01}; t_1; l_1) \times F_{(}M_{02}; t_2; l_2) = \mathrm{e}^{-2kd} \tag{2.25}$$

where the functions $F_{\pm}(M_{0i};t_i;l_i)$ are given by

$$\begin{aligned}F_{\pm}(M_{0i};t_i;l_i) = {}& [{}^{(i)}\beta^{(i)}\mu_{22}(\tanh(kl_i)+1)\tanh^{-1}({}^{(i)}\beta kt_i) \\ & + {}^{(i)}\mu_{11}{}^{(i)}\mu_{22} - {}^{(i)}\mu_{12}^2 \pm s^{(i)}\mu_{12}(\tanh(kl_i)-1)+\tanh(kl_i)] \\ & \times[{}^{(i)}\beta^{(i)}\mu_{22}(\tanh(kl_i)-1)\tanh^{-1}({}^{(i)}\beta kl_i) - {}^{(i)}\mu_{12}^2 \pm s^{(i)}\mu_{12} \\ & (\tanh(kl_i)+1) + {}^{(i)}\mu_{11}{}^{(i)}\mu_{22} - \tanh(kl_i)]^* \quad i=1,2\end{aligned} \tag{2.26}$$

From the formulae (2.25) and (2.26) follow all the special cases of surface and volume magnetostatic wave propagation which have been investigated in [1, 8] up to now.

2.2 Summary of basic dispersion formulae for magnetostatic waves in layers

Using formulae (2.25) and (2.26) and considering (2.11)–(2.13) for different special cases, the following equations will be obtained:

Surface waves in coupled structures loaded by metal (Fig. 2.2a)

In this case $H = H_z$, ${}^{(i)}\mu_{11} = {}^{(i)}\mu_{22} = \mu_{1,2}$, ${}^{(i)}\beta = 1$; ${}^{(i)}\mu_{12} = \mu_{21} = \mu_{a1,2}$ and from (2.25)–(2.26) there follows

$$\begin{aligned}& \frac{\mu_1(1+\tanh(kl_1)\tanh^{-1}((kt_1)+\mu_1^2-\mu_{a1}^2+s\mu_{a1}(1-\tanh(kl_1))+\tanh(kl_1)}{\mu_1^2-\mu_{a1}^2+s\mu_{a1}(1+\tanh(kl_1))-\mu_1(1-\tanh(kl_1))\tanh^{-1}(kt_1)-\tanh(kl_1)} \\ & \times\frac{\mu_2(1+\tanh(kl_2))\tanh^{-1}(kt_2)+\mu_2^2-\mu_{a2}^2-s\mu_{a2}(1-\tanh(kl_2))+\tanh(kl_2)}{\mu_2^2-\mu_{a2}^2+s\mu_{a2}(1+\tanh(kl_2))-\mu_2(1-\tanh(kl_2))\tanh^{-1}(kt_2)-\tanh(kl_2)} \\ & = \mathrm{e}^{-2kd}\end{aligned} \tag{2.27}$$

Coupled surface waves in a free structure (Fig. 2.2b)

In this case it is necessary to put $l_1 = l_2 \to \infty$, $\tanh(kl_1) = \tanh(kl_2) \to 1$ and then from (2.27) there follows

$$\begin{aligned}& \frac{2\mu_1\tanh^{-1}(kt_1)+\mu_1^2-\mu_{a1}^2+1}{\mu_1^2+2s\mu_1-\mu_{a1}^2-1}\times\frac{2\mu_2\tanh^{-1}(kt_2)+\mu_2^2-\mu_{a2}^2+1}{\mu_2^2+2s\mu_2-\mu_{a2}^2-1} \\ & = \mathrm{e}^{-2kd}\end{aligned} \tag{2.28}$$

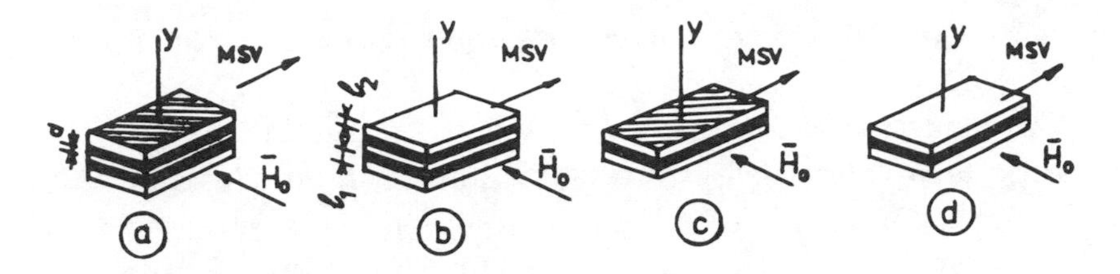

Fig. 2.2. Surface magnetostatic waves in various coupled structures, a — coupled layers loaded by a metallic interface, b — in a free structure, c — one ferrite layer loaded by a metallic interface, d — free ferrite layer.

Surface waves in an independent layer loaded by metallic planes (Fig. 2.2c)

In this case in (2.28) it is necessary to put $d \to 0$, $t_1 \to 0$ (or $t_2 \to 0$), $\mu_1 = \mu_2 = \mu$, $\mu_{a1} = \mu_{a2} = \mu_a$. Then

$$\tanh(kl_1).\tanh(kl_2) + \mu(\tanh(kl_1) + \tanh(kl_2))\tanh^{-1}(kt) + s\mu_a(\tanh(kl_1) - \tanh(kl_2)) + \mu^2 - \mu_a^2 = 0 \tag{2.29}$$

Surfacc waves in a free ferrite layer (Fig. 2.2d)

Now $l_1 = l_2 \to \infty$, $\tanh(kl_1) = \tanh(kl_2) \to 1$ and from (2.29) we have

$$1 + 2\mu \cdot \tanh^{-1}(kt) + \mu^2 - \mu_a^2 = 0 \tag{2.30}$$

Surface waves in a metal-coated layer

Let $l_1 = l_2 \to 0$, $\tanh(kl_1) = \tanh(kl_2) \to 0$ and then from (2.29) there follows

$$\mu^2 - \mu_a^2 = 0 \tag{2.31}$$

Formulae (2.27)–(2.31) represent all possible variations of dispersion equations for surface magnetostatic waves known from the literature.

Forward volume waves in coupled structures loaded by metal layers

The structure is represented in Fig. 2.2a, but in this case the magnetic field is oriented in the y-axis direction, i.e. $\vec{H} = \vec{H}_y$. According to (2.2), in this case ${}^{(i)}\mu_{11} = \mu_{1,2}; {}^{(i)}\mu_{22} = 1; {}^{(i)}\mu_{1,2} = 0$. It can be shown that in the case of a volume wave propagating in a structure $\mu_{1,2} < 0$ and to keep the propagation constant as a real variable in equations (2.25–2.26) it is necessary to do the following substitution:

$$ {}^{(i)}\beta = j\xi_{1,2} \quad \text{where } \xi_{1,2} = \sqrt{(-\mu_{12})} $$

After substitution in (2.26) the result is

$$ \frac{\xi_1(1+\tanh(kl_1)) + (\tanh(kl_1)+\mu_1)\cdot\tan(\xi_1 kt_1)}{\xi_1(1-\tanh(kl_1)) + (\tanh(kl_1)-\mu_1)\cdot\tan(\xi_1 kt_1)} \times $$
$$ \times \frac{\xi_2(1+\tanh(kl_2)) + (\tanh(kl_2)+\mu_2)\cdot\tan(\xi_2 kt_2)}{\xi_2(1-\tanh(kl_2)) + (\tanh(kl_2)-\mu_2)\cdot\tan(\xi_2 kt_2)} = \mathrm{e}^{-2kd} \tag{2.32} $$

Coupled forward volume waves in non-loaded structures

In this case $l_1 = l_2 \to \infty$, $\tanh(kl_1) = \tanh(kl_2) \to 1$ and from (2.32) it follows that

$$ \frac{2\xi_1 + (1+\mu_1)\cdot\tan(\xi_1 kt_1)}{(1-\mu_1)\cdot\tan(\xi_1 kt_1)} \times \frac{2\xi_2 + (1+\mu_2)\cdot\tan(\xi_2 kt_2)}{(1-\mu_2)\tan(\xi_2 kt_2)} = \mathrm{e}^{-2kd} \tag{2.33} $$

Coupled forward volume waves in a ferrite layer loaded by conductive planes

For this case it is necessary to put in (2.33) $d \to 0$, $t_1 \to 0$ (or $t_2 \to 0$) $\mu_1 = \mu_2 = \mu$, giving

$$ \mu + (\tanh(kl_1)+\tanh(kl_2))\xi\cdot\tan^{-1}(\xi kt) + \tanh(kl_1)\cdot\tanh(kl_2) = 0 \tag{2.34} $$

Forward volume waves in a free layer

In this case $l_1 = l_2 \to \infty$, $\tanh(kl_1) = \tanh(kl_2) \to 1$ and from (2.34) follows

$$\frac{2\xi}{\xi^2 - 1} = \tan(\xi kt) \tag{2.35}$$

Forward volume waves in a metal-coated ferrite slab

In (2.34) we put $l_1 = l_2 \to 0$, $\tanh(kl_1) = \tanh(kl_2) \to 0$, which results in

$$\mu = -\left(\frac{m\pi}{kt}\right)^2 \tag{2.36}$$

Backward volume waves in coupled structures loaded by metal layers

In this case $\vec{H} = \vec{H}_x$, i.e. the field is oriented along the x-axis. Then ${}^{(i)}\mu_{11} = 1$; ${}^{(i)}\mu_{22} = \mu_{1,2}$; ${}^{(i)}\mu_{1,2} = 0$. Since for volume waves $\mu_{1,2} < 0$ holds, in order that the propagation constant be a real number it is necessary to put

$${}^{(i)}\beta = \frac{1}{j\xi_{1,2}} \qquad \text{where} \quad \xi_{1,2} = \sqrt{(-\mu_{1,2})} \tag{2.37}$$

The substitution of (2.37) into (2.26) results in

$$\frac{(1 + \tanh(kl_1))\xi_1 \tan^{-1}(kt_1/\xi_1) + \mu_1 + \tanh(kl_1)}{\mu_1 - (1 - \tanh(kl_1))\xi_1 \tan^{-1}(kt_1/\xi_1) - \tanh(kl_1)} \times \frac{(1 + \tanh(kl_2))\xi_2 \tan^{-1}(kt_2/\xi_2) + \mu_2 + \tanh(kl_2)}{\mu_2 - (1 - \tanh(kl_2))\xi_2 \tan^{-1}(kt_2/\xi_2) - \tanh(kl_2)} = \mathrm{e}^{-2kd} \tag{2.38}$$

Coupled backward volume waves in free structures

In this case $l_1 = l_2 \to \infty$ and then $\tanh(kl_1) = \tanh(kl_2) \to 0$ and from (2.38) follows

$$\frac{2\xi_1 + (1 + \mu_1)\tan(kt_1/\xi_1)}{(1 - \mu_1)\tan(kt_1/\xi_1)} \times \frac{2\xi_2 + (1 + \mu_2)\tan(kt_2/\xi_2)}{(1 - \mu_2)\tan(kt_2/\xi_2)} = \mathrm{e}^{-2kd} \tag{2.39}$$

Backward volume waves in a ferrite layer loaded by conductive planes

In (2.38) we put $d \to 0$, $t_1 \to 0$ (or $t_2 \to 0$), $\mu_1 = \mu_2 = \mu$ and obtain

$$\mu - (\tanh(kl_1) + \tanh(kl_2))\xi \tan^{-1}(kt/\xi) + \tanh(kl_1)\tanh(kl_2) = 0; \tag{2.40}$$

Backward volume waves in a free layer

$l_1 = l_2 \to \infty$, $\tanh(kl_1) = \tanh(kl_2) \to 1$ and from (2.40) one gets

$$\frac{2\xi}{\xi^2 - 1} = \tan(kt/\xi) \tag{2.41a}$$

Backward volume waves in a metal-coated layer

The substitution $l_1 = l_2 \to 0$, $\tanh(kl_1) = \tanh(kl_2) \to 0$ into (2.40) gives

$$\frac{1}{\mu} = -(\frac{m\pi}{kt})^2 \tag{2.41b}$$

Now the properties of the volume and surface magnetostatic waves propagating in ferrite layers will be investigated separately. First, forward volume magnetostatic waves which propagate in perpendicularly to the layer of magnetized ferrites will be investigated.

2.3 Forward volume magnetostatic waves

Dispersion properties, i.e. the relation between the propagation constant (wave number) and frequency ω of the forward volume magnetostatic waves are described by the dispersion formulae (2.32)–(2.36). First, the frequency band in which forward volume magnetostatic waves can propagate will be determined. Let us consider the formula (2.36) for a metal-coated layer (as the simplest case). Since

μ is given by (1.43), after substitution and modification from (2.36) there follows

$$\omega^2 = \omega_H \cdot \left[\omega_H + \frac{\omega_M}{1 + \left(\frac{m\pi}{kt}\right)^2} \right] \tag{2.42}$$

If $k \to 0$, then $\omega_1 \to \omega_H$; if $k \to \infty$, then $\omega_2 \to \sqrt{(\omega_H(\omega_H + \omega_M))}$. This means that this system represents a MSW band filter with the frequency band

$$\omega_H \leq \omega \leq \sqrt{(\omega_H(\omega_H + \omega_M))} \tag{2.43}$$

The analysis shows that pass band of volume MSW at $n \neq 0$ does not depend on external conditions (metal coating, free layer, metal layers) and is totally determined by the ferrite parameters and by the external magnetic field. As follows from (2.43), volume waves propagate at higher frequencies than the frequency of the ferromagnetic resonance. Moreover, the frequency of the ferromagnetic resonance is the bottom frequency of the volume magnetostatic wave pass band. The dispersion curves of volume MSW in a metal-coated layer for various values of n using (2.42) are depicted in Fig. 2.3. The variable

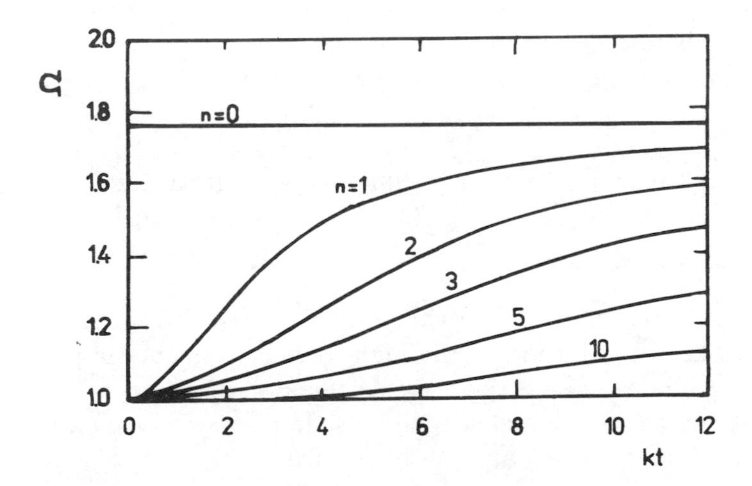

Fig. 2.3. Dispersion curves of volume MSW in a metal-coated layer $l_1 = l_2 = 0$, $\omega_M/\omega_H = 2$. Parameter $n = m$.

n here is the number of the order of excited volume magnetostatic waves, i.e. it is equal to the number of half-waves along the layer thickness (along the y-axis). For $n = 0$ the dispersion curve degenerates into a line and the wave stops propagating. There is uniform precession in the whole layer. From Fig. 2.3 it follows that with increasing frequency k grows, i.e. the dispersion is normal and propagating waves are forward. The group velocity of forward volume waves $v_g = \mathrm{d}\omega/\mathrm{d}k$ is oriented in the same direction as the phase velocity. So an infinite number of volume waves with different values of n can exist in a metal-coated layer, which depend on the conditions of excitation. It is noteworthy that the values of the variable μ in the pass band of forward volume magnetostatic waves are within the interval

$$-\infty \leq \mu \leq 0$$

i.e. μ is always negative (see Fig. 1.4). The same follows also from Walker's formula. Considering (2.12) and (2.15) for forward volume MSW, the formula $\mu k_x^2 + k_y^2 = 0$ can be obtained. Since k_x and k_y have to be real (propagating waves), μ can be negative only.

The question of why these waves are called volume magnetostatic waves may be raised. In the general case, solutions of Walker's equation have the following form:

$$\psi = (A\mathrm{e}^{-\beta ky} + B\mathrm{e}^{\beta ky}) \cdot \mathrm{e}^{jskx} \tag{2.15}$$

where $\beta = (\mu_{11}/\mu_{22})^{1/2}$, $s = +1$ (forward wave). For magnetization which is perpendicular to the layer plane the diagonal components of the high-frequency permeability are given by (2.2), i.e.

$$\beta = \sqrt{\mu}$$

and since in the frequency band of volume wave excitation $\mu < 0$, the variable β is imaginary (i.e. $\beta = j\xi$). Then (2.15) changes into

$$\psi = (A\mathrm{e}^{-j\xi ky} + B\mathrm{e}^{j\xi ky}) \cdot \mathrm{e}^{jkx} \tag{2.44}$$

The function in parentheses (often called also the membrane function) describes the distribution of the high-frequency magnetic potential across the layer (along the y-axis). This distribution is described in this case by trigonometric functions. In the case of a metal-coated layer at $y = 0$, $\psi = 0$ and as follows from (2.44) $A = -B$. Then the function ψ will be

$$\psi(x, y) = -2jA\sin(\xi ky)\mathrm{e}^{jkx} \tag{2.45}$$

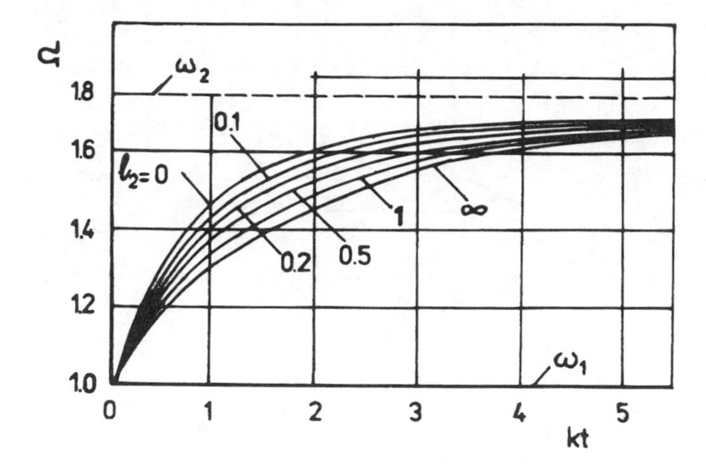

Fig. 2.4. Dispersion curves for forward volume MSW. $l_1 = \infty$, $n = 0$, $\Omega_M = 2$. Parameter l_2/t.

Then the transverse distribution of high-frequency magnetic potential (high-frequency magnetic field) is described by a harmonic function, i.e. it is a periodic function of the y co-ordinate. The high-frequency magnetic field is located in the whole volume of the layer and has nodes on the top and bottom surfaces ($y = 0$, $y = t$). The high-frequency field has a resonance character and across the layer a standing MSW is distributed. The MSW energy flux includes the whole layer thickness and for this reason the propagating wave is a volume wave. Up to now the simplest case of a metal-coated YIG layer has been investigated. The analysis shows that if the metal planes are taken apart from the YIG layer the MSW dispersion properties are changed very slightly. In Figs. 2.4 and Fig. 2.5 the dispersion characteristics of forward volume MSW for $n = 0$, 1 and various values of l_1 and l_2 are depicted in accordance with the formulae (2.34), (2.35). As can be seen, the presence of metallic surfaces does not change the character of dispersion, the dispersion is normal and the waves are forward. Even a big change of boundary conditions (from a metal-coated to a free plate) changes the dispersion of a forward volume MSW only very slightly. It is also noteworthy that the boundary frequencies ω_1 and ω_2 of the forward volume wave excitation band do not change (with the exception of the basic zone in a metal-coated plate).

From this follows an important practical consequence. The excitation band of forward volume MSW in a normally magnetized layer

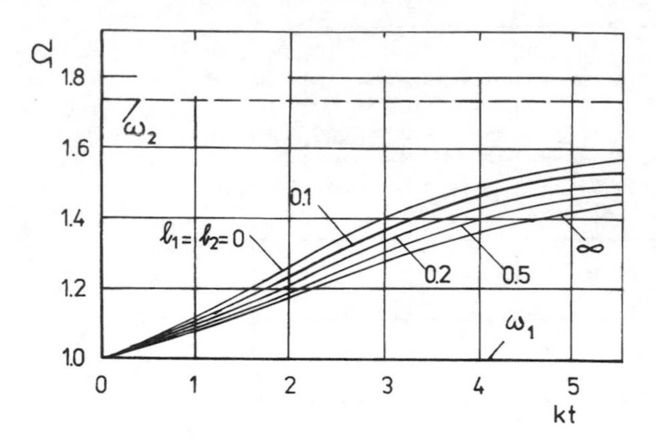

Fig. 2.5. Dispersion curves for forward volume MSW. $n = 1$, $\Omega_M = 2$. Parameter $l_1/t = l_2/t$.

does not depend on the layer load (sputtered metal, metallic layers separated by a dielectric gap, free layer) and the dispersion changes slightly. The excitation band and the dispersion are determined only by internal parameters of ferromagnetic materials $\omega_M = \gamma\mu_0 M_0$ and by an external magnetic field $\omega_M = \gamma\mu_0 H_0$. An important factor here is the ratio between M_0 and H_0 or ω_M and ω_H. Also with the rise of ω_M/ω_H the relative pass band for forward volume waves increases. For a given material the rise of ω_M/ω_H means a decrease of operating frequency (external magnetic field and frequency of FMR). On the contrary, with a rise in the operating frequency (for the same operating material) the relative excitation band of volume MSW narrows. For the most common operating material in MSW technology — YIG — saturation magnetization M_0 is about $140\,\mathrm{kA\,m^{-1}}$. In this case, at low frequencies (up to 1–2 GHz) the pass bands are one octave or even more, and at high frequencies, $\simeq$ 30–40 GHz, they are of the order of a few tenths of one percent.

Using the simple formula (2.42) the group velocity of forward volume MSW in a normally magnetized metal-coated layer can be simply computed in an analytical way. In the simplest variant the absolute value of the group velocity can be computed in the following way:

$$v_{gx} = d\omega/dk \tag{2.46}$$

(a more exact definition will be given in Chapter 3). After diferentiation of (2.42) by k and elimination of k by means of (2.42) the following formula for the group velocity of forward volume MSW can be found:

$$v_{gx} = \left(\frac{\omega_H t}{n\pi}\right) \cdot \frac{\omega_H^2}{\omega.\omega_M} \cdot \left[\frac{\omega^2}{\omega_H^2} - 1\right]^{1/2} \cdot \left[\left(1 + \frac{\omega_M}{\omega_H}\right) - \frac{\omega^2}{\omega_H^2}\right]^{3/2} \quad (2.47)$$

From (2.47) it follows that the group velocity is equal to zero on the boundaries of the pass band for $\omega_1 = \omega_2$ $(k = 0)$ and at $\omega_2 = \sqrt{(\omega_H(\omega_H + \omega_M))}$. The dependence of v_g on frequency is represented in Fig. 2.6: the maximum value of v_g for forward volume

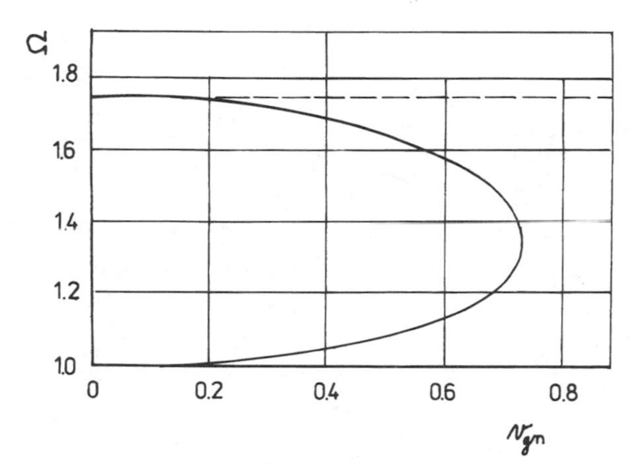

Fig. 2.6. The dependence of normalized group velocity of a forward volume wave on frequency Ω. $\Omega_M = 2.v_{gn} = (n\pi/\omega_H t)v_g$.

waves is around the middle of the pass band. The presence of conductive planes changes (increases) the steepness of dispersion, which can lead to a substantial increase of the group velocity of forward volume MSW in the middle of the pass band. It should be noted that the absolute values of dispersion, phase and group velocities depend strongly on the layer thickness. The thinner the YIG layer, the slower the magnetostatic wave, the weaker the dispersion, and the less the group velocity (if other conditions do not change). For the practically used YIG layer thickness of about 10–20 μm at $k = 10^2$–10^3, the delay is $c/v_{\text{gr}} \simeq 10^2$–$10^3$, i.e. it is big enough (in comparison with e.g. periodic delay systems used in high-frequency technology).

In practical applications forward volume MSW are used much less frequently than surface waves. One of the reasons is the multimode excitation of volume MSW, which may cause substantial worsening of the spectrum of volume MSW in a pass band. There are also other reasons that will be discussed later.

2.4 Backward volume magnetostatic waves

Backward volume magnetostatic waves propagate along a magnetic field in a longitudinal magnetized layer. The dispersion equations for various specific cases are described by formulae (2.37)–(2.41a,b). The last formula is the simplest one and therefore will be discussed in more detail. In the case of a metal-coated layer, from (2.41b) there follows (considering (1.43))

$$\omega^2 = \omega_H \left[\omega_H + \frac{\omega_M}{1 + \left(\dfrac{kt}{m\pi} \right)^2} \right] \tag{2.48}$$

Again, if it is supposed that $k = 0$, $k = \infty$, respectively, the pass bands of the backward volume waves can be found. These are identical with (2.43). However, the dispersion of these waves is different from the case of forward volume waves. In Fig. 2.7 the dispersion dependences of backward volume MSW for a metal-coated layer and various zones of excitation according to (2.48) are depicted. As can be seen, for arbitrary n the propagation constant k decreases with increasing frequency, i.e. the dispersion is anomalous. The group velocity of these waves is negative, i.e. oriented against the phase velocity (from Fig. 2.7 it follows that $\mathrm{d}\omega/\mathrm{d}k < 0$). Waves of this

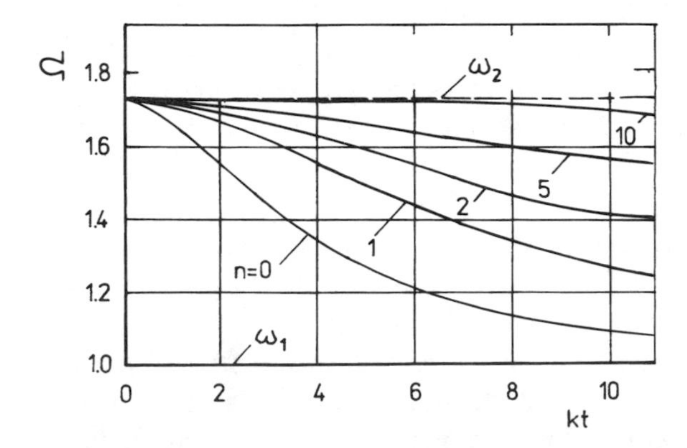

Fig. 2.7. Dispersion curves for backward volume waves. $\Omega_M = 2$, $l_1 = l_2 = 0$. Parameter $n = m - 1$, $\omega_1 = \omega_H$, $\omega_2 = \sqrt{\omega_H(\omega_H + \omega_M)}$.

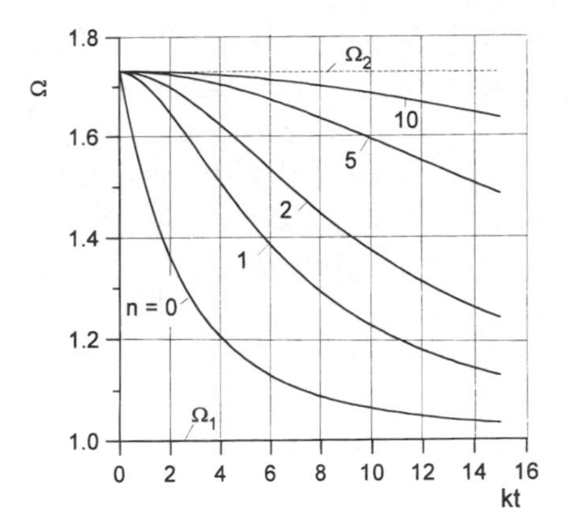

Fig. 2.8. Dispersion curves for backward volume waves. $\Omega_M = 2$, $l_1 = l_2 = \infty$. Parameter n.

kind are called backward.

As for forward volume waves, μ is negative in the pass band of backward waves. From (2.12) and (2.15) it follows that the relationship between k_x and k_y is in this case given by $\mu k_y^2 + k_x^2 = 0$. Since k_x and k_y are both real, μ is negative ($\mu < 0$). By the use of (2.15) it can be shown that for backward volume waves the membrane function $\psi(y)$ is harmonic, i.e. the waves are volume waves. MSW energy flux fills up the cross volume of the layer.

In Figs. 2.7 and 2.8 the dispersion dependences of backward volume MSW in metal-coated and free layers are depicted. As can be seen, the influence of metallic planes, as in the case of forward

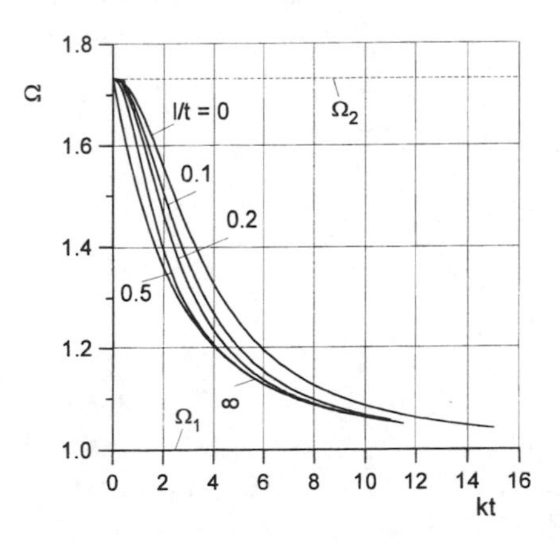

Fig. 2.9. Dispersion curves for backward volume waves. $\Omega_M = 2$, $n = 0$. Parameter $l_1/t = l_2/t = l/t$.

volume waves, is small. This can be seen well in Fig. 2.9, in which dispersion dependences $\omega(k)$ for the basic zone $n = 0$ and the case of symmetrically located metallic plates ($l_1 = l_2$) for various distances of metallic interfaces from the ferrite plate are depicted. As follows from Fig. 2.9, changing l from 0 to ∞, i.e. from the metal-coated to the free layer, the dispersion dependence changes only slightly. The same fact follows also from Fig. 2.10, in which the influence of asymmetrically located metallic plates on the dispersion dependences of backward volume MSW is investigated.

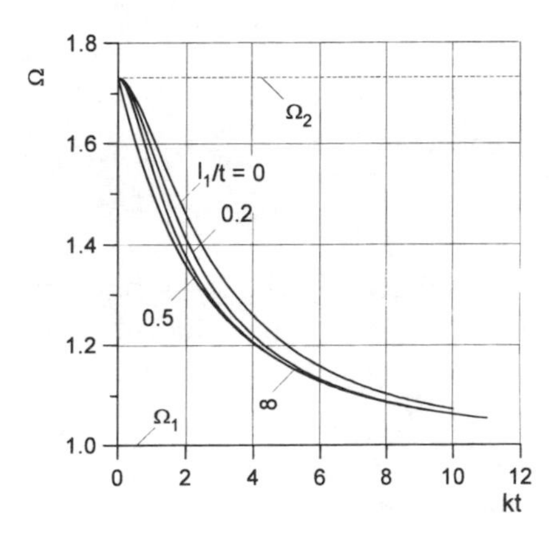

Fig. 2.10. Dispersion curves for backward volume waves. $\Omega_M = 2$, $n = 0$, $l_1 = \infty$. Parameter l_2/t.

Since in the simplest case of a metal-coated layer the dispersion dependence is relatively simple (2.48), the formula for group velocity will be simple as well. Using (2.48) for the amplitude of the group velocity v_g we have

$$v_g = -\left(\frac{\omega_H t}{n\pi}\right) \cdot \frac{\omega_H^2}{\omega_M \omega} \cdot \left[\frac{\omega^2}{\omega_H^2} - 1\right]^{3/2} \cdot \left[\left(1 + \frac{\omega_M}{\omega_H}\right) - \frac{\omega^2}{\omega_H^2}\right]^{1/2} \quad (2.49)$$

The value of k is always positive, i.e. it is oriented in the positive direction of the x-axis. From (2.49) it follows that $v_g < 0$, i.e. it is oriented in the negative direction of the x-axis. The wave is backward volume and it has anomalous (negative) dispersion. The formula (2.49) has the same structure as (2.47) for forward waves. For $\omega = \omega_H$ and $\omega = \sqrt{(\omega_H(\omega_H + \omega_M))}$, v_g limits to zero and the dependence $v_g(\omega)$ has an analogous shape to that depicted in Fig. 2.6. As for forward volume waves the absolute values of dispersion and group velocity depend on the layer thickness. The thinner the YIG layer, the smaller the dispersion and group velocity (if other conditions do not change). It has been mentioned above this is a general property of all types of MSW in ferrite layers.

To finish the investigation of volume magnetostatic waves it is necessary to remember once more their important practical property. Volume waves are influenced slightly by external metallic planes. This is connected with the fact that since the energy flow of volume waves is in fact concentrated inside the slab [1], the field does not "feel" the metal. This can have a positive as well as negative influence, depending on the function that MSW elements are supposed to perform. This fact will be analyzed in Chapter 8 in more detail.

2.5 Surface magnetostatic waves

Surface magnetostatic waves are the most common and well investigated class of magnetostatic waves. As follows from (2.27)–(2.31), these waves propagate in ferromagnetic materials magnetized in the layer plane perpendicularly to the direction of the magnetic field. Using our description it means that the waves propagate in the x-axis direction and the magnetic field at the excitation of surface magnetostatic forces is oriented in the z-axis direction. The dependence (2.30) for surface waves in a free slab will be investigated later with the aim of explaining the character of these waves. After substitution for μ and μ_a from (1.43) and (1.44), the following dispersion equation will be obtained:

$$\omega^2 = \omega_H(\omega_H + \omega_M) + \frac{\omega_M^2}{2(1 + \tanh^{-1}(kt))} \qquad (2.50)$$

This dispersion equation is a classical one and was published for the first time by Damon and Eshbach in 1961 [11]. For this reason the waves obtained from equation (2.50) are called Damon–Eshbach waves. From (2.50) it is easy to obtain frequency limits for propagation of these waves. The equation (2.50) after substitution for limit values of k results in

at $k \to 0$ $\quad \omega_2 = \sqrt{(\omega_H(\omega_H + \omega_M))}$ $\qquad$ lower limit at $k \to \infty$ $\quad \omega_3 = \omega_M + (1/2)\omega_M$ upper limit

i.e. the lower limit of the surface MSW pass band is identical with the upper limit of volume MSW. This limit is determined by the zero crossing of the diagonal component of the high-frequency permeability tensor μ (see Fig. 1.4). For frequencies $\omega < \omega_2$ if $\mu < 0$, only volume waves can propagate in the slab. For frequencies $\omega > \omega_2$ only surface waves are possible, for which $\mu > 0$.

In Fig. 2.11 the dispersion dependence of surface MSW in a free independent slab for $\omega_M/\omega_H = 0.15$ is represented. As can be seen the dispersion is in this case normal and surface waves in a free slab are forward waves, i.e. the orientations of group and phase velocities are the same. This can be proved using (2.50) and it can be seen also from Fig. 2.11. From (2.50) for the group velocity we have

$$v_g = \frac{\omega_H t}{4} \cdot \frac{(\omega_M/\omega_H)^2}{(\omega/\omega_H)} \cdot \left[1 - 4 \cdot \frac{(\omega/\omega_H)^2 - (1 + \omega_M/\omega_H)}{(\omega_M/\omega_H)^2}\right] \qquad (2.51)$$

In contrast to the volume MSW for which the group velocity is equal

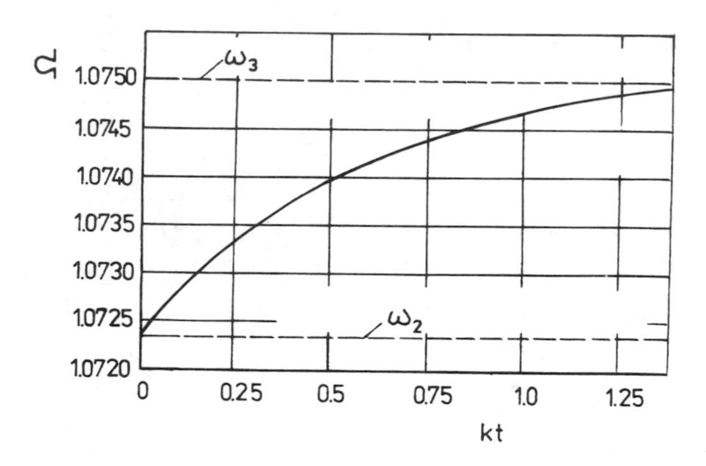

Fig. 2.11. Dispersion curve for surface MSW in a free slab. $\Omega_M = 0.15$, $l_1 = l_2 = \infty$, $\omega_3 = \omega_H + \omega_M/2$.

to zero both on the upper and the lower pass band limits ($k = 0, \infty$), the group velocity for surface MSW on the upper boundary ($k = \infty$) is equal to zero and on the lower boundary approaches the value

$$v_g = \omega_H t(\omega_M/\omega_H)^2/(4\sqrt{(1 + \omega_M/\omega_H)})$$

i.e. it remains finite. From (2.51) it follows that the smaller the slab thickness the less the group velocity; this is the same as in the case of volume waves. Now, the reason why these waves are called "surface" waves will be treated. For this reason the formula (2.15) for the distribution of high-frequency potential $\psi(x, y)$ will be investigated. For the surface waves $\beta = \sqrt{(\mu_{11}/\mu_{22})} = 1$ [(2.11)] and the transverse distribution of potential is described by exponential functions of the following type:

$$\psi(x, y) = (A \cdot \mathrm{e}^{-ky} + B \cdot \mathrm{e}^{ky}) \cdot \mathrm{e}^{\pm jkx}$$

i.e. a high-frequency potential has its maximum at one interface and exponentially decreases with distance from the interface, both into and out of the layer. Electromagnetic energy spreads near to one of the interfaces and the wave seems to be tied to this layer interface. The slower the wave, i.e. the bigger the propagation constant k, the more distinct is the "surface" character of the magnetostatic wave. The next question is on which surface the wave propagates. The fact is that it depends on the direction of the wave propagation. If $\vec{n}$ is the vector of an external normal to the slab, then the wave propagates on the surface [8], on which the vectors $\vec{H}_0$, $\vec{k}$ and $\vec{n}$ create a right-handed co-ordinate system. Therefore at excitation of surface MSW into the forward direction the wave propagates on one surface and at excitation in the backward direction on the other one. Changing the direction of the external magnetic field moves the MSW from one surface to the other. It is said that surface waves are nonreciprocal, i.e. in the forward and backward directions they propagate on different interfaces of the ferrite layer. The fact that for the surface waves the electromagnetic energy flux is concentrated close to one of the interfaces results in a considerable rise of high-frequency potential (high-frequency magnetic field) outside the layer, and therefore a wave of this kind has to "feel" well (in contrast to volume waves) the presence of metal on the interface. Experiments have shown that surface MSW are exceptionally sensitive to the presence of conductive planes and that by means of these conductive planes both the band width of surface MSW and their dispersion can be changed over a wide range. The dispersion can change from normal (positive) to anomalous (negative). The waves change from forward to backward, i.e. the direction of group velocity changes.

The possibility of regulating the dispersion and characteristics of surface MSW by means of a metallic interface can be illustrated by

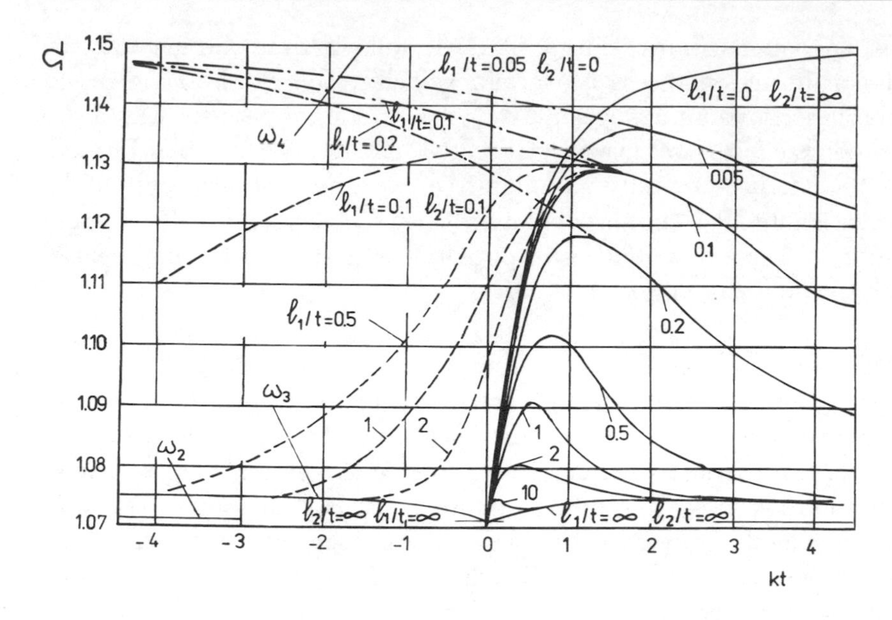

Fig. 2.12. The influence of metallic interface on the dispersion characteristics of surface MSW for various directions of propagation. l_1/t, l_2/t parameters. $\Omega_M = 0.15$, $\omega_4 = \omega_H + \omega_M$.

the generalized dependence in Fig. 2.12, which represents dispersion dependences $\omega(k)$ for various directions of MSW propagation (because of nonreciprocity the propagating waves in the direction $\pm k$ will have different characteristics) and for various distances l_1 and l_2 from the metallic interfaces. The general dispersion equation (2.29) was used for computation. It is worth treating Fig. 2.12 in more detail. Three groups of dispersion curves are depicted.

The full curves represent the dispersion of surface MSW for layers with one metallic interface. The dashed curves represent the case of two metallic interfaces (symmetrical and asymmetrical case) and, finally, the third group of curves, dashed-dot, correspond to the layer metal-coated on one side.

The first group of curves represent the most frequently investigated case of surface MSW in a slab with a one-sided metallic interface. From the point of view of practical realization it is the simplest case because thin a YIG layer is created by epitaxial growth on the GGG substrate 500–1000 μm thick, the ratio l_1/t is 50–100 and more, and $l_2/t \to \infty$. A control metallic plane is placed on the other operating side of the layer, the distance to which varies in a range from

0 to ∞. Let us start with the lowest full curve. It describes the dispersion of surface MSW in a free slab $l_1/t = l_2/t \rightarrow \infty$, i.e. the metallic interfaces are infinitely far away. This is the simplest case, investigated also by Damon and Eshbach (see also Fig. 2.11). For the value $\omega_M/\omega_H = 0.15$ (for YIG it corresponds to the frequency band 30–40 GHz) the pass band in a free slab is

$$\sqrt{(\omega_H(\omega_H + \omega_M))} \leq \omega \leq \omega_H + (\omega_M/2)$$

The wave dispersion is normal and the waves are forward. Let us start to bring the metallic plane from the operating side closer. A characteristic "beak" appears on dispersion curves at small values of kt, which increases when kt decreases. An extremum on the dispersion curve appears where the derivative $\mathrm{d}\omega/\mathrm{d}k$ changes its sign. Left of the extremum v_g is positive, the wave is forward and it has a normal dispersion. Right of the extremum, group velocity is negative, the wave will be backward and dispersion is anomalous. Bringing the metallic plane nearer, the pass band of the system broadens and the extremum moves to the right into the region of high values of kt. In a limit case for $l_1/t \rightarrow 0$, $l_2/t \rightarrow \infty$ the pass band is maximal and given by

$$\sqrt{(\omega_H(\omega_H + \omega_M))} \leq \omega \leq \omega_H + \omega_M$$

It is noteworthy also that for $kt \rightarrow \infty$ all dispersion curves coalesce with the Damon–Eshbach curve for a free layer. In the case that l_1 and l_2 swap the whole graph is displayed on the left half of the figure in a mirror image. It can be said then that the already one-sided presence of the metallic interface enables a change of the dispersion and pass band of surface MSW in an effective way. As will be shown in Chapter 8 this fact is frequently used in practical applications — in MSW transmission lines with the specified dispersion properties.

Now the second group of curves (dashed) will be investigated. These are constructed for one of the possible cases of a YIG layer loaded on two sides by metallic plates. This case was investigated by one of the authors in [8]. It should be noted that in the case of symmetrically placed metallic planes only backward waves exist in the system. The picture is then mirror symmetrical and the waves have anomalous dispersion; the group velocity is oriented in the opposite direction to the phase velocity. Bringing the metallic planes closer to the ferrite slab the frequency band of surface MSW becomes wider and on the boundary of the pass band the group velocity is equal to zero. If the metallic interfaces are unsymmetrically placed, curve symmetry is also disturbed. Simultaneously, the parts of curves with normal and anomalous dispersion appear. The extrema move to the right or to the left, depending on the ratio l_1/l_2. The realization of this case is almost impossible for thin YIG layers on a substrate but it is possible in the case of a free YIG slab carved out of a monocrystal [12].

The third group of curves (dash-dot) represent dispersion dependences of surface MSW in a one-sided metal-coated layer. Here there are also many possible variants, but from the practical point of view this case is difficult to realize to the same measure as the second one. In spite of this from the theoretical point of view the investigation of all variants of the influence of metallic planes is important for the design of MSW transmission lines with prescribed dispersion properties.

One of the main advantages of MSW in comparison with acoustic waves is, as already mentioned above, the possibility of influencing the dispersion of MSW and consequently the signal delay in a wide range and in a wide frequency band. As results from the analysis in [1], depending on the requirements placed on an MSW device, a sufficiently great delay $\tau(\omega)$ can be reached. It is determined only by the losses in the magnetostatic line and by the device sensitivity. In addition to the delay, the operating frequency band and the character of the $\tau(\omega)$ dependence in this operating band are also important requirements for practical applications. The signal delay is given by the group velocity and per length L is equal to

$$\tau(\omega) = L/v_g(\omega) \tag{2.52}$$

The group velocity on the right side of (2.52) is defined by the already mentioned well-known formula

$$v_g = \frac{\mathrm{d}\omega}{\mathrm{d}k} \tag{2.53}$$

In practical applications either dispersionless delay lines with $\tau(\omega) =$ const. in the operating frequency band, or delay lines with a linear dependence $\tau(\omega)$ on frequency, i.e. $\tau(\omega) = C_1\omega$, are the most frequently required. The first case occurs if in the operating frequency

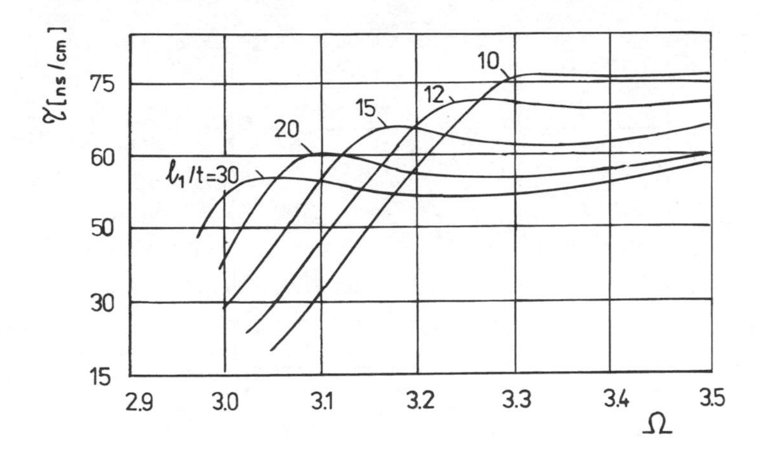

Fig. 2.13. The dependence of a delay for surface MSW as a function of frequency. l_1/t — parameter. Regions with $\tau(\omega) = \text{const.}$ $\omega_M/\omega_H = 7$, $l_2/t = \infty$, $t = 10^{-4}$ m are obvious.

interval the dispersion dependence $\omega(k)$ is linear, i.e. $\omega(k) = C_2 k$ and $v_g = C_2 = \text{const.}$ The second case occurs when the dispersion dependence is quadratic, i.e. $\omega = C_3\sqrt{k}$ and then $v_g = \mathrm{d}\omega/\mathrm{d}k = C_3/(2\sqrt{k}) = C_3^2/(2\omega)$, i.e. $\tau(\omega) = (2L/C_3^2)\omega$ depends linearly on frequency. The smaller the absolute value of coefficients C_2 and C_3 the bigger the delay. The delay characteristics for surface and volume magnetostatic waves were investigated in detail in [1] and [8] and two important cases are represented in Fig. 2.13 and Fig. 2.14.

To end this section the dispersion dependences of coupled MSWs will be investigated, i.e. waves in the system consisting of two (or more) thin YIG layers placed next to each other (Fig. 2.2a, b) and coupled together by means of high-frequency electromagnetic fields. In a similar way, coupled magnetostatic waves were investigated in [8]. The dispersion dependences of surface MSW in free structures ($l_1 = l_2 = \infty$) coupled through the gap d (in accordance with the formula (2.28)) for various values of the width d/t are represented in Fig. 2.15. As follows from Fig. 2.15 the solution is, in this case, described by two groups of waves, the dispersion dependences of which are above and below the dispersion curve of surface MSW in a simple layer. In agreement with the usual terminology, one group of waves (with lower values of k for the given frequency) is called fast waves (the upper curves in the figure) and the second one slow waves (the lower curves in the figure). Both dispersion branches

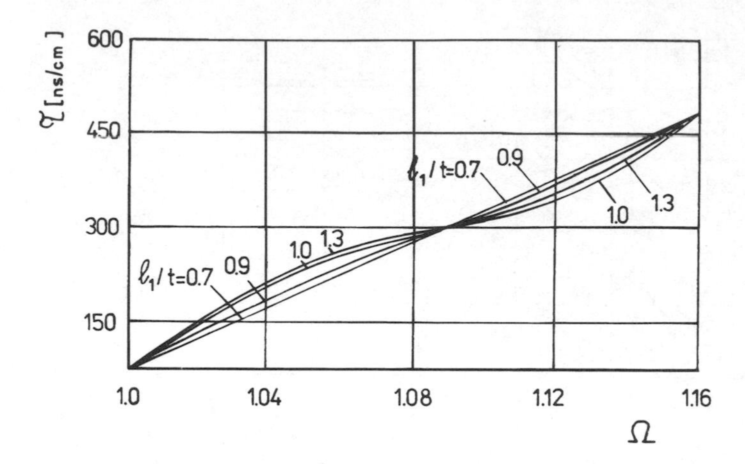

Fig. 2.14. The dependence of a delay for forward volume MSW on frequency. l_1/t — parameter. Linear dependence in some areas is obvious. $\omega_M/\omega_H = 0.5$, $l_2/t = \infty$, $t = 10^{-4}$m.

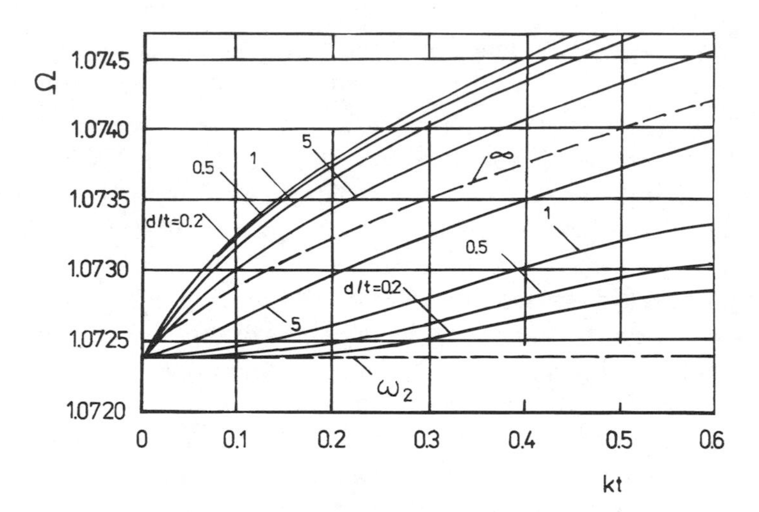

Fig. 2.15. Dispersion dependences of surface MSW in a free coupled structure. $t_1 = t_2$, $\Omega_M = 0.15$. Parameter d/t, d — distance between layers.

differ considerably for small values of d/t and converge when the gap between layers increases. In the case that layers are infinitely far away from each other then both dispersion branches converge to the ordinary Damon–Eshbach dispersion dependence for a single slab (the dashed line in Fig. 2.15). The structure of electromagnetic field in coupled layers is interesting. Detailed analysis can be found in

[8]. Since the fast and slow waves are excited simultaneously (at the same frequency) and have different values of k (i.e. different phase velocities), then, as a consequence of the interference, a summary high-frequency field will be characterized by alternating maxima and minima (impacts of modes). Moreover, the maximum value of the high-frequency electromagnetic field occurs alternatively once in the first and once in the second layer. This corresponds to the periodic "pouring" of energy from one layer to the other [8]. This kind of coupling between interacting wave processes is called "passive" in high-frequency electronics.

2.6 Excitation of magnetostatic waves

One of the serious problems in the production of MSW elements is the design of elements for the transformation of the electromagnetic wave into MSW. A transformer of this type has to fulfil two conditions: (1) the creation of a high-frequency magnetic field with components perpendicular to the constant magnetizing field H_0 and (2) very effective transformation of the input electromagnetic power into the MSW power and vice versa. Transformation elements can be implemented in different variants — coaxial, microstrip or waveguide. Nowadays, microstrip transducers are the most commonly used for this reason and therefore we will pay them specific attention. The simplest way of exciting a ferrite structure by means of a microstrip line is represented in Fig. 2.16. A ferromagnetic layer is located a distance d above the microstrip line, l_1 and $l_2 + d_h$ are distances to the conductive planes and w is the microstrip width. Electromagnetic power propagates in the microstrip line in the z-axis direction. The excited MSW propagates in the x-axis direction. If the constant magnetic field is oriented in the y-axis direction, the volume MSW is excited in the structure. If the constant magnetic field is oriented in the z-axis direction, the surface MSW is excited. A high-frequency current flows through the microstrip; this current is considered to be given, and in the first approximation uniformly distributed over the microstrip surface and equal to zero outside it [9]. It will be supposed that the microstrip is infinitively thin in the y-axis direction, the microstrip width is much less than the MSW wavelength and the electromagnetic wavelength in the microstrip line is much less than the ferrite structure width in the z-axis direction. The solution of

MSW excitation will be decomposed into two stages. Firstly, the connection between the distribution of high-frequency fields and the given current $I = I_0 e^{-j\omega t}$ in the microstrip will be found. Then the MSW power flux for the given high-frequency fields will be com-

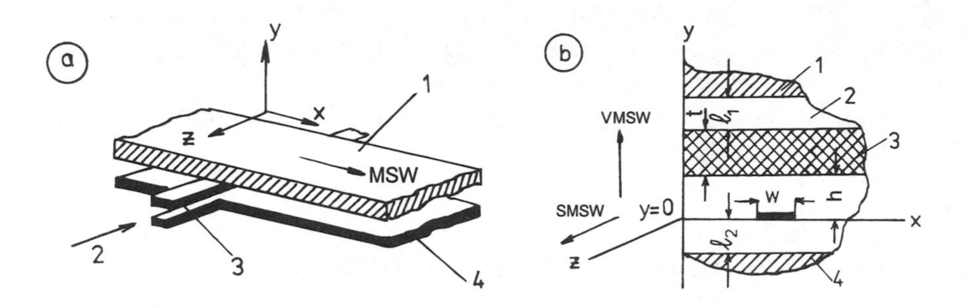

Fig. 2.16. a) Possible excitation structure 1 — YIG layer, 2 — input electromagnetic wave, 3 — microstrip, 4 — metallic interface. b) Geometry of the structure 1 — metal, 2 — dielectric, 3 — ferrite, 4 — metal.

puted. When the power propagating in ferrite (the Poynting vector) is known, the so-called radiation resistance of the microstrip line loaded by a ferrite structure can be introduced, which represents the variable crucial for the solution of the given problem. The input microstrip line has to be adjusted to this resistance.

Then it is supposed that current flows in the microstrip of width w in the z-axis direction. This current excites high-frequency fields with components h_x, h_y, e_z (TE mode). The components of the vector $\vec{b}$ are coupled with the components of the vector $\vec{h}$ by the formula (2.5) in which the components of the tensor $\overline{\overline{\mu}}$ are given by (2.6)–(2.8). It will be supposed that all components of the MSW high-frequency field change in accordance with the formula $e^{-j(\omega t - kx)}$ (i.e. $s = +1$). From Maxwell's equations (1.65.(1)) the relationship between e_z and b_y is found to be as follows:

$$e_z = -\frac{\omega}{k} \cdot b_y \tag{2.54}$$

where

$$b_y = \mu_0(\overline{\overline{\mu}} \cdot h)_y$$

On the metallic strip w (in the plane $y = 0$) the tangential component of the magnetic field h_x is discontinuous and it holds that

$$h_x^+ - h_x^- = -j_z(x) \tag{2.55}$$

where $h_x^{\pm}$ are the values of the high-frequency fields above and below the microstrip, respectively. The function $j_z(x)$ is in the form

$$\begin{aligned} &j_z(x) = j_0 = I_0/w && \text{on the strip } (0 \leq x \leq w) \\ &j_z(x) = 0 && \text{outside the strip } (x < 0; x > w) \end{aligned} \tag{2.56}$$

Since inhomogeneous boundary conditions (2.56) hold on the microstrip, the components b_y, $h_x^{\pm}$ and $j_z(x)$ can be expressed by means of the Fourier integrals

$$b_y = \frac{1}{2\pi} \int b_k e^{jkx}\, dk \tag{2.57.1}$$

$$h_x^{\pm} = \frac{1}{2\pi} \int h_k^{\pm} e^{jkx}\, dk \tag{2.57.2}$$

$$j_z(x) = \frac{1}{2\pi} \int j_k e^{jkx}\, dk \tag{2.57.3}$$

where b_k, $h_k^{\pm}$, j_k are Fourier components of the high-frequency magnetic induction, magnetic field intensity and current density.

If the formula (2.1) is used for surface magnetic permeability for the kth Fourier component, the boundary conditions (2.55) can be written as follows

$$jb_k \left(\frac{1}{\mu_s^+} - \frac{1}{\mu_s^-} \right) = j_k I_0$$

Then, the y component of magnetic induction from (2.57.1) is determined (for $y = 0$) by the formula

$$b_y = -j \frac{I_0}{2\pi} \int j_k \left(\frac{1}{\mu_s^+} - \frac{1}{\mu_s^-} \right)^{-1} e^{jkx}\, dk \tag{2.58}$$

the computation of this integral gives [1, 9]:

$$b_y = I_o j_k \frac{\mu_s^{+2}}{\dfrac{\partial}{\partial k_+}(\mu_s^- - \mu_s^+)} e^{jk_+ x} \tag{2.59}$$

and the components $h_x(0)$ in the plane $y = 0$ (for $x \gg w$) are given by

$$h_x(0) = -jb_y/\mu_s^- \tag{2.60}$$

The high-frequency energy flux at the MSW propagation in the ferrite structure for the given distribution of high-frequency magnetic field in the plane $y = 0$ will be investigated now. At first, the losses in ferrite will be neglected and the propagation of MSW without attenuation will be considered. The mean value of power flux through the closed area S is given by [14]

$$P = (1/2) \int_S (\vec{e} \times \vec{h}^*) \cdot \mathrm{d}\vec{S} \tag{2.61}$$

The magnetostatic wave (Fig. 2.16) has the components of high-frequency fields h_x, h_y, e_z. Taking this fact into account the vector product $\vec{e} \times \vec{h}$ can be written as

$$\vec{e} \times \vec{h}^* = -\vec{i}(e_z h_y^*) + \vec{j}(e_z h_x) \tag{2.62}$$

where $\vec{i}$, $\vec{j}$ are unit vectors in the x-axis and y-axis directions, respectively. The area S in (2.61) will be the area bounded in the y-axis direction by the planes $y = -l_2$ and $y = t + l_1$ (for the sake of simplicity we will put $h = 0$, Fig. 2.16b), in the x-axis direction by the planes $x = \pm x$, and in the z-axis direction by the planes $z = 0$, $z = \Delta z$. It can be shown easily that transverse energy fluxes through the planes $y =$ const. are equal to zero (because $e_z = 0$ on metal for $y = -l_2;\ t + l_1$); then only longitudinal energy fluxes through transverse $x =$ const. planes of the width Δz remain. To compute them it is necessary to find high-frequency field intensities and inductions in the area above the ferrite, in it, and over it. If the general solution for magnetostatic potential (2.15) and corresponding boundary conditions are used, the following formulae for the y-components of high-frequency fields and inductions will be obtained in every investigated area. Only the y-components are interesting for us because only they contribute to the longitudinal energy fluxes. Then, in the area $-t_2 \leq y \leq 0$:

$$\begin{aligned} h_y &= -jsh_x(0)\sinh[k(y + l_2)]/\cosh(kl_2) \\ b_y &= \mu_0 h_y \end{aligned} \tag{2.63}$$

In the area $0 \leq y \leq t$:

$$h_y = \frac{-jsh_x(0)\beta}{1+D}(-De^{-\beta ky} + Ue^{\beta ky})$$
$$b_y = \frac{-sh_x(0)}{1+D}(DUe^{-\beta ky} + Ve^{\beta ky}) \tag{2.64}$$

where

$$D = -\frac{1+V}{1-V}e^{2\beta kt}; \quad U = s\mu_{12} - \beta\mu_{22}; \quad V = s\mu_{12} + \beta\mu_{22}; s = \pm 1$$

In the area $t \leq y \leq t + l_1$:

$$h_y = -jsh_x(t)\sinh(k[y - (t + l_1)])/\cosh(kl_1) \tag{2.65}$$

where

$$h_x(t) = h_x(0)\frac{\beta\mu_{22}}{(1 + s\mu_{12})\sinh(\beta kt) + \mu_{22}\beta\cosh(\beta kt)}.$$

In the case that the ferrite layer is free from the upper side ($l \to \infty$), the formula (2.65) will be as follows

$$h_y = jsh_x(t)e^{k(t-y)} \tag{2.66}$$

It should be noted that all formulae for the magnetic field intensities and inductions are defined by means of the component $h_x(0)$, which at the excitation of the structure by microstrip line is given by the formulae (2.59) and (2.60).

If the results obtained for high-frequency fields are used, the longitudinal MSW power flows propagating in, above and below the ferrite, in the positive direction of the x-axis, from the microstrip line can be computed. Adding together we have

$$P^+ = \frac{|h_x(0)|^2}{4} \cdot \frac{\partial}{\partial k}(\mu_s^- - \mu_s^+) + \frac{|h_x(t)|^2}{4} \cdot \left[\frac{\mu_s(+t) - 1}{k} - \frac{\partial\mu_s(+t)}{\partial k}\right] \tag{2.67}$$

In the case that the dielectric layer is free from the upper side ($l_1 \to \infty$), then (2.67) will be as follows

$$P^+ = \frac{|h_x(0)|^2}{4} \cdot \frac{\partial}{\partial k}(\mu_s^- - \mu_s^+) \tag{2.68}$$

An analogous formula is also obtained for the energy propagating in the negative direction of the x-axis. If $h_x(0)$ from (2.60) and (2.59) is substituted into (2.68) the following formula will be obtained:

$$P^{\pm} = \frac{1}{4} I_0^2 j_k^2 (\mu_s^-)^2 \left[\frac{\partial}{\partial k} (\mu_s^- - \mu_s^+)^{-1} \right]_{s=\pm 1} \tag{2.69}$$

where the surface permeabilities μ_s^+ and μ_s^- are defined by (2.19), i.e.

$$\mu_s^- = -\tanh(kl_1).$$

$$\mu_s^+ = \frac{s\beta\mu_{22} + (\mu_{11}\mu_{22} - \mu_{12}^2 - s\mu_{12})\tanh(\beta kt)}{(\beta\mu_{22} + s\mu_{12} + s)\tanh(\beta kt)} \tag{2.70}$$

The formula (2.69) for the energy flux transmitted by MSW can also be given in the following form:

$$P^{\pm} = (1/2) R^{\pm} I_0^2 \tag{2.71}$$

$$R = R^+ + R^- = j_k{}^2 \left[C_a^+ + C_a^- \right] \tag{2.72}$$

where the coefficients $C_a^{\pm}$ are given by

$$C_a^{\pm} = \frac{(\mu_s^-)^2}{2} \left[\frac{\partial}{\partial k} (\mu_s^- - \mu_s^+)^{-1} \right]_{s=\pm 1} \tag{2.73}$$

Now, $(\partial \mu_s / \partial k)$ will be computed. Taking into account (2.70) and after some algebra the formula (2.73) can be expressed as follows

$$C_a^{\pm} = \frac{1}{2} \cdot \frac{\tanh^2(kl_2)}{k[(t/\mu_{22})[(\mu_{12} \mp \tanh(kl_2))^2 - \mu_{11}\mu_{22}] - l_2 \cdot \mathrm{sech}^2(kl_2)]} \tag{2.74}$$

If the microstrip is placed in the distance d_h from the ferrite layer (Fig. 2.16b) then for the coefficients $C_a^{\pm}$ holds:

$$C_a^{\pm}(d_h) = \frac{\sinh^2[k(l_2 - d_h)]}{\sinh^2(kl_2)} \cdot C_a^{\pm} \tag{2.75}$$

If the lower ground plane is located far enough away, the formula (2.74) for $C_a^{\pm}$ will be simpler:

$$C_a^{\pm} = \frac{\mu_{22} \mathrm{e}^{-2kd_h}}{2kt[(1 \mp \mu_{12})^2 - \mu_{11}\mu_{22}]} \tag{2.76}$$

The values of μ_{11}, μ_{12}, μ_{22} in the formulae (2.74)–(2.76) are given by the formula (2.13) for the backward volume MSW and by the formula (2.11) for the surface MSW. Using (2.72)–(2.76) radiation resistances for various special cases can be computed.

For forward volume MSW the total radiation resistance per unit length is given by

$$R = 79f \cdot \frac{\sin^2\left(\frac{kw}{2}\right)}{\left(\frac{kw}{2}\right)^2} \times \frac{\sinh^2[k(l_2 - d_h)]}{\sinh^2(kl_2)} \times \frac{\tanh^2(kl_2)}{kt[\tanh^2(kl_2) - \mu_{11}^2 - (l_2/t)\mathrm{sech}^2(kl_2)]} \quad [\Omega\,\mathrm{cm}^{-1}] \tag{2.77}$$

and for surface MSW we obtain

$$R_{\pm} = 79f \cdot \frac{\sin^2\left(\frac{kw}{2}\right)}{\left(\frac{kw}{2}\right)^2} \times \frac{\sinh^2([k(l_2 - d_h)]}{\sinh^2(kl_2)} \times \frac{\mu_{11}\tanh^2(kl_2)}{2kt[(\mu_{12} \mp \tanh(kl_2))^2 - \mu_{11}^2 - (l_2/t)\mathrm{sech}^2(kl_2)]}; \quad [\Omega\,\mathrm{cm}^{-1}] \tag{2.78}$$

The signal frequency f in (2.77) and (2.78) is in GHz. For the computation of the dependence k on the frequency it is necessary to use the dispersion relation obtained in Section 2. Further, we will treat some special cases.

The results of the radiation resistance R computation for volume MSW using the formula (2.77) for the thin YIG layer ($M_0 = 140\,\mathrm{kA\,m^{-1}}$) for $d_h = 0$ are presented in Figs. 2.17–2.19. The dependence of the radiation resistance on frequency for various values of the strip width (w/t) is shown in Fig. 2.17. It can be seen that the excitation band is getting wider when the strip is getting narrower and vice versa. By altering the microstrip width the excitation band of the MSW transmission line can be changed over a wide range. This fact is used in the regulation of the pass band of MSW filters.

The influence of the gap between the ferrite and metal l_2/t for $t =$ const. on the magnitude of R is shown in Fig. 2.18. As can be seen, on varying l_2/t the excitation band does not change; this is

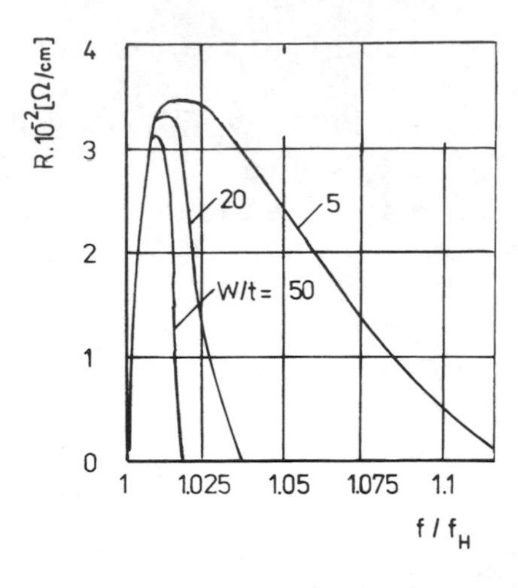

Fig. 2.17. The dependence of radiation resistance of forward volume waves on frequency for various widths of microstrip, $f_H = 9.8\,\text{GHz}$, $l_2/t = 50$.

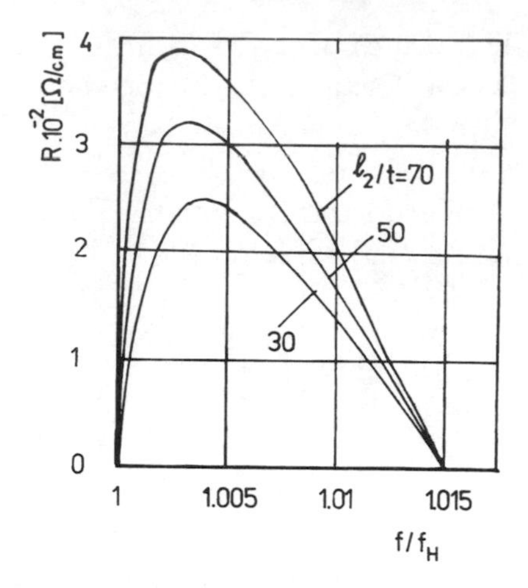

Fig. 2.18. The influence of the distance of metal plane on the radiation resistance of forward volume MSW, $f_H = 9.8\,\text{GHz}$, $w/t = 50$.

given by the strip width w/t. On changing l_2 the maximum value R_{max} changes, and with l_2 increasing, R_{max} increases and the maximum is shifted towards the bottom frequency of the forward volume MSW pass band. For $l_2 \to \infty$ R has the maximum value at the frequency $f/f_H = 1$.

The dependences of the radiation resistance on the frequency for various frequency bands are represented in Fig. 2.19. As can be seen, on moving to the short-wave band the relative excitation band of volume MSW is getting narrower. Simultaneously also the maximum value of the radiation resistance decreases.

Analogous curves for surface MSW are shown in Figs. 2.20–2.22. As follows from these figures, almost all conclusions concerning the behaviour of the radiation resistance in the frequency band for forward volume MSW are valid also for surface MSW. It is noteworthy that for surface MSW the absolute values of the radiation resistance are about 10 times bigger than for volume MSW.

The losses in the ferrite cause the change of excitation conditions. This fact, as will be shown later, plays an important role in the investigation of processes in nonlinear MSW elements, in which losses can increase at high input signals levels as a result of parametric effects. The losses in the ferrite at the excitation of MSW by a

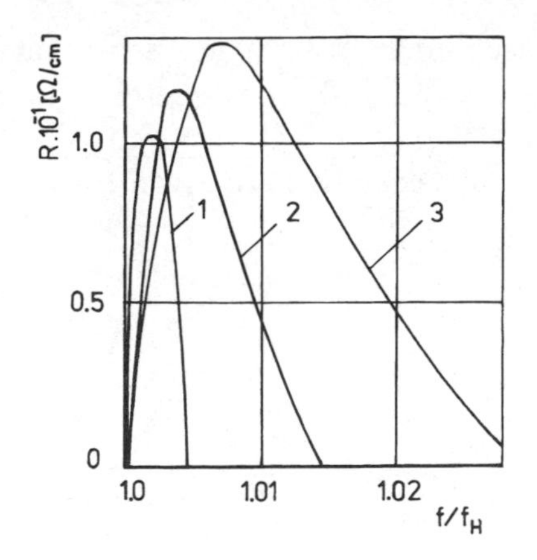

Fig. 2.19. The dependence of radiation resistance for forward volume MSW on frequency for various frequency bands. $\omega_H = 2\pi . f_H$ GHz, $w/t = 20$, $l_2/t = 50$, 1 — $R.10^{-3}$, $f_H = 35.5$ GHz, 2 — $R.5 \times 10^{-3}$, $f_H = 9.8$ GHz, 3 — $R.10^{-2}$, $f_H = 4.9$ GHz.

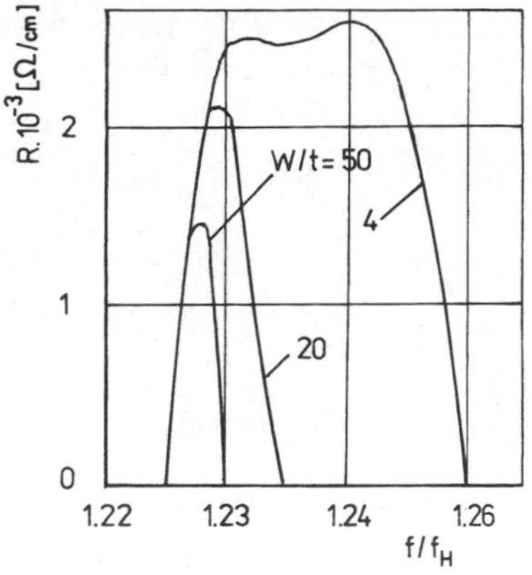

Fig. 2.20. The dependence of radiation resistance for surface MSW on frequency for various microstrip widths. $f_H = 9.8$ GHz, $l_1/t = 50$.

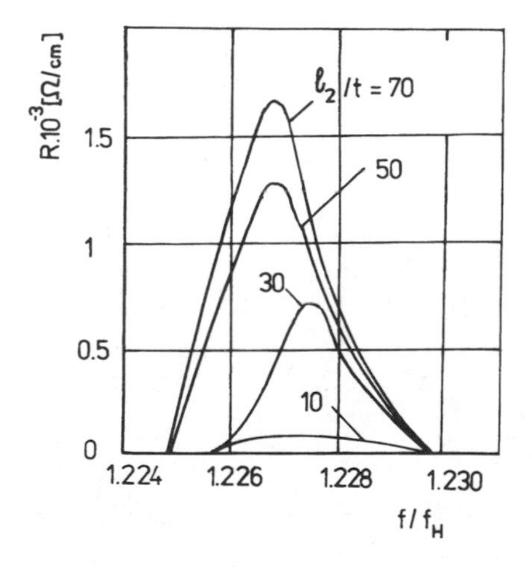

Fig. 2.21. The influence of metallic interface on the dependence of radiation resistance for surface MSW. $f_H = 9.8$ GHz, $w/t = 50$.

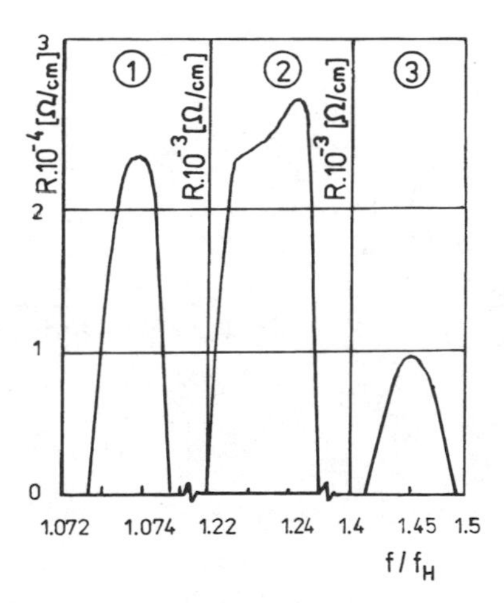

Fig. 2.22. The dependence of radiation resistance of surface MSW on frequency for various frequency bands, $l_1/t = 50$, $w/t = 4$, 1 — $f_H = 31.5$ GHz, 2 — $f_H = 9.8$ GHz, 3 — $f_H = 4.9$ GHz.

microstrip antenna were taken into account in [13]. The losses will be characterized in the same way as above by the parameter α in the Landau–Lifshitz equation (1.22). This means that (1.49) applies and the components of high-frequency permeability are given by

$$\mu = 1 + \frac{\gamma\mu_0 M_0(\omega_H + j\omega\alpha)}{\omega_H^2 - (1+\alpha^2)\omega^2 + 2j\omega\omega_H\alpha}$$
$$\mu_a = \frac{\gamma\mu_0 M_0\omega}{\omega_H^2 - (1+\alpha^2)\omega^2 + 2j\omega\omega_H\alpha} \tag{2.79}$$

Taking losses into consideration the components of the permeability tensor become complex numbers. The propagation constant in this case also will be a complex number defined as [8]:

$$k = k' + jk''$$

The complex propagation constant k satisfies a complex dispersion equation (e.g. of the type (2.19)) where μ_{s0} and μ_{s1} are now complex variables given by the formulae (2.21), (2.23) and (2.79). Using this approach the complex radiation impedance $Z = R + jX$ in the plane $x = 0$ was computed in [1]. It is given by

$$Z_{\pm} = \frac{1}{2}\left|\frac{\sin\dfrac{kw}{2}}{\left(\dfrac{kw}{2}\right)}\right|^2 \times (I_1 + I_2 + I)$$
$$\times \frac{|\tanh^2(kt)|^2}{|(t/\mu_{22})[(\mu_{12} \mp s\tanh(kt))^2 - \mu_{11}\mu_{22}] - l_2\mathrm{sech}^2(kt)|^2}; \tag{2.80}$$

where

$$I_1 = \frac{1}{k't}\cdot\frac{\sinh(2k'l_2) - (k'/k'')\sin(2k''t)}{\cosh(2k'l_2) + \cos(2k't)}$$

$$I_2 = \frac{\beta^*}{|1+D|^2}\left[U|D|^2\cdot\frac{e^{-2\gamma'} - 1}{\gamma'} + V\frac{e^{-2\gamma'} - 1}{\gamma'}\right.$$
$$\left. + j\left[DU\cdot\frac{e^{-2j\gamma''} - 1}{\gamma''} + VD^*\cdot\frac{e^{-2j\gamma''} - 1}{\gamma''}\right]\right]$$
$$\gamma' = \mathrm{Re}(\beta kt), \qquad \gamma'' = \mathrm{Im}(\beta kt)$$

$$I_3 = \frac{1}{k't}\left|\frac{\beta\mu_{22}}{(1+s\mu_{12})\sinh(\beta kt) + \beta\mu_{22}\cosh(\beta kt)}\right|^2 \tag{2.81}$$

For $\alpha \to 0$, $k'' \to 0$ and the formulae (2.80) and (2.81) are reduced to the formulae (2.72)–(2.76) for the real component of the radiation impedance. The components of the permeability tensor in the formulae (2.80) and (2.81) are for the forward volume MSW given by (2.12), for the surface MSW by (2.11) and for the backward volume waves by (2.13).

The real component of the radiation impedance R in (2.80) is connected with the power transmitted by MSW and the imaginary component jX characterizes the power localized in the structure. In the investigated model at $\alpha \to 0$, $k'' \to 0$, $jX \to 0$. When losses in a ferrite layer increase (the parameter α increases), the imaginary component of the radiation impedance also increases, i.e. the power localized in the structure increases. It can be said that high-frequency power localizes essentially around the microstrip and the component of propagating power decreases. In other words, the increase of losses in ferrite decreases the efficiency of the magnetostatic wave excitation in ferrite, and high-frequency power propagates in the microstrip without attenuation.

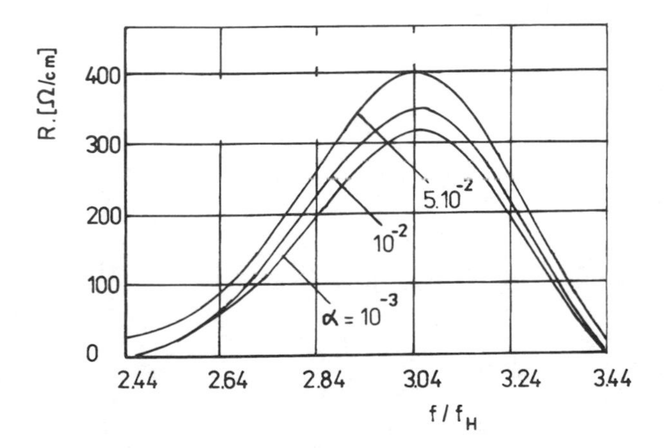

Fig. 2.23. The frequency dependence of the radiation resistance for surface MSW for various values of the loss parameter α, $f_H = 1\,\text{GHz}$, $l_2/t = 5$, $w/t = 3$.

The dependences $R(\omega)$ and $X(\omega)$ for various values of the parameter α corresponding to the excited surface waves are presented in Fig. 2.23 and Fig. 2.24. These dependences well illustrate what has been said. If the loss parameter changes in the range $\alpha = 10^{-3}$–10^{-2}, the imaginary component of the radiation impedance jX increases

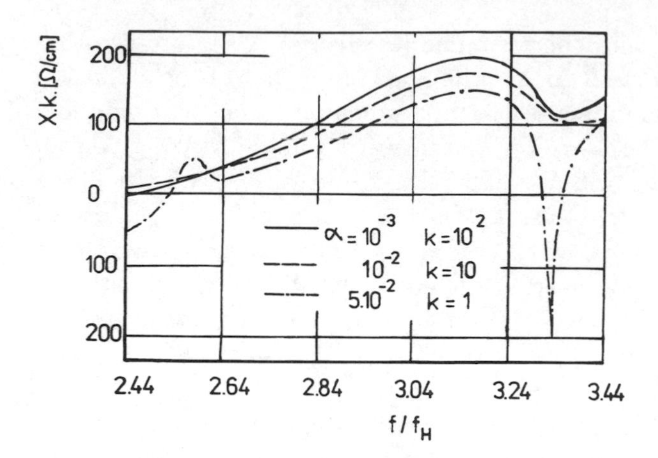

Fig. 2.24. The frequency dependence of the reactance of radiation impedance for various values of the loss parameter α. Other parameters as in Fig. 2.23.

100-fold. The high-frequency power does not penetrate into the ferrite and periodic exchange of reactive power between the microstrip and the ferrite occurs.

Chapter 3

ANISOTROPY OF MAGNETOSTATIC WAVES*

3.1 Magnetostatic wave propagation at an arbitrary angle with regard to the magnetic field

In Chapter 1 magnetostatic waves propagating in a three-dimensional coordinate system (x, y, z) only in the x-axis direction were investigated. In this case the magnetic field could be oriented in the y-axis direction (forward volume waves), in the z-axis direction (surface MSW), and in the x-axis direction (backward volume MSW) (Fig. 2.1, Fig. 2.2). The relationship between the components $\vec{b}$ and $\vec{h}$ is defined for all given cases by the formulae (2.2) – (2.4) and the corresponding components of the magnetic permeability tensor by the formulae (2.6) – (2.8). The propagation constant $\vec{k}$ introduced in (2.15) was in principle the propagation constant k_x in the x-axis direction.

The general case when magnetostatic waves can propagate at an arbitrary angle to the magnetic field will be investigated now. In this case, as will be shown later, the characteristics depend on the given angle, i.e. propagation of magnetostatic waves is anisotropic with regard to the direction of magnetic field H_0. The case when the magnetic field is perpendicular to the layer plane will be investigated first (see Fig. 3.1).

Waves in a perpendicularly magnetized layer

In the general case of a ferrite layer magnetized perpendicularly to the layer plane (at corresponding excitation) MSW can propagate

---

*Substantial parts of the results in this chapters were obtained by A. Stalmachov and A. V. Vaškovský (USSR) and are published in this book with their kind permission.

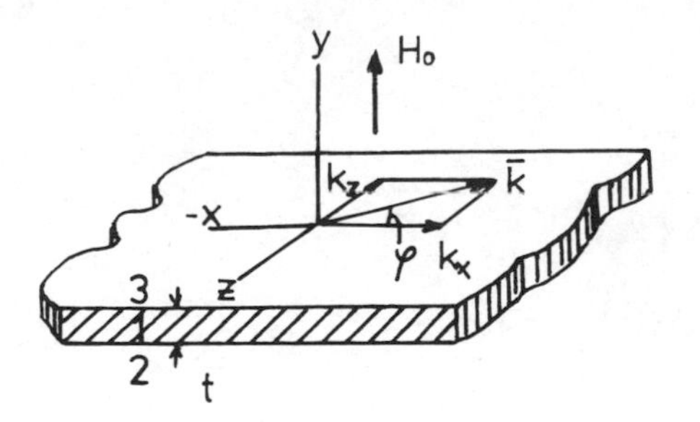

Fig. 3.1. Geometry for volume MSW, 1 — ferrite, 2 — substrate, 3 — dielectric, t — layer thickness.

in any direction in the xz plane. This can be provided for example by rotating the plane of an excitation antenna around the y-axis. In this case the wave vector $\vec{k}$ can be expressed as follows:

$$k_x = k\cos\varphi; \quad k_z = k\sin\varphi; \quad k_x^2 + k_z^2 = k^2 \tag{3.1}$$

where $k = |\vec{k}|$. Let k_y denote the wave number of MSW in the y-axis direction. The relationship between the components of the propagation constant k_x, k_y and k_z is given by the formulae (1.72) or (1.73) considering also (1.74). For a perpendicularly magnetized layer:

$$\begin{aligned} &\mu k_x^2 + \mu k_z^2 + k_y^2 = 0 && \text{inside the ferrite layer} \\ &k_x^2 + k_z^2 + k_y^2 = 0 && \text{outside the layer} \end{aligned} \tag{3.2}$$

In the monograph [8] the dispersion equation was obtained for MSW in a perpendicularly magnetized layer 1 in the case that arbitrary dielectric layers 2 and 3 are around the ferrite. The corresponding equation is as follows:

$$\frac{k_{y1}k_{y2} + k_{y1}k_{y3}}{k_{y1}^2 - k_{y2}k_{y3}} = \tan(k_{y1}t) \tag{3.3}$$

If it is supposed that the dielectric above and below the ferrite layer is the same, after substituting (3.2) into (3.3) we have

$$\frac{2\xi}{\xi^2 - 1} = \tan(\xi k t) \tag{3.4}$$

where $\xi = \sqrt{(-\mu)}$.

The equation (3.4) is formally the same as the dispersion equation (2.35) for forward volume magnetostatic waves. Now, however, the wave vector $\vec{k}$ lies in the xz plane and is oriented at arbitrary angle φ to the x-axis. In this case the formula (3.4) does not describe the dispersion curve but the dispersion surface in the co-ordinate system ω, k_x, k_z. The section of this dispersion surface through the plane (ω, k) represents the ordinary dispersion curves of direct volume magnetostatic waves in a perpendicularly magnetized layer what was investigated in the previous chapter.

The dispersion surfaces of forward volume waves obtained by means of (3.4) for $n = 1, 2$, $\omega_M/\omega_H = 2$ (see also Fig. 2.3) are depicted in Fig. 3.2. Since (3.4) does not depend on an angle, dis-

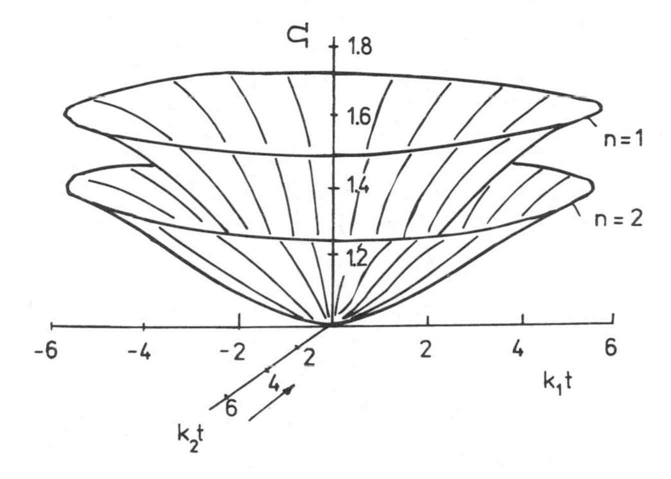

Fig. 3.2. Dispersion surfaces of forward volume MSW for various n.

persion surfaces of forward volume MSW are axially symmetrical to the ω axis and are funnel-shaped with complex cross-section. For the above given modes the dispersion surfaces also represent funnel-shaped cups symmetrical around the ω axis and placed one in the other. From the point of view of the energy-wave properties the ferrite layer magnetized perpendicularly to the layer plane represents for MSW in the xz plane an isotropic medium in which all directions of the orientation of the wave vector $\vec{k}$ are equivalent and its amplitude $|\vec{k}|$ does not depend on φ. As has been already mentioned above, the direction of the wave vector is determined by the method of MSW excitation.

Waves in a tangentially magnetized layer

The magnetic field applied tangentially to the ferrite slab (i.e. it lies in the slab plane) will be investigated now. Furthermore, it will be supposed that the magnetic field is oriented in the z-axis direction (Fig. 3.3).

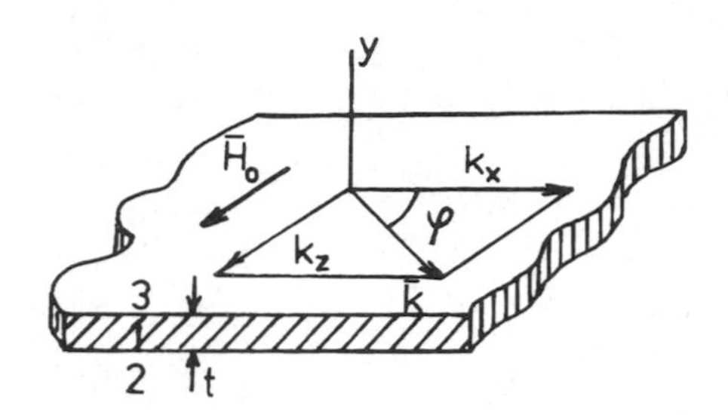

Fig. 3.3. Geometry of the task for the in-plane magnetized layer.

The ferrite slab is loaded by the dielectric layers 2 and 3. This case, at present already a classical one, was for the first time investigated by R. Damon and D. Eshbach [11]. If the magnetic field is oriented in the z-axis direction, the relations between the wave vector components k_x, k_y, and k_z are as follows [1, 8]:

$$\begin{aligned} k_{y1}^2 &= -(k_x^2 + \frac{k_z^2}{\mu}) && \text{inside the slab} \\ k_{y2,3}^2 &= k_x^2 + k_z^2 && \text{outside the slab} \end{aligned} \tag{3.5}$$

The general dispersion equation for magnetostatic waves in a free tangentially magnetized ferrite layer, propagating in the xz plane at arbitrary angle φ to the x-axis (Fig. 3.3) was obtained in [8]. Similarly, as in the previous case, the orientation of $\vec{k}$ is given by the orientation of an excitation antenna. The corresponding equation is as follows (the formula (6.26) in [8]):

$$\begin{aligned} &k_{y2}k_{y3} + (k_{y2} + k_{y3})\mu k_{y1}\tan^{-1}(k_{y1}t) + (k_{y2} - k_{y3})\mu_a k_x \\ &-(\mu k_{y1})^2 - (\mu_a k_x)^2 = 0 \end{aligned} \tag{3.6}$$

If it is supposed that the ferrite is loaded by the same dielectric, i.e.

$k_{y2} = k_{y3}$, then (3.6) simplifies to

$$(k_x^2 + k_z^2) + 2\mu\sqrt{(k_x^2 + k_z^2)} \cdot \sqrt{\left[-\left(k_x^2 + \frac{k_z^2}{\mu}\right)\right]} \cdot$$
$$\tan^{-1}\left[\sqrt{\left[-\left(k_x^2 + \frac{k_z^2}{\mu}\right)\right]}t\right] + \mu^2\left(k_x^2 + \frac{k_z^2}{\mu}\right) - \mu_a k_x^2 = 0 \qquad (3.7)$$

This formula is identical with that obtained in [11]. The equation (3.7) relates three variable k_x, k_y, ω and it represents the equation of a dispersion surface in the co-ordinate system k_x, k_y, ω. Introducing the angle φ and using notation

$$k_x = k \cdot \cos\varphi; \quad k_z = k \cdot \sin\varphi; \quad k_z/k_x = \tan\varphi$$

from (3.7) follows:

$$1 + 2\mu\cos\varphi\sqrt{(1 + \tan^2\varphi/\mu)} \cdot \tanh^{-1}\left[kt\cos\varphi\sqrt{(1 + \tan^2\varphi/\mu)}\right]$$
$$+\mu^2\cos^2\varphi(1 + \tan^2\varphi/\mu) - \mu_a^2\cos^2\varphi = 0 \qquad (3.8)$$

This is also an equation of a dispersion surface, but in k, φ, ω space. Solution of any of the equations (3.7) and (3.8) must give the same result. Before doing this, some general notes concerning the character of the solutions obtained will be presented.

Volume MSW propagating in a ferrite slab magnetized perpendicularly to the slab plane do not depend on the propagation direction (in equation (3.4) the angle φ is not present). It means that propagation of forward volume waves is isotropic. These is a quite different situation in the case of a tangentially magnetized slab. As can be seen in equation (3.8) the angle φ, which characterizes the orientation of the wave vector, does appear. For different values of the angle φ the k and ω, i.e. the dispersion dependence, will be different. It means that propagation of MSW in a tangentially magnetized layer is anisotropic. The question raises which MSW modes can propagate in a tangentially magnetized layer. It was shown in Chapter 2 that surface and backward volume MSW propagate in the plane of the magnetized layer. They differ from each other by the shape of the membrane function $\tan^{-1}\left[t\sqrt{[-(k_x^2 + k_z^2/\mu)]}\right]$ in (3.7) which describes transverse distribution of the high-frequency field.

If the membrane function has a harmonic (periodic) character, then the solution describes the system of volume waves. In the case that the membrane function has an exponential character, the solution represents surface waves. The shape of the functions depends on the sign of the formula $\left(k_x^2 + k_z^2/\mu\right)$ which occurs under the root. The frequency dependences of the diagonal component of the high-frequency permeability tensor for various values of $\Omega_M = \omega_M/\omega_H$ are depicted in Fig. 1.4. As was presented in Chapter 2, the operating band of volume and also surface waves is in the interval

$$1 \leq \Omega \leq 1 + \Omega_M/2$$

The interface between them is the point $\Omega = \sqrt{(1+\Omega_M)}$. This point corresponds to the value $\mu = 0$ on the curves depicted in Fig. 1.4. Then, within the MSW pass band the value of μ crosses zero and changes its sign.

The membrane function for volume waves has the form of ctg, i.e. $(\mu k_x^2 + k_z^2)/\mu < 0$. Since always $k_x^2, k_y^2 > 0$, theoretically two cases are possible:

$$\begin{aligned} &(1) \quad \mu \leq 0 \qquad \mu k_x^2 + k_z^2 \geq 0 \quad \text{or} \quad -\mu k_x^2 \leq k_z^2 \\ &\qquad \tan^2\varphi = (k_z^2/k_x^2) \geq |\mu| \end{aligned}$$

for any angle φ from $\varphi = \pi/2$ (backward volume waves propagating in the direction of magnetic field) to $\varphi = 0$, if $\mu = 0$.

$$(2) \quad \mu \geq 0; \qquad \mu k_x^2 + k_z^2 \leq 0;$$

because now $\mu : k_x^2 : k_z^2 > 0$, this case cannot be implemented.

From this analysis, it follows that volume MSW exist only for $\mu \leq 0$. The membrane function for surface waves has the form of coth, i.e. $(\mu k_x^2 + k_z^2)/\mu > 0$. Here too there are theoretically two possibilities:

$$(1) \quad \mu \geq 0 \qquad \mu k_x^2 + k_z^2 \geq 0$$

holds for defined values of k_x and k_y, i.e. for defined values of the angle φ from $\varphi = 0$ to φ_c which is given by the condition $\mu = 0$ from (3.7). In this case

$$\varphi_c = \operatorname{atan}(\sqrt{\Omega_M}) \tag{3.9}$$

and therefore the angle φ of permitted directions of wave vector will be in the interval

$$0 \leq \varphi \leq \varphi_c$$

(2) Theoretically also the second situation is possible:

$$\mu \leq 0; \qquad \mu k_x^2 + k_z^2 \leq 0 \quad \text{or} \quad \mu \leq -\tan^2\varphi,$$

which can be fulfilled within the frequency band

$$1 \leq \Omega \leq \sqrt{(1 + \Omega_M \cos^2 \varphi)} \tag{3.10}$$

or for angles

$$0 \leq \varphi \leq \arccos\left[\sqrt{[(\Omega^2 - 1)/\Omega_M]}\right]$$

Therefore in the frequency band given by the formula (3.10), the existence of a solution of Walker's equation in the form of surface waves for $\mu < 0$ is also theoretically possible. However, for a solitary slab the surface solutions of the dispersion equation (3.7) in the frequency band given by (3.10) do not exist. These waves will exist in coupled structures with various magnetizations of individual layers. The estimation made shows that the spectrum of magnetostatic waves in a tangentially magnetized layer in the case of the general direction of MSW propagation with regard to the magnetic field direction may have a complex shape.

As follows from equation (3.7), on changing the angle φ (this angle is given only by the location of the plane of an excitation antenna with regard to the magnetic field direction), the shape of the dispersion dependence changes, and the width of the pass band may change as well as the location of the upper and lower band frequencies of surface and volume MSW, etc. To answer these questions, equation (3.8) has to be solved for given values of Ω_M at $\varphi = \text{const.}$ Series of values of $\varphi = \text{const.}$ give series of dispersion curves representing the cross-sections of dispersion surfaces through the planes $\varphi = \text{const.}$ All given calculations hereinafter are for $\Omega_M = 2$. For this value of Ω_M μ crosses zero at the frequency $\Omega = 1.7320$ (Fig. 1.4).

Volume MSW will be investigated first. For $\mu \leq 0$; $1 \leq \Omega \leq \sqrt{(1 + \Omega_M)} = 1.7320$ these waves exist in two adjacent quadrants

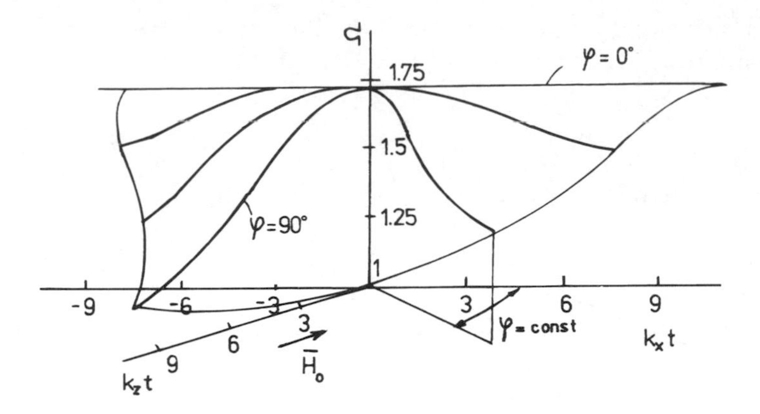

Fig. 3.4. Dispersion surface of a basic mode of volume MSW in a tangentially magnetized layer; $n = 1$, $\Omega_M = 2$.

and the dispersion surface represents in principle a saddle-shaped structure. The dispersion surface of volume MSW for the basic mode ($n = 1$) in a tangentially magnetized plate is depicted in Fig. 3.4.

In the same figure the sections of dispersion surface are depicted which in fact represent dispersion curves for various orientations of the vector $\vec{k}$. On decreasing the angle φ dispersion curves gradually disappear. For the plane perpendicular to H_0 the dispersion curve degenerates into the line $\Omega = \sqrt{(1 + \Omega_M)}$. On decreasing φ the band of existence of volume MSW gets narrower because the lower boundary of the pass band increases. Dispersion curves of backward volume MSW in a free ferrite slab for various orientations of MSW propagation (various φ) are presented in Fig. 3.5. Backward volume MSW exist with the frequency band $|\mu| \leq \tan^2\varphi$ which is given by the inequality

$$\sqrt{(1 + \Omega_M \cos^2 \varphi)} \leq \Omega \leq \sqrt{(1 + \Omega_M)}$$

For $\varphi = 0$ the band width is equal to zero and for $\varphi = 90°$ the band width is maximum.

Let us investigate surface MSW. In this case $\mu \geq 0$ and $\Omega \geq \sqrt{(1 + \Omega_M)}$. The upper limit of surface MSW propagating perpendicularly to the magnetic field direction ($\varphi = 0$) is $\Omega = 1 + \Omega_M/2$. Considering the formula (3.8) for $\varphi =$ const., the dependence $k(\omega)$ can be found and we get a dispersion surface for surface MSW (Fig. 3.6). As can be seen, this dispersion surface lies in four quadrants along the x-axis and represents a complex strongly saddle-

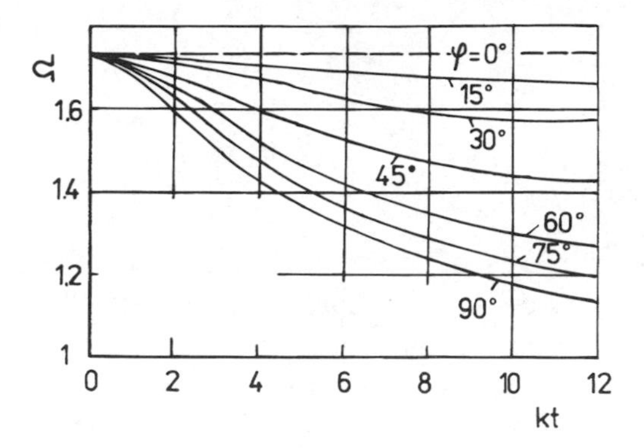

Fig. 3.5. Dispersion curves of backward volume MSW in a free ferrite slab for various angles of propagation, $\Omega_M = 2$, $l_1 = l_2 = \infty$, $n = 1$.

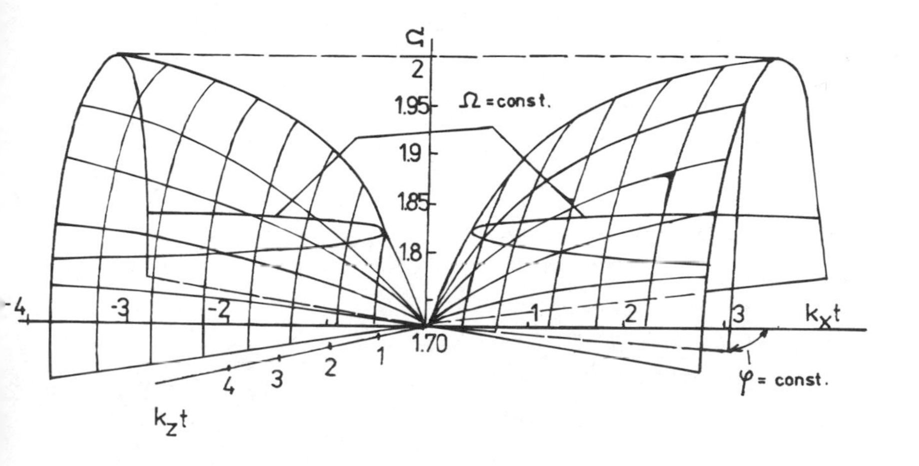

Fig. 3.6. Dispersion surfaces of surface MSW, $\Omega_M = 2$, $l_1 = l_2 = \infty$. The thick line represents the cross-section for constant frequency.

shaped space structure. In Fig. 3.6 the critical angle $\varphi_c = \text{atan}\sqrt{\Omega_M}$ is indicated at which the surface waves cease to exist. As in the case of backward volume waves approaching the critical angle φ_c the frequency band of existence of surface MSW narrows. This is a consequence of the decrease of the upper band frequency, which in general is given by

$$\frac{\mu_a^2 - \mu - 1}{2\mu} \cos^2 \varphi - \frac{1 + \mu}{2\mu} \sin^2 \varphi - \sqrt{\left(\frac{\sin^2 \varphi}{\mu} + \cos^2 \varphi \right)} = 0$$

Cross-sections of dispersion surfaces of surface MSW in the first quadrant through the plane $\varphi = \text{const.}$ are represented in Fig. 3.7. Actually, as can be seen, if the angle φ increases, the band of existence of MSW narrows and for $\varphi = \varphi_c$ is equal to zero. The solution of Walker's equation in the form of surface waves is theoretically possible also for $\mu \leq 0$. This is a frequency area from which backward volume waves "go away" because the bottom band frequency of backward volume MSW increases (the angle φ changes from 90° to 0°). This area should be "filled in" with surface waves. However, as can be seen from the analysis [15, 16], it cannot happen for a free ferrite slab and the dispersion equations (3.7) and (3.8) in this frequency band do not have a solution in the form of surface waves. However, this case may happen for a coupled structure.

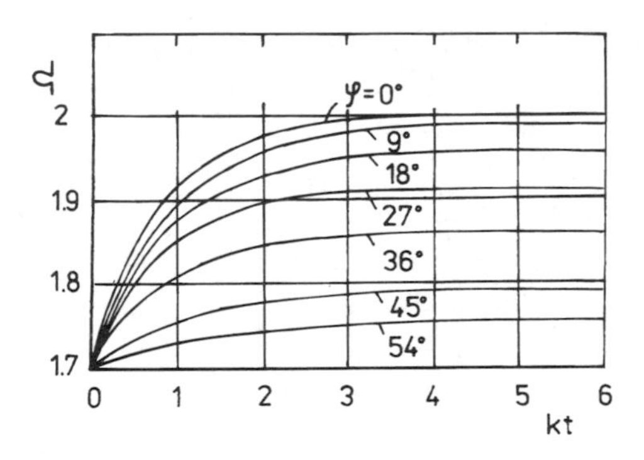

Fig. 3.7. Cross-sections of the dispersion surface of surface MSW through planes $\varphi = \text{const.}$, $\Omega_M = 2$, $l_1 = l_2 = \infty$.

To end of this section we will investigate the dispersion surfaces of surface MSW in a more complex structure, namely in a ferrite layer loaded by a metallic plane at some distance from the ferrite. These surfaces are depicted in Fig. 3.8 for $\Omega_M = 0.15$, $l_1/t = 0.15$, $l_2/t = \infty$, $\varphi_c = 22°$. The general structure of dispersion surface remains the same; however, its cross-sections are more complex.

From the above analysis it follows that in a layer magnetized perpendicularly to its plane propagation of MSW is isotropic, i.e

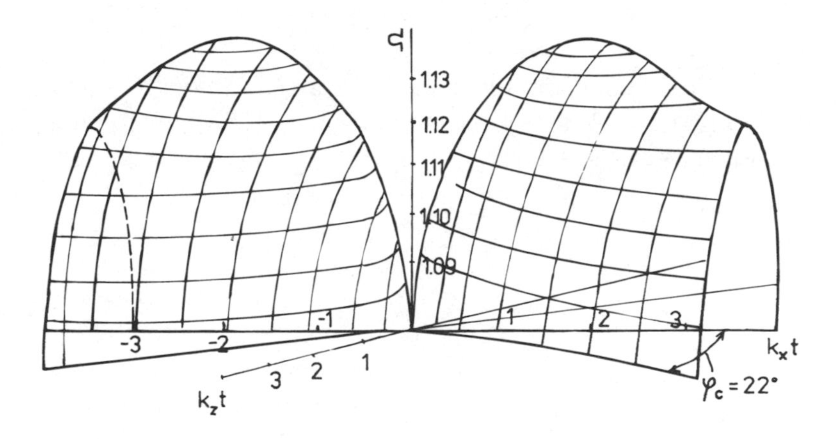

Fig. 3.8. Dispersion surfaces of surface MSW for a metallized layer. $\Omega_M = 0.15$, $l_1/t = 0.15$, $l_2/t = \infty$.

the amplitude of the wave vector $\vec{k}$ is the same in all directions and does not depend on the angle φ. In a tangentially magnetized layer the situation is different. In this case the wave vector $\vec{k}$ is a function of the angle of propagation φ, and the dependence $k(\varphi)$ is different in different frequency bands and is determined by the type of wave excited. Volume as well as surface waves can be excited. It can be said that in a tangentially magnetized layer MSW *propagate anisotropically.*

Let us proceed, now, with the investigation of questions of energy characteristics of MSW and especially the question of energy "flux" or group velocity when MSW propagate at a certain angle with regard to magnetic field direction. As we will see later, this question is extraordinarily interesting and goes beyond the framework of magnetostatic waves.

3.2 Group velocity and anisotropy of group velocity of magnetostatic waves

The orientation of group velocity in anisotropic media (in this case in a gyrotropic medium) in general is not identical with the orientation of propagation of a wave front, i.e. the group velocity vector $\vec{v}_g$ and the wave vector $\vec{k}$ are not collinear. By the term anisotropy of group velocities we will understand noncollinearity of the vectors $\vec{v}_g$ and $\vec{k}$ and also the dependence of $\vec{v}_g$ (amplitude and orientation) on frequency for fixed $\vec{k}$. The group velocity of the wave propagating in an anisotropic medium is given by the frequency gradient in wave vector space [15], i.e.

$$v_g = \text{grad}_k \omega \tag{3.11}$$

or

$$\vec{v}_g = \vec{i}_x \frac{\partial \omega}{\partial k_x} + \vec{i}_y \frac{\partial \omega}{\partial k_y} + \vec{i}_z \frac{\partial \omega}{\partial k_z}$$

where $\vec{i}_z$, $\vec{i}_y$, $\vec{i}_z$ are unit vectors in the direction of the x, y and z-axes respectively. In the case investigated, MSW propagate in the xz plane and for this reason the wave and group velocities have only two components

$$\begin{aligned} k_x &= k \cos \varphi \qquad & v_{gx} &= \partial \omega / \partial k_x \\ k_z &= k \sin \varphi \qquad & v_{gz} &= \partial \omega / \partial k_z \end{aligned} \tag{3.12}$$

The left side of the dispersion equation (3.7) [or (3.6), (3.8)] is a function of the frequency ω and the wave numbers k_x and k_y, i.e. $F(\Omega, k_x, k_z)$. If the rules for the derivation of implicit functions are used, for the components of group velocity vector we have

$$\begin{aligned} v_{gx} &= \frac{\partial \omega}{\partial k_x} = -\omega_H \frac{\partial F}{\partial k_x} \left(\frac{\partial F}{\partial \Omega} \right)^{-1} \\ v_{gz} &= \frac{\partial \omega}{\partial k_z} = -\omega_H \frac{\partial F}{\partial k_z} \left(\frac{\partial F}{\partial \Omega} \right)^{-1} \end{aligned} \tag{3.13}$$

where $F(\Omega, k_x, k_z)$ is given by the left-hand side of (3.7).

We can compute $\partial F / \partial k_x$, $\partial F / \partial k_z$ and $\partial F / \partial \Omega$ from (3.7) and after substituting into (3.13) for the components of group velocity in a tangentially magnetized ferrite layer we have

$$
\begin{aligned}
\frac{v_{gx}}{\omega_H t} &= -\frac{\cos\varphi}{B(\Omega,k,\varphi)}\Bigg((\mu^2-\mu_a^2-1)+\mu\tanh^{-1}[A(\Omega,\varphi)kt] \\
&\qquad \times\left[A(\Omega,\varphi)+\frac{1}{A(\Omega,\varphi)}-\frac{2kt}{\sinh[2A(\Omega,\varphi)kt]}\right]\Bigg) \\
\frac{v_{gz}}{\omega_H t} &= -\frac{\sin\varphi}{B(\Omega,k,\varphi)}\Bigg((\mu+1)+\tanh^{-1}[A(\Omega,\varphi)kt] \\
&\qquad \times\left[\mu A(\Omega,\varphi)+\frac{1}{A(\Omega,\varphi)}-\frac{2kt}{\sinh[2A(\Omega,\varphi)kt]}\right]\Bigg)
\end{aligned}
\tag{3.14}
$$

where

$$
\begin{aligned}
A(\Omega,\varphi) &= \sqrt{\left(1+\left(\frac{1}{\mu}-1\right)\sin^2\varphi\right)} \\
B(\Omega,k,\varphi) &= \frac{\mu_a kt}{1-\Omega^2}\Bigg(1+(1+\Omega_M)\cos^2\varphi+\tanh^{-1}[A(\Omega,\varphi)kt] \\
&\qquad \times\sin^2\varphi\cdot\left[\frac{2A(\Omega,\varphi)}{\sin^2\varphi}-\frac{1}{\mu A(\Omega,\varphi)}-\frac{2kt}{\mu\sinh[2A(\Omega,\varphi)kt]}\right]\Bigg)
\end{aligned}
\tag{3.15}
$$

For a given angle φ, the dispersion equation (3.8) can be solved, and the substitution of found values of Ω, k, μ, μ_a for given φ into (3.15) gives $A(\Omega,\varphi)$ and $B(\Omega,k,\varphi)$. After substituting $A(\Omega,\varphi)$ and $B(\Omega,k,\varphi)$ into (3.14) we can find $v_{gx}/\omega_H t$ and $v_{gz}/\omega_H t$ as functions of frequency or wave number. It is obvious that the dependence of the orientation and amplitude of group velocity on the orientation and amplitude of wave number (frequency) is rather complicated.

As follows from (3.14) for $\varphi = 0$ only one component of group velocity v_{gx} exists and the vector $\vec{v}_g$ is collinear with the wave vector $\vec{k}$, i.e. when the excitation of surface MSW is perpendicular to the orientation of magnetic field ($\vec{k}\perp\vec{H}_0$) the directions of phase and group velocities are identical. Practically this case is investigated in most papers on surface MSW. However, if the phase velocity begins to deviate from perpendicular to the direction of $\vec{H}_0$, interesting peculiarities begin to appear in the behaviour of the vector $\vec{v}_g$ (Fig. 3.9). If the vector $\vec{k}$ (antenna) deviates to the right from the normal with

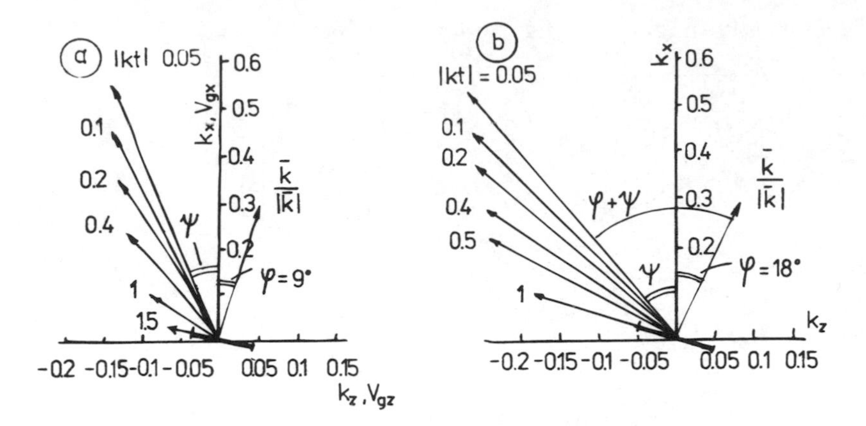

Fig. 3.9. The dependence of the orientation of group velocity on wave number for a constant deviation of the antenna a) $\varphi = 9°$ and b) $\varphi = 18°$, $\Omega_M = 2$.

regard to the magnetic field, then the vector $\vec{v}_g$ deviates from the normal but to the left and vectors $\vec{k}$ and $\vec{v}_g$ become noncollinear. As can be seen from Fig. 3.9a for the fixed direction of an antenna (orientation of $\vec{k}$ or angle φ) we observe a strong dependence of the vector $\vec{v}_g$ amplitude and orientation on the amplitude of the wave vector $\vec{k}$ (or frequency). With increasing frequency (or kt) the angle ψ between the vector $\vec{v}_g$ and the normal to the magnetic field direction increases and the amplitude of group velocity decreases. If the deviation of k from the normal increases to the right, the deviation of the vector $\vec{v}_g$ from the normal to the left increases as well (Fig. 3.9b). If the angle φ approaches the critical angle φ_c, the group velocity vector lies almost in the plane of the antenna. In this case MSW does not "go away" from the antenna but propagates along it. In this way it can be explained that surface MSW do not propagate at certain deviation angles φ.

The behaviour of the vector $\vec{v}_g$ at the change of the direction and the amplitude of the vector $\vec{k}$ (frequency) can be judged on the basis of the shapes of cross-sections of the dispersion surface at a constant frequency (horizontal cross-sections of Fig. 3.6). The curves of this type for two quadrants of the plane k_x, k_z are represented in Fig. 3.10. The normals at every point of such a curve, oriented to the direction in which the function $\Omega(k_x, k_z)$ increases, represent the orientations of group velocities (they are depicted by arrows in Fig. 3.10). This is the orientation of "flowing down" of

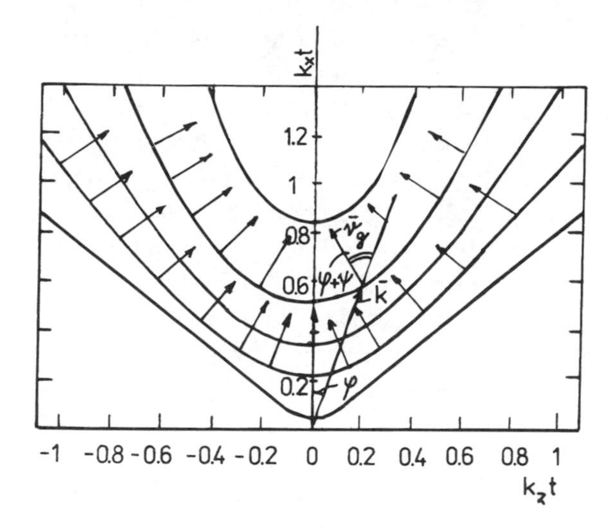

Fig. 3.10. Cross-sections of constant frequency of dispersion surface of surface MSW for two quadrants k_x, k_z of the plane $\Omega_M = 2$. Forward directions of group velocity are indicated by arrows.

energy from a "dispersion" hill. The density of constant frequency curves gives information about the amplitude of the group velocity (the steeper the slope, i.e. the larger the gradient, the larger the group velocity). As can be seen, the existence of anisotropy of the group velocity directly follows from the complex shape of the MSW dispersion surfaces $\Omega(k_x, k_z)$. The same anisotropy exists in the case of backward volume waves (Fig. 3.4). Only forward volume waves propagate isotropically. In this case the constant frequency curves are concentric circles with the centres on the Ω-axis and the orientations of $\vec{k}$ and $\vec{v}_g$ are identical for any angle φ.

3.3 Accumulation of energy and energy flux of magnetostatic waves Anisotropy of MSW energy flux

As was already noted in Chapter 1, magnetostatic waves in magnetized bodies (thin layers) in principle represent slow electromagnetic waves in anisotropic (gyrotropic) media. Since these waves are electromagnetic, they have to have a magnetic as well as an electric component of high-frequency field. These fields can be computed either using the complex system of Maxwell's equations (1.65) or from the reduced, so-called magnetostatic system of Maxwell's equations (1.70). These solutions and their comparison were made in the monograph [1]. The fundamental conclusions are the following. In a slow magnetostatic wave the dominant component of the high-frequency field is the magnetic field, i.e. the amplitude h of the high-frequency magnetic field is much bigger than the high-frequency component of the electric field e (see, e.g. the formula (2.25) in [1]). The electromagnetic invariant for these waves is $e^2 - h^2 < 0$, i.e. it is negative. Waves of this type are called in the literature predominantly magnetic. Let us investigate now the energy carried by such a wave as well as the problems concerning the energy flux (this question was partially solved in Chapter 2). First of all, the time average of MSW energy will be computed. For the anisotropic medium when losses are neglected, the electromagnetic energy density is given by [14]

$$W = \frac{1}{4}\mathrm{Re}\left[\vec{h}\cdot\frac{\partial(\omega\overline{\overline{\mu}}^*)}{\partial\omega}\cdot\vec{h}^* + \vec{e}\cdot\frac{\partial(\omega\overline{\overline{\varepsilon}}^*)}{\partial\omega}\cdot\vec{e}^*\right] \tag{3.16}$$

Since for MSW and $h^2 \gg e^2$ the second term in (3.16) can be neglected, (3.16) can be written in the form

$$W = \frac{1}{4}\mathrm{Re}\left[\vec{h}\cdot\frac{\partial(\omega\overline{\overline{\mu}}^*)}{\partial\omega}\cdot\vec{h}^*\right] \tag{3.17}$$

where $\overline{\overline{\mu}}^*$ and $\vec{h}^*$ are the complex conjugates of $\overline{\overline{\mu}}$ and $\vec{h}$. Since $\vec{h} = \nabla\psi$ (Chapter 1), for the energy density we can write

$$W = \frac{1}{4}\mathrm{Re}\left[\nabla\psi\cdot\frac{\partial(\omega\overline{\overline{\mu}}^*)}{\partial\omega}\cdot\nabla\psi^*\right] \tag{3.18}$$

For the time average of the energy flux density (the Poynting vector) holds (Chapter 2):

$$\overrightarrow{\overline{P}} = \frac{1}{2}\mathrm{Re}(\vec{e}\times\vec{h}^*) \tag{3.19}$$

This formula will be applied to magnetostatic waves, while the following formula from vector algebra will be used:

$$\mathrm{div}(\vec{e}\times\vec{h}^*) = \vec{h}^*\cdot\mathrm{rot}\vec{e} - \vec{e}\cdot\mathrm{rot}\vec{h}^*$$

If (1.70) is substituted for rot$\vec{e}$ and rot$\vec{h}^*$, we obtain

$$\mathrm{div}(\vec{e} \times \vec{h}^*) = -j\omega\vec{b} \cdot \nabla\psi^*$$

where $\vec{b} = \mu_0\overline{\overline{\mu}} \cdot \vec{h}$. Considering (1.70.3) the last formula can be rewritten as

$$\mathrm{div}(\vec{e} \times \vec{h}^*) = -j\omega\mathrm{div}(\psi^*\vec{b})$$

from what follows (neglecting the constant)

$$(\vec{e} \times \vec{h}^*) = -j\omega(\psi^*\vec{b}) \tag{3.20}$$

If (3.20) is substituted into (3.19), for the Poynting vector of the magnetostatic wave we have

$$\overrightarrow{\overline{P}} = -\frac{\omega}{2}\mathrm{Re}(j\psi^*\vec{b}) \tag{3.21}$$

The formulae (3.18) and (3.21) are fundamental formulae for the computation of MSW energy characteristics. To do so we have to know the scalar magnetostatic potential ψ, $\nabla\psi$, $\overline{\overline{\mu}}$ and the complex conjugate variables. To make this task more concrete a tangentially magnetized structure (Fig. 3.3) will bc investigated. For the scalar potential $\psi(x, y, z)$ in three areas of the structure (ferrite — 1, dielectric — 2, 3) we can write

$$\begin{aligned} \psi_1 &= \left(A\mathrm{e}^{-k_{y1}y} + B\mathrm{e}^{k_{y1}y}\right)\mathrm{e}^{j(k_x x + k_z z)} \\ \psi_2 &= C\mathrm{e}^{k_{y2}y} \cdot \mathrm{e}^{j(k_x x + k_z z)} \\ \psi_3 &= D\mathrm{e}^{k_{y3}y} \cdot \mathrm{e}^{j(k_x x + k_z z)} \end{aligned} \tag{3.22}$$

Let $k_{y2} = k_{y3} = k_y$ and use the boundary conditions (1.75). For the interfaces $y = 0$ and $y = t$ we will obtain a system of equations from which we will compute the unknown constants A, B, D (the constant C is supposed to be equal to 1). In general

$$\begin{aligned} A &= \frac{k_y + \mu k_{y1} - \mu_a k_x}{2\mu k_{y1}}; \qquad B = \frac{\mu k_{y1} + \mu_a k_x - k_y}{2\mu k_{y1}} \\ D &= (A\mathrm{e}^{k_{y1}t} + B\mathrm{e}^{-k_{y1}t})\mathrm{e}^{k_y t} \end{aligned} \tag{3.23}$$

From the formulae (3.18), (3.21) and (3.33) the energy density w and the components of the Poynting vector P_x, P_z of magnetostatic wave for each area of the investigated structure can be computed. By integration along the transverse co-ordinate of each area within corresponding boundaries (area 1: $0 \leq y \leq t$; area 2: $-\infty \leq y \leq 0$; area 3: $t \leq y \leq \infty$) and taking into account that $k_z = k \sin\varphi$, $k_x = k\cos\varphi$ and by introducing $\sqrt{[\cos^2\varphi + (\sin^2\varphi/\mu)]} = \alpha$ for the energy in the ferrite layer ($0 \leq y \leq t$) we will obtain

$$
\begin{aligned}
W_1 = -\frac{k}{8}\Bigg[&\left(1 + \Omega_M \frac{1+\Omega^2}{(1-\Omega^2)^2}\right) \times \left(\frac{\alpha^2 + \cos^2\varphi}{\alpha}\right) \\
&\times \left(A^2(\mathrm{e}^{2\alpha kt} - 1) - B^2(\mathrm{e}^{-2\alpha kt} - 1) - 4AB(\alpha^2 - \cos^2\alpha)kt\right) \\
&+ \frac{\sin^2\varphi}{\alpha}\left(A^2(\mathrm{e}^{2\alpha kt} - 1) - B^2(\mathrm{e}^{-2\alpha kt} - 1) + 4AB\alpha kt\right)\Bigg]
\end{aligned} \tag{3.24.1}
$$

and for the Poynting vector components

$$
\begin{aligned}
P_{x1} = -\frac{\omega}{4}\Bigg(&A^2 \frac{\alpha\mu_a + \mu\cos\varphi}{\alpha}(\mathrm{e}^{2\alpha kt} - 1) + 4AB\mu kt\cos\varphi \\
&+ B^2 \frac{\alpha\mu_a - \mu\cos\varphi}{\alpha}(\mathrm{e}^{-2\alpha kt} - 1)\Bigg)
\end{aligned} \tag{3.24.2}
$$

$$
P_{z1} = -\frac{\omega}{4}\frac{\sin\varphi}{\alpha}\left(A^2(\mathrm{e}^{2\alpha kt} - 1) + B^2(\mathrm{e}^{-2\alpha kt} - 1) + 4AB\alpha kt\right) \tag{3.24.3}
$$

For the areas under the ferrite layer ($-\infty \leq y \leq 0$)

$$
W_2 = -\frac{k}{4} \tag{3.25.1}
$$

$$
P_{x2} = -\frac{\omega}{4}\cos\varphi \tag{3.25.2}
$$

$$
P_{z2} = -\frac{\omega}{4}\sin\varphi \tag{3.25.3}
$$

and for the area over the ferrite layer ($t \leq y \leq \infty$)

$$
W_3 = \frac{k}{4}\left(A\mathrm{e}^{\alpha kt} + B\mathrm{e}^{-\alpha kt}\right)^2 \tag{3.26.1}
$$

$$
P_{x3} = -\frac{\omega}{4}\left(A\mathrm{e}^{\alpha kt} + B\mathrm{e}^{-\alpha kt}\right)\cos\varphi \tag{3.26.2}
$$

$$
P_{z3} = -\frac{\omega}{4}\left(A\mathrm{e}^{\alpha kt} + B\mathrm{e}^{-\alpha kt}\right)\sin\varphi \tag{3.26.3}
$$

The total energy accumulated in the total transverse (infinite) volume with the base of unit area (specific energy density) will be the sum of (3.24.1), (3.25.1) and (3.26.1):

$$W = W_1 + W_2 + W_3 \tag{3.27}$$

The components of the resulting Poynting vector (the energy flux through the transverse area of unit width) will be

$$\begin{aligned} P_x &= P_{x1} + P_{x2} + P_{x3} \\ P_z &= P_{z1} + P_{z2} + P_{z3} \end{aligned} \tag{3.28}$$

The amplitude and direction of the resulting Poynting vector are given by the formulae

$$|\vec{P}| = \sqrt{(P_x^2 + P_z^2)}; \qquad \Theta = \mathrm{atan}(P_z/P_x) \tag{3.29}$$

It should be noted that partial energy fluxes P_{x1}, P_{x2}, P_{x3} and P_{z1}, P_{z2}, P_{z3} do not have a physical meaning because they do not realize themselves in an independent way and they have a purely mathematical character. Only the components of total MSW power flow (energy flux) (3.28) through an infinite transverse area ($-\infty \leq y \leq \infty$), P_x, P_z and $\vec{P}$ have a physical meaning.

If the total Poynting vector $|\vec{P}|$ and accumulated energy W are known, the velocity of energy transmission (velocity of energy flux) can be found

$$v_e = \frac{|\vec{P}|}{W} = \frac{\sqrt{(P_x^2 + P_z^2)}}{W} \tag{3.30}$$

The behaviour of the total energy flux will be analyzed now and the results obtained will be compared with the results of the previous section. As above, anisotropy of MSW energy flux will mean the dependence of the $\vec{P}$ direction on frequency for fixed k and non-collinearity of the vectors $\vec{P}$ and $\vec{k}$ in general.

First of all the case $\varphi = 0°$ will be investigated, i.e. the surface MSW is excited in a magnetic crystal, the wave front is parallel to the magnetic field direction. From the formulae (3.24.3), (3.25.3) and (3.26.3) it can be seen that all components of total energy flux are equal to zero and the MSW energy propagates in the x-axis direction, i.e. it is identical with the direction of the vector $\vec{k}$. When the vector $\vec{k}$ deviates from the normal to the magnetic field direction, the situation changes substantially.

Directions of the total MSW energy flux $\vec{P}$ through an infinite transverse plane for the various values of frequency Ω in the pass band of the system of surface magnetostatic waves ($\Omega_M = 2; 1.73 \leq \Omega \leq 2$) are depicted in Fig. 3.11 for various directions of $\vec{k}$.

As was expected, in accordance with the results of the previous section, if the vector $\vec{k}$ deviates from the normal to the magnetic field direction (the x-axis), the total energy flux deviates to the left. In addition, the higher the signal frequency the bigger the deviation. For every value of φ a defined sector of the permitted orientations of the total energy flux vector exists dependent on the frequency at constant k. As follows from Fig. 3.11, if the angle φ increases, this sector narrows and in the limit $\varphi \to \varphi_c$ the width of this sector goes to zero. In this case the energy flux is oriented along the antenna, which does not radiate. In Fig. 3.11 the right limit of the sector

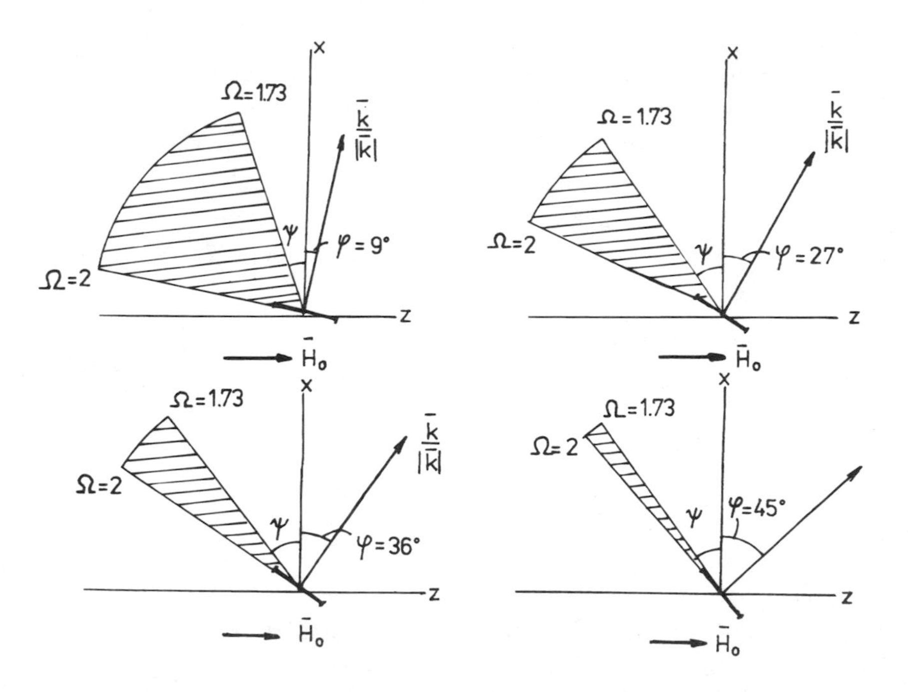

Fig. 3.11. Forward flow of the total power for various orientations of wave vector (exciting antenna). Hatched areas correspond to the areas of energy flux of the propagating wave.

of the permitted flux orientations corresponds to the lower limit of the pass band of surface MSW ($\Omega_M = 2$; $\Omega = 1.73$) and the left limit corresponds to the upper limit ($\Omega = 2$). As also follows from Fig. 3.11, if the angle φ increases, the angle ψ between the normal to $\vec{H}_0$ (the x-axis) and the direction of the Poynting vector (for the right sector limit in Fig. 3.11) increases. From calculations it follows that for the given direction $\vec{k}$ and given frequency Ω the direction of the Poynting vector P is identical with the orientation of the group velocity $\vec{v}_g$ [1].

The dependences of the orientation of the total energy flux of surface MSW (the area characterized by the angle φ measured to the left from the normal to the magnetic field direction) on the location of an excitation antenna (characterized by the angle φ measured to the right from the normal) for various frequencies in the pass band are presented in Fig. 3.12. All curves have one source point in which the direction of k is identical with the normal to the magnetic field and they end on the limit depicted in Fig. 3.12 by the dashed curve. This curve corresponds to the limit case when $|\vec{k}| \to \infty$ and the angle $(\varphi + \psi) = 90°$. In this case the total energy flux is oriented along the phase front of surface MSW. Below this curve the sum of angles $(\varphi + \psi) < 90°$, i.e. the angle between the direction of MSW energy flux and the wave vector is less than 90°. It is noteworthy that if the antenna deviates to the right, the angle ψ increases to the left.

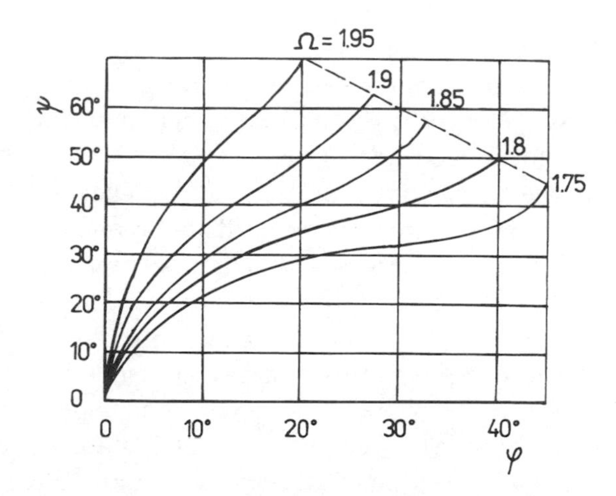

Fig. 3.12. The dependence of total energy flux orientation on the position of an exciting antenna for surface MSW. $\Omega_M = 2$, $l_1 = l_2 = \infty$. Parameter Ω.

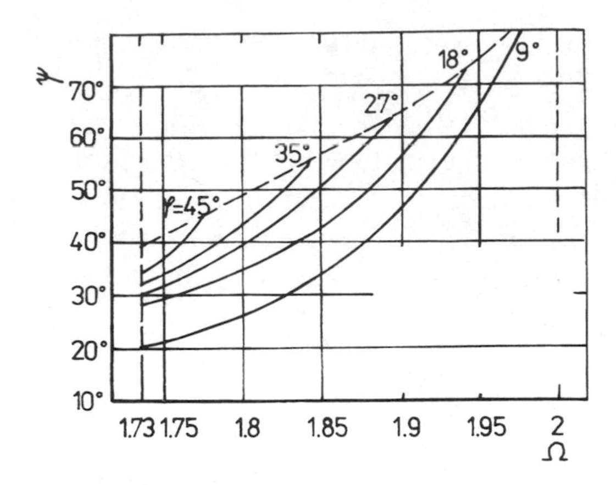

Fig. 3.13. The dependence of the Poynting vector on frequency. The orientation of an exciting antenna φ — parameter. Dashed lines represent the boundaries of the pass band.

If the frequency in the pass band increases, the dependence of the angle ψ on the angle φ is more explicit.

Fig. 3.13 represents the dependence of the Poynting vector orientation on the increase of frequency at a fixed antenna location (the angle φ). As can be seen, with increasing frequency the angle ψ increases as well, i.e. the orientation of P deviates more explicitly from the normal. Approaching the upper frequency limit, the steepness of the dependence $\psi(\Omega)$ increases as well. When the direction of the vector $\vec{k}$ approaches the critical angle ($\varphi = 54°$), the width of the sector of possible power flow orientations tends to zero.

Let us now investigate the question of MSW energy concentrated in a transverse volume ($-\infty \leq y \leq \infty$), the base of which is a unit area. Energy is given by the formulae (3.24.1), (3.25.1), (3.26.1) and (3.27). The dependences of the energy of the system on frequency for various φ are depicted in Fig. 3.14. As can be seen, if the frequency in the pass band increases, the energy increases, and approaching the upper band frequency (for a given angle φ) it tends infinity. This is correct, because in this case the group velocity tends to zero.

Finally, the results of the computation of energy transmission velocity given by the formula (3.30) will be investigated. The dependences of the energy flux velocity on the frequency for various angles φ are depicted in Fig. 3.15. The fundamental conclusions are the following. The orientation of the total MSW energy flux coalesces with the orientation of MSW group velocity. The amplitude of en-

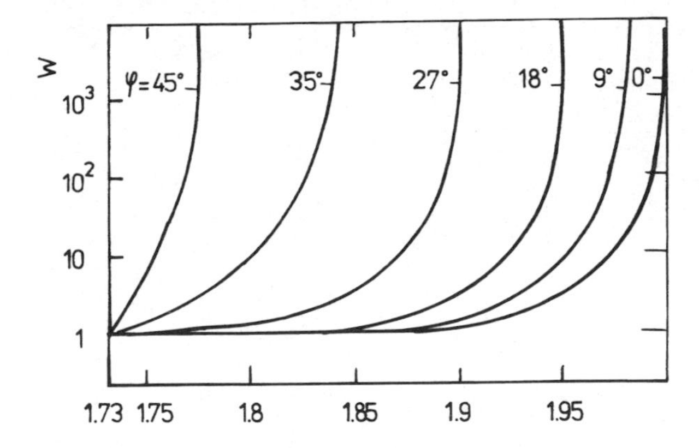

Fig. 3.14. The dependence of energy W on frequency for various values of φ, $\Omega_M = 2$, $l_1 = l_2 = \infty$.

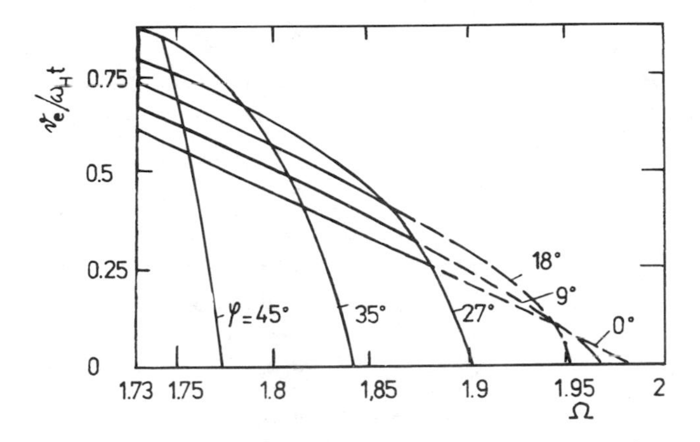

Fig. 3.15. The dependence of energy flux velocity on frequency for various values of φ, $\Omega_M = 2$, $l_1 = l_2 = \infty$.

ergy transmission velocity $v_e = |P|/W$ is equal to the amplitude of group velocity $|v_g|$. It can be said that in the case of MSW these notions are almost identical. From Fig. 3.15 it follows that at small angles φ the velocity of energy transmission with frequency decreases almost linearly in a wide frequency band. On approaching the upper band frequency the transmission velocity steeply decreases, the delay increases, and the accumulated energy increases as well. For a constant frequency the velocity of the energy transmission increases with the rise of the angle φ. At the lower band frequency ($\Omega = 1.732$ for $\Omega_M = 2$) the velocity of surface MSW transmission is finite for

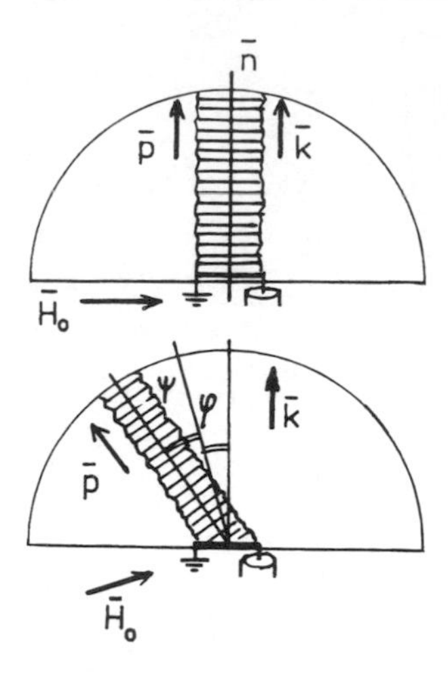

Fig. 3.16. Schematic representation of MSW beam excitation and propagation at various angles with regard to magnetic field.

any angle φ.

In [17] an experimental investigation of anisotropic properties of surface MSW in the frequency band 4 – 4.5 GHz is described. Thin YIG films about 10 μm thick in the shape of half-discs with diameter 40 mm were used in the experiment. Excitation of MSW was realized by a microstrip antenna with aperture 4 mm, which enabled the excitation of a flat wave front of MSW parallel to a transducer (Fig. 3.16). The angle φ between the vector $\vec{k}$ and the normal to $\vec{H}_0$ was given by the rotation of the whole system with regard to the magnetic field direction. Reception of surface MSW was realized by the loop inductive probe with a very small aperture which could be located at any point of the thin YIG layer and the orientation of the loop plane to the input transducer could be changed. The behaviour of the energy flux of the group velocity at the deviation of wave vector from the normal to the magnetic field direction is schematically depicted in Fig. 3.16. If the wave vector deviates to the right of the normal, the wave beam deviates to the left. Measurements by means of a probe show that at $\varphi = 0°$ the orientation of energy flux of surface MSW coincides with the orientation of $\vec{k}$, i.e. $\psi = 0°$. The maximum signal in the probe was observed in the case when the plane of the probe was parallel to the plane of the excitation antenna (phase wave front) and decreased to zero on the rotation of the probe by 90°. When the angle between the wave vector $\vec{k}$ and the magnetic field direction $\vec{H}_0$ began to increase, the measurements by the probe showed that, in accordance with the theory, the orien-

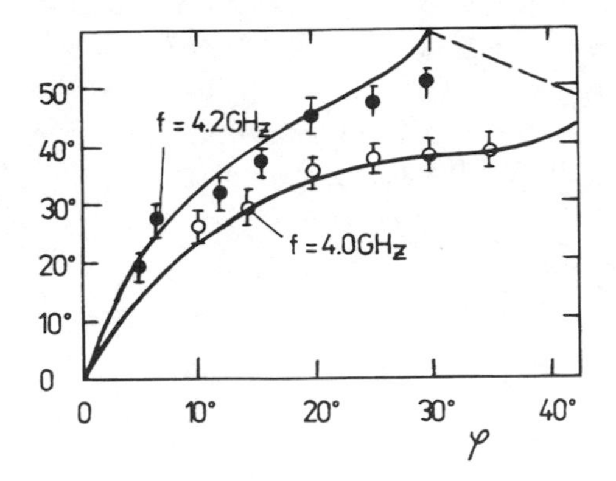

Fig. 3.17. The comparison of theoretical and experimental dependences $\psi(\varphi)$ for two frequencies 4.0 and 4.2 GHz at $H_0 = 62.4\,\mathrm{kA\,m^{-1}}$, $l_1 = l_2 = \infty$.

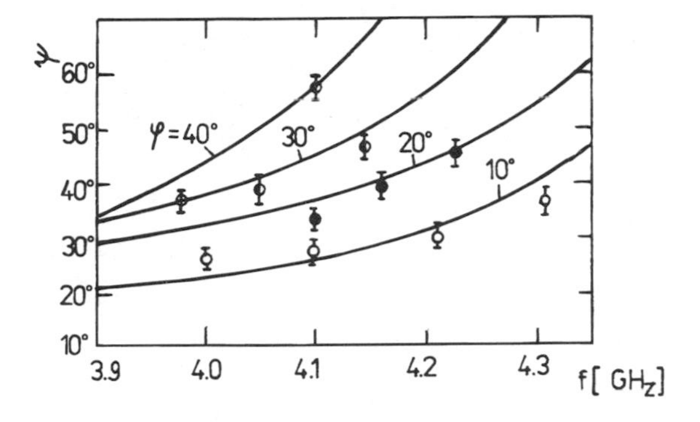

Fig. 3.18. The dependence of the change of the forward direction of energy flux on frequency for various orientations of the vector $\vec{k}$, $H_0 = 62.4\,\mathrm{kA\,m^{-1}}$, $l_1 - l_2 - \infty$.

tation of $\vec{P}$ began to deviate from the direction of $\vec{k}$ to the left. The theoretical curves $\psi(\varphi)$ for the special case $\omega/2\pi = 4; 4.2\,\mathrm{GHz}$ and $H_0 = 62.4\,\mathrm{kA\,m^{-1}}$ are presented in Fig. 3.17; experimental points are also depicted. As can be seen, good qualitative agreement exists between experiment and theory. In Fig. 3.18 the dependence of the direction of energy flux on the frequency for various orientations of the vector $\vec{k}$ is shown. It is obvious that for a constant value of $\vec{k}$ with increasing amplitude the angle ψ increases as well (compare with Fig. 3.13). Again, as above there is a good qualitative agreement between theory and experiment.

It is noteworthy that if the magnetostatic beam deviates from the normal to the magnetic field direction, the beam becomes narrower and at the critical angle its width tends to zero. In this way a very narrow magnetostatic beam ("magnetostatic needle") can be formed, the direction of which depends on the signal frequency. On this principle frequency-spatial MSW filters can be constructed.

Chapter 4

QUASIOPTICS OF MAGNETOSTATIC WAVES

4.1 Beams of magnetostatic waves Reflection and refraction of magnetostatic waves

As a result of a large deceleration, magnetostatic waves are characterized by a relatively small wavelength in comparison with the wavelength of electromagnetic waves of the same frequency in vacuum. In typical cases in thin YIG layers the wavelength of magnetostatic waves (MSW) is from one to a hundred micrometres, i.e. much smaller than the linear size of waveguides of magnetostatic structures: 10^3 – 10^4 times smaller. This condition is even more appropriate if we do not use magnetostatic waveguides, but instead thin YIG films on the substrate with diameters of 50 – 100 mm and more, produced in series. In such structures magnetostatic waves can exist in the form of magnetostatic beams which are similar to light beams propagating in a transparent medium. Such a beam propagates "slowly" in a structure (ten thousand times more slowly than a light beam) — either on the surface of a thin film (surface wave) or in the volume of a film (volume wave). A strict mathematical definition of this beam does not exist. We can only say that in optics a beam is a finite rather narrow bundle of electromagnetic waves which can exist independently of other beams [18]. A beam formed by some source (antenna) cannot be infinitely narrow because as a result of diffraction wave propagation will be accompanied by an increase of the transverse size of the beam, i.e. its widening at the sides. The wider a beam is, the wider diffraction broadening will be. We can use the term beams or bundle when this widening is much smaller than the transverse size of the beam. From the theory of diffraction we find that the term beam is usable for the bundle of electromagnetic waves on the condition that the following inequality is valid

$$L \ll D^2/\lambda \tag{4.1}$$

where L is the length of the beam, D is the minimum size of a diaphragm (antenna), λ is the wavelength in the medium we are considering. As a result of (4.1), under the same conditions, the larger the delay, i.e. the smaller the wavelength, the bigger L will be, i.e. the waves will propagate a longer distance as beams. All these considerations are known in connection with light (optical) beams. In full they can be used for slow magnetostatic waves in magnetically ordered matter. Difficulties will be caused only by the fact that unlike ordinary light propagation the propagation of magnetostatic waves can be anisotropic in many cases. In addition to this, magnetically ordered matter is characterized by strong dispersion of magnetostatic waves.

For an examination of the properties of an MSW beam we will use the approximation of geometrical optics, i.e. we will neglect diffractive phenomena in the magnetostatic beam. This is why the wavelength of a beam will be limited by (4.1). Then we will broaden the task and consider inevitable diffractive corrections. Four laws which are valid for linear media, i.e. properties of these media do not depend on the amplitude of a wave, are the basis for geometrical optics. They are:

1. The law of rectilinear propagation of light;
2. The law of independent propagation of light beams;
3. The laws of reflection of light;
4. The laws of refraction of light.

In our case there is no reason to assume that the propagation of magnetostatic waves in magnetically ordered matter will not obey the experimental laws of geometrical optics. A magnetostatic wave is a very slow electromagnetic wave propagating in a medium and it should differ from monochromatic light beams in optical media only quantitatively. It is necessary to add that the energy flux of a magnetostatic wave has a pure electromagnetic character, unlike e.g. acoustic waves in elastic media. But only an experiment can answer the question of whether it is possible to apply and use methods of geometrical optics for the investigation of the optical properties of magnetostatic waves and the computation of various quasioptical component elements for MSW — magnetostatic lenses, prisms, mirrors, etc. That is why we will proceed as follows: we will present the basic theory of reflection and refraction of magnetostatic waves using results from the previous chapters and we will compare the results with an experiment [1].

The investigation of the properties of MSW, which we are interested in, is illustrated by the example of surface MSW propagating in a free ferrite layer. The knowledge acquired about optical properties of surface MSW will be used in the investigation of peculiarities of the behaviour of volume MSW. We will base our argument on materials presented in Chapter 3.

Let us return to the dispersion equation for surface MSW propagating at an arbitrary angle to the magnetic field (e.g. equation (3.8)). We can write it in the form

$$F = \tanh\left[k\cos\varphi\sqrt{\left(1+\frac{\tan\varphi^2}{\mu}\right)}\,t\right] + 2\frac{\sqrt{[\mu(\mu+\tan\varphi^2)(1+\tan\varphi^2)]}}{(1+\mu^2-\mu_a^2)+(1+\mu)\tan^2\varphi} \equiv 0 \quad (4.2)$$

Dispersion surfaces corresponding to equation (4.2) are shown in Fig. 3.10. By analyzing curves of a constant frequency we can understand easily that for $\varphi = 0$ the orientations of phase and group velocities are the same and for $\varphi = \varphi_c$ they will be perpendicular to one another. As has been mentioned above, this is the physical essence of the critical angle and that is why in a wave process no energy transmission can occur in the direction of a plane wave front propagation. The direction at right angles to the magnetizing field, i.e. the perpendicular to the magnetic field has the meaning of an optical axis in a uniaxial crystal. Hence the direction in which vectors $\vec{k}$ and $\vec{v}_g$ are collinear will be called the optical axis of a thin layer. Obviously, the direction of the optical axis is determined only by the orientation of the magnetic field. Let angle φ determine the direction k for the optical axis and angle ψ the direction of the group velocity for the same axis. As was shown in the previous chapter, the relationship between the direction of MSW front propagation (angle φ) and the energy of MSW (the direction of group velocity, angle ψ) is nonlinear (Fig. 3.12). We will examine this issue in more detail. From (3.13), angle ψ is given by the equation

$$\tan\psi = \frac{v_{gz}}{v_{gx}} = \frac{\partial F/\partial k_z}{\partial F/\partial k_x} \quad (4.3)$$

where F is given by equations (4.2) or (3.8). After differentiation we get

$$\tan\psi = \tan\varphi \frac{2A(1+\mu)\sinh^2(Akt) + (1+\mu A^2)\sinh(2Akt) + 2Akt}{2A(1+\mu^2-\mu_a^2)\sinh^2(Akt) + \mu(1+A^2)\sinh(2Akt) + 2\mu Akt} \tag{4.4}$$

where $A = \sqrt{[\cos^2\varphi + (\sin^2\varphi)/\mu)]}$.

The results of the calculations for positive and negative values are plotted in Fig. 4.1. These curves are analogous to the ones in

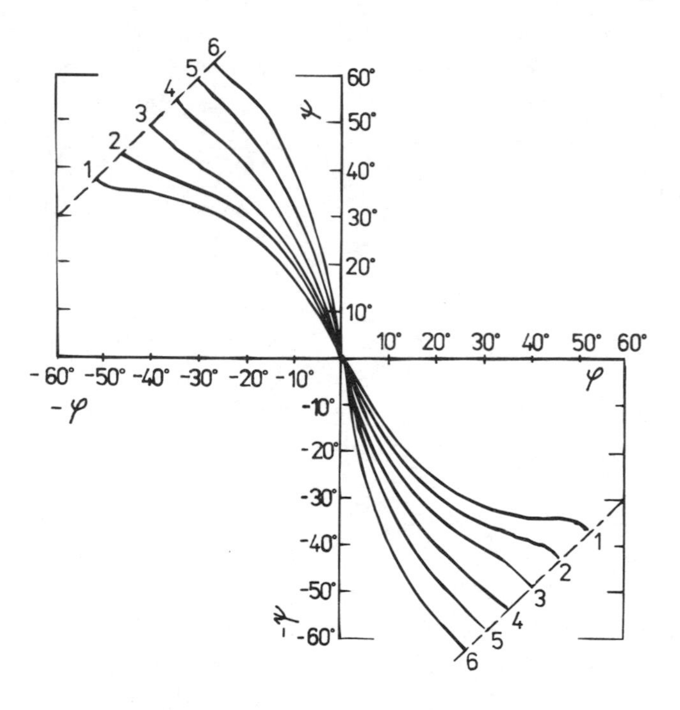

Fig. 4.1. Mutual dependence of the angle ψ on the antenna rotation angle φ for surface MSW and various frequencies. 1 — 3.4, 2 — 3.5, 3 — 3.6, 4 — 3.7, 5 — 3.8, 6 — 3.9 GHz, $M_0 = 140\,\text{kA m}^{-1}$, $H_0 = 54.8\,\text{kA m}^{-1}$.

Fig. 3.12. Notice that there is a complicated character of dependences $\psi(\varphi)$ on the frequency. For frequencies in the vicinity of a lower boundary frequency (close to $\sqrt{[\omega_H(\omega_H+\omega_M)]}$) the dependence $\psi(\varphi)$ is considerably nonlinear and the curves have a distinct flat part parallel to the co-ordinate axis. For frequencies in the vicinity of the upper boundary frequency (close to $(\omega_H + 1/2\omega_M)$) the dependence $\psi(\varphi)$ straightens up and is almost a straight line. All the curves finish in the cut-off lines of a surface MSW whilst the equations of these lines are given by $\varphi = \varphi_c + 90°$. We can see again that if $\varphi = \varphi_c$ then the orientation of the group velocity is perpendicular to $\vec{k}$. Let us notice another peculiarity of surface MSW propagation. Obviously, it is valid that

$$\varphi_c = \arccos\left[\frac{\omega + \sqrt{[\omega^2 - \omega_H(\omega_H + \omega_M)]}}{\omega_H + \omega_M}\right] \tag{4.5}$$

i.e. the higher the frequency is inside the existence interval of surface MSW the narrower the wave propagation spectrum will be, and the wider the energy transmission sector will be. In other words, the more wave fronts are oriented in the direction of the magnetic field (vector $\vec{k}$ is parallel to the optical axis) the wider sector in which the energy of a wave "gets blurred".

Now let us examine the laws of reflection and refraction of magnetostatic waves. We will begin with reflection. Let a thin film, in which wave propagation is in progress, have a plane mirror boundary from which an incident wave reflects completely, as from an ideal mirror. At the same time the basic condition of the geometrical law of reflection — *the projections of wave vectors to the mirror plane are equal* — must be fulfilled. We will use this condition in order to find the direction of a reflected wave. In an anisotropic medium this is a considerably complicated task as can be seen from an examination of the constant frequency curves for surface MSW in Fig. 3.10. If the mirror position is oblique to the orientation of an optical axis, then it is obvious that the angle of reflection must differ considerably from the angle of incidence. This results from the fact that for a single frequency (for a single curve of a constant frequency) the magnitude of a wave vector $\vec{k}$ is determined by its direction $\vec{k}/k$ with regard to the optical axis. Since after the reflection of MSW the direction of the vector $\vec{k}$ and its magnitude are changing and the equality of tangential projections must be kept, it is obvious that the angle of reflection cannot be equal to the angle of incidence. At the same time there is a question of what will happen if a wave after reflection from an ideal mirror is propagating in a direction which lies in the sector where wave propagation is not possible ($\varphi > \varphi_c$, Fig. 3.10). Let a surface wave propagate in the positive direction of the x-axis (Fig. 4.2). The magnetizing field $\vec{H}_0$ is oriented in the direction of

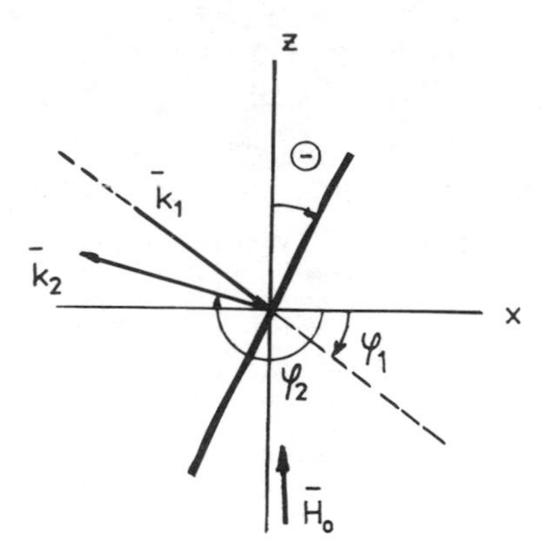

Fig. 4.2. The geometry of the problem.

the z-xis, i.e. the x-axis is an optical axis. The incident waves inside an allowed propagation sector $\varphi < \varphi_c$. In this case the magnetic potential contains the coefficient $\exp(\pm jk_z z)\cdot\exp(jk_x x)(k_x > 0)$ and

$$\varphi = \arctan\left|\frac{k_z}{k_x}\right| \leq \varphi_c \tag{4.6}$$

Let an ideal mirror whose plane has an angle Θ with the z-axis be placed in the path of this beam, as depicted in Fig. 4.2. Let us denote the quantities of the incident and reflected waves with the indices 1 and 2, respectively. Then, the condition that the wave vector projections to the mirror plane are equal to each other results in

$$k_1(\tan\varphi_1 - \tan\Theta)\cos\varphi_1 = k_2(\tan\varphi_2 - \tan\Theta)\cdot\cos\varphi_2 \tag{4.7}$$

The angles φ_1 and φ_2 open clockwise from the positive direction of the x-axis (optical axis). Hence if the angle φ_1 is in the first or fourth quadrant (as is shown in Fig. 4.2 and the condition (4.6) is fulfilled), then, depending on the mirror location (angle Θ) the angle φ_2 can appear in an arbitrary quadrant. In the simplest case if $\Theta = 0$ and $\varphi_1 = 0$, then $\varphi_2 = 180°$. Let us note that in our case it is not advantageous if angles open from the mirror plane to the different sides.

In equation (4.7) $k^2 = k_x^2 + k_z^2$ and depends on the propagation direction, i.e. on the angle φ which is given (for the constant frequency) by the dispersion equation (4.2). In order to find the dependence φ_2 on φ_1 it is necessary to solve a system of equations consisting of equation (4.7) and two dispersion equations for the incident and reflected waves in a form similar to equation (4.2)

$$k_1(\tan\varphi_1 - \tan\Theta)\cos\varphi_1 - k_2(\tan\varphi_2 - \tan\Theta)\cos\varphi_2 = 0$$

$$\tanh\left[k_1\cos\varphi_1\sqrt{[1+(\tan^2\varphi_1/\mu)]t}\right] + 2\frac{\sqrt{[\mu(\mu+\tan^2\varphi_1)(1+\tan^2\varphi_1)]}}{(1+\mu^2-\mu_a^2)(1+\mu)\tan^2\varphi_1} = 0$$

$$\tanh\left[k_2\cos\varphi_2\sqrt{[1+(\tan^2\varphi_2/\mu)]t}\right] + 2\frac{\sqrt{[\mu(\mu+\tan^2\varphi_2)(1+\tan^2\varphi_2)]}}{(1+\mu^2-\mu_a^2)(1+\mu)\tan\varphi_2} = 0 \tag{4.8}$$

These three equations can be reduced to one which gives the connection between the angle of incidence φ_1 and the angle of reflection φ_2 for a given mirror slope angle Θ and a given frequency, i.e. the connection between μ and μ_a for a selected magnetization value ω_M. This equation is as follows:

$$\frac{\tan\varphi_1 - \tan\Theta}{\sqrt{(\mu+\tan^2\varphi_1)}}\text{atanh}\left[2\frac{\sqrt{[\mu(\mu+\tan^2\varphi_1)(1+\tan^2\varphi_1)]}}{(1+\mu^2-\mu_a^2)(1+\mu)\tan^2\varphi_1}\right] =$$

$$= \frac{\tan\varphi_2 - \tan\Theta}{\sqrt{(\mu+\tan^2\varphi_2)}}\text{atanh}\left[2\frac{\sqrt{[\mu(\mu+\tanh^2\varphi_2)(1+\tan^2\varphi_2)]}}{(1+\mu^2-\mu^2)(1+\mu)\tan^2\varphi_2}\right] \tag{4.9}$$

Fig. 4.3a shows a typical dependence of φ_2 on φ_1 for various values of angle Θ, for the case $\Omega_M = 3$, $\Omega = 2$. The incident wave is propagating in the positive direction of the x-axis in the angle sector $-51^\circ \leq \varphi \leq 51^\circ$ (i.e. the critical angle $\varphi_c = \pm 51^\circ$). As follows from Fig. 4.3, the corresponding values φ_2 dependent on the mirror slope (angle Θ) can be in the two regions $-51^\circ \leq \varphi_2 \leq 51^\circ$ and $129^\circ \leq \varphi_2 \leq 231^\circ$. In the first region of φ_2 the projections of the wave vectors of the incident and reflected waves to the optical axis have the same signs and in the second region they have opposite signs (Fig. 4.4a, b).

If $\Theta = 0^\circ$, then the dependence $\varphi_2(\varphi_1)$ is a straight line going through the point $(0^\circ, 180^\circ)$. In this case the normal of the mirror is oriented in the direction of the optical axis and divides the layer into two half-planes which are symmetrical with regard to their properties. In this case the laws of mirror reflection will be fulfilled i.e.

the angle of incidence is equal to the angle of reflection. So, e.g. if $\varphi_1 = 10°$ then the angle of reflection is equal 170° or 10°, depending on whether it moves from the normal counterclockwise or clockwise, respectively. We can get similar straight lines if the angle of the mirror orientation Θ is equal to 90° or 270°.

The analysis of the calculated dependences $\varphi_2(\varphi_1)$ plotted in Fig. 4.3 shows that by turning the mirror (changing the angle Θ) we can find four intervals of Θ values in which a reflected wave can exist. The first interval includes values Θ from 0° to 51°. In this angle interval a reflected wave propagates to the left in the negative direction of the x-axis ($k_x < 0$, the upper part of Fig. 4.3). All curves describing reflections in this interval converge to the two points corresponding to the critical angles $\varphi_c = \pm 51°$. In the second interval of angles of reflection ($51° \leq \Theta \leq 129°$) a reflected wave propagates to the right in the positive direction of the x-axis ($k_x > 0$, the lower part of Fig. 4.3). In this case the curves describing the reflection are limited by the two points again. One of them represents the critical angle and the second one is an angle at which the reflected wave is parallel to the mirror plane. The point with co-ordinates (0,0) in Fig. 4.3, at which the curves for $\Theta = 90°$ and 270° converge, is empty. Then the angle interval $120° \leq \Theta \leq 180°$ follows, in which MSW incident on the mirror and propagating in the positive direction of the x-axis cannot exist. Further, everything is repeating with mirror symmetry with regard to the point $\Theta = 180°$.

Special attention must be devoted to the situations in which $\Theta = 51°$ and 309°. These are the angles at which, for all possible incidence directions of a surface MSW, the boundary conditions require the existence of a reflected wave propagating outside the possible propagation directions sector of surface MSW, and this is not possible due to the anisotropic properties of a thin layer. There is evidence [1] that in this case an edge wave arises propagating along the mirror plane (the edge of a layer). The front of this wave is perpendicular to the mirror plane and its amplitude decreases in the perpendicular direction to the mirror plane. A reflected wave propagating into the sample does not exist. This phenomenon was called "total nonreflection" [1]. It is a very important phenomenon for the implementation of component elements in a forward wave regime.

Up to now everything mentioned has been related to the wave vector orientation. However, it is more important to know how the group velocity orientation will change after reflection from a mirror.

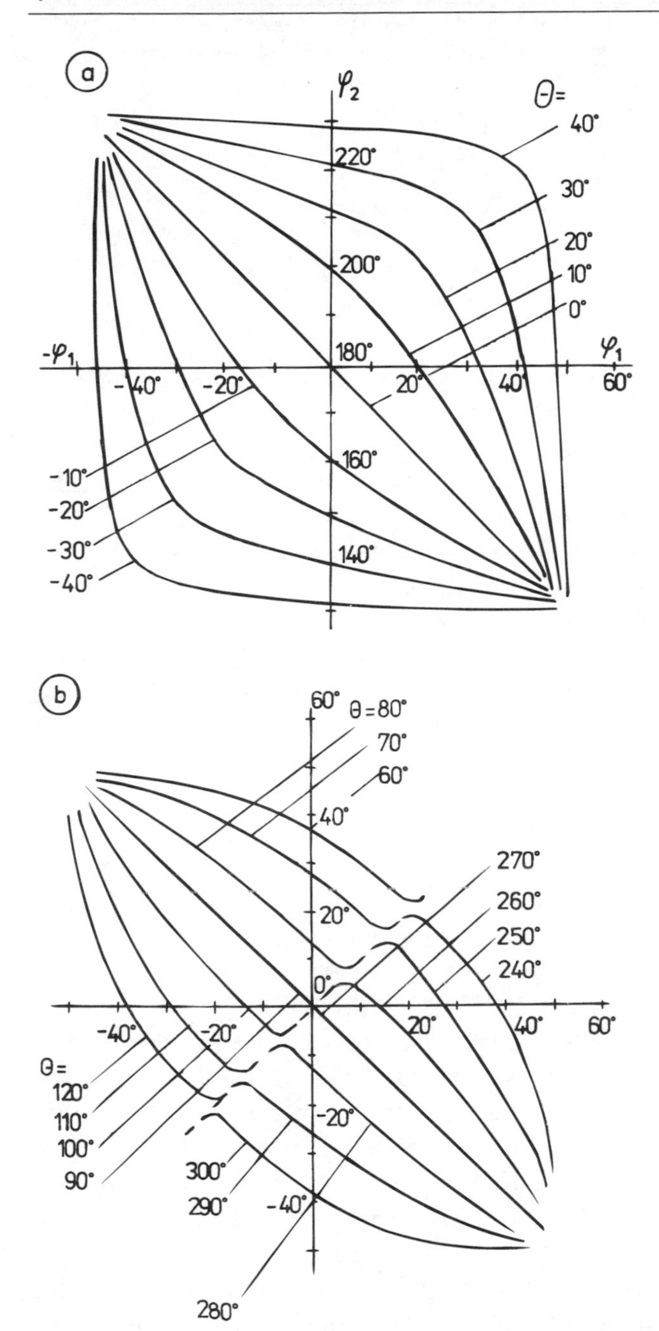

Fig. 4.3. Typical dependence of φ_2 on φ_1 for various values of the angle Θ. $\Omega_M = 3$, $\Omega = 2$ ($M_0 = 140\,\mathrm{kA\,m^{-1}}$; $H_0 = 70\,\mathrm{kA\,m^{-1}}$, $f = 3400\,\mathrm{MHz}$).

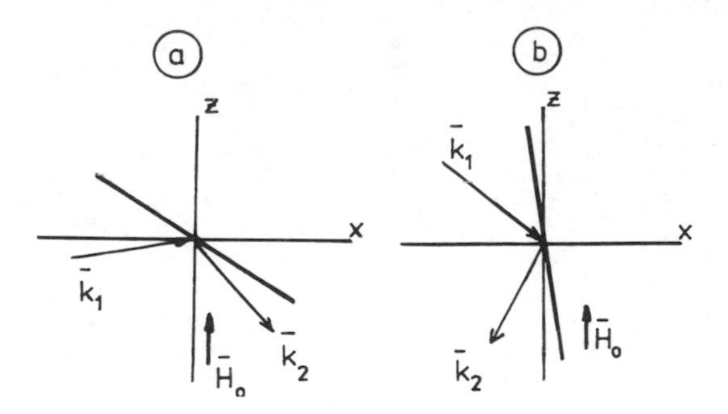

Fig. 4.4. Schematic representation of the wave vector of an incident and reflected wave for two orientations of magnetic field.

The relationship between group velocities and wave vector directions is given by (4.4). It is possible using (4.4) to transform the dependences in Fig. 4.3 to corresponding dependences for orientations of group velocities. The results are presented in Fig. 4.5. Dependences in Fig. 4.3 and 4.5 enable us to find the directions of a reflected beam and also the orientation of a wave front inside the beam.

Let us turn our attention to the interesting peculiarities of Fig. 4.5. If a mirror and the optical axis contain the angle (20° – 30°), then all beams incident on the mirror at various angles are reflected in the same direction for the whole sector of existence of surface MSW. Only in the vicinity of the critical angle is the situation changing considerably. In this case a very small change of the angle of incidence ψ_1 causes a big change of the angle of reflection ψ_2. If, e.g. $\Theta = 20°$, then all beams incident on the mirror in the sector $\pm 35°$ are reflected at the one angle $-216°$ (more exactly in the narrow interval $216° \pm 2°$). On the other hand, beams incident at an angle $36 \pm 2°$ reflect in a wide interval $(180° \pm 35°)$. Reflected beams falling at big angles Θ either converge in a narrow interval (at $\Theta = 300°$) or diverge in a wide interval ψ_2 (at $\Theta = 120°$). Experimental results for the investigations of surface magnetostatic wave reflection from the planar interface of a layer are presented in [1]. The results obtained fit with the theory, and this confirms the theoretical views which have been presented.

Now, the refraction of magnetostatic waves will be investigated. All waves refract at an interface between two media if on the interface between the media the phase velocities are changed. The dispersal dependences described in Chapter 2 enable us to propose various realizations for the interface between the media, for example in the

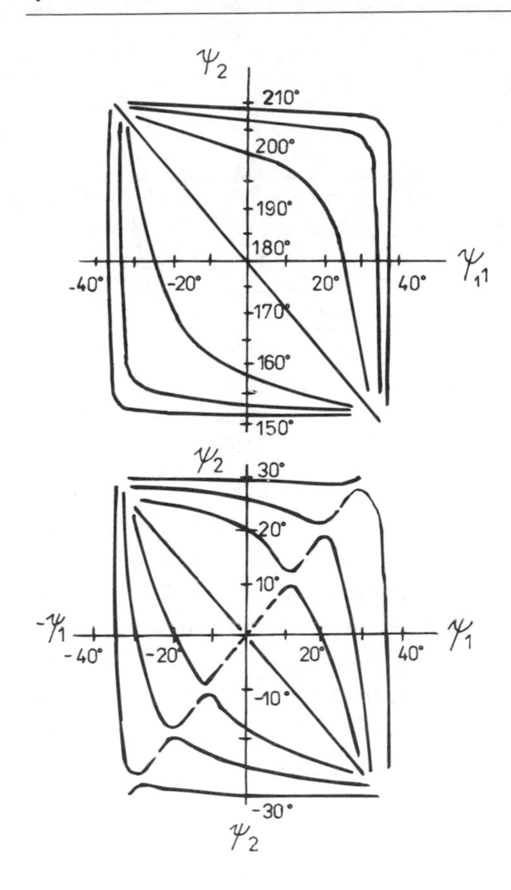

Fig. 4.5. Re-computed dependences from Fig. 4.3 to corresponding dependences of group velocities.

following ways: (1) to change the saturation magnetization of a thin film (e.g. implanting ions changing magnetization); (2) to change the thickness of a layer t (e.g. with a step produced by chemical etching of one part of a thin layer surface); (3) locating close to some part of the film a thin metallic layer. In the following text, an ideal (infinitely thin) interface will be supposed.

In agreement with the conditions of geometrical optics on the interface of two media the following condition must be fulfilled: *wave vector projections of incident, reflected and transmitted waves into the plane of the interface must be equal.* The reflection of a wave has already been investigated and now the laws of refraction of MSW will be treated. Using the condition given above an equation analogous to (4.7) can be found. To compute the direction of a refracted

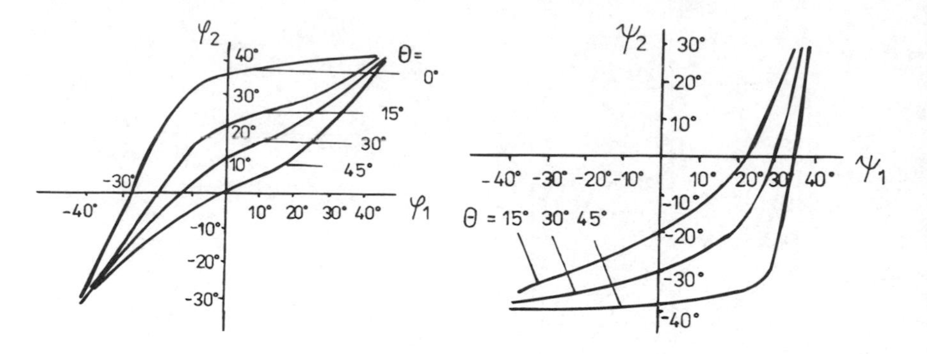

Fig. 4.6. The dependence of the wave vector of the refracted beam φ_2 on the direction of an incident wave φ_1 for various angles Θ.

Fig. 4.7. Dependences of orientations of group velocities of incident (ψ_1) and refracted (ψ_2) MSW for various angles Θ.

wave propagation a system of equations, analogous to (4.8), has to be solved. The results of the computation of the refracted surface MSW characteristics for the parameters given in Fig. 4.3 are presented in [1]. The interface is in the form of a small step through a layer thickness in which the step size is $(t_1 - t_2)/t_1 = 0.5$. The dependence of the wave vector orientation of a refracted beam φ_2 on the direction of the incident wave φ_1, for various values of the interface orientation Θ with reference to the z-axis, is depicted in Fig. 4.6. The dependences of the directions of the group velocities ψ_1 and ψ_2 of the incident and refracted MSW are depicted in Fig. 4.7. As can be seen, the dependences $\varphi_2(\varphi_1)$ and $\psi_1(\psi_2)$ for the refracted waves are analogous to the dependences for reflected MSW and they are similarly complex. For small deviation angles between the interface and the normal to the optical axis these dependences are almost linear. If Θ increases this "linearity" disappears and a visible bend appears. For $\Theta = 40°$ the characteristics have a "rectangular" shape. For these angles of the interface to the optical axis a relatively big change in the angle of the incident wave front φ_1 and with it of the refracted beam ψ_1 almost does not change the location of the refracted beam. Only close to critical angles do these dependences not hold. Here, small changes of the "input parameters" change very visibly the characteristics of the refracted beam. The results of an experimental investigation of MSW refraction are presented in [1]. They are in satisfactory agreement with the theory.

The quasioptical properties of volume MSW will be investigated to end this section. Volume waves in a perpendicularly magnetized layer are characterized by isotropic propagation. The curves of constant frequencies are concentric circles (Fig. 3.2). For this reason quasioptics of forward volume waves is isotropic, the directions of group and phase velocities are the same, the wave front is always perpendicular to the beam, and mirror reflection occurs, i.e. there is a full analogy with light beam behaviour in an isotropic medium.

In the tangentially magnetized layers, in which the existence of a backward volume MSW is possible, the situation is even more complicated. These waves are characterized by a complex dispersive surface (Fig. 3.4) and curves of constant frequencies have equally complex shapes also. The dependences of group and phase velocities for backward volume MSW are given in Fig. 4.8: they are analogous

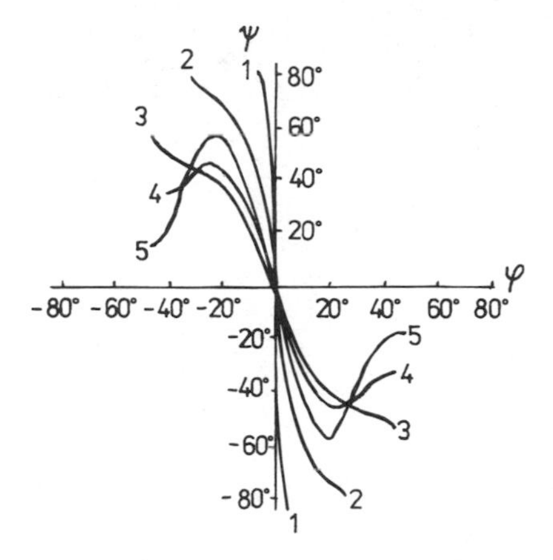

Fig. 4.8. Dependences of phase and group velocities of backward volume waves for various frequencies. $M_0 = 140\,\mathrm{kA\,m^{-1}}$, $H_0 = 46.8\,\mathrm{kA\,m^{-1}}$, f =: 1 — 1.6; 2 — 2.0; 3 — 2.4; 4 — 2.8; 5 — 3.2 GHz.

to the dependences for the surface waves in Fig. 4.1. As follows from this figure, the dependences $\psi(\varphi)$ for backward volume waves are much more complicated than for surface waves. The basic difficulty lies in the fact that a bend appears on the curves $\psi(\varphi)$, which results in an ambiguous dependence of the phase and group velocity direction. From this fact big difficulties ensue in the construction of quasioptical elements using backward volume MSW. For this reason they are not very popular in practical applications.

4.2 Quasioptical analogues of MSW microwave elements

Some possible applications of the quasioptical properties of surface MSW in the construction of quasioptical elements with MSW will be considered.

Positive lens

The dependence of the direction of a refracted beam on angle Θ (the orientation of the refraction plane with regard to the normal line of the optical axis) enables us to find, in the framework of geometric optics methods, a profile of the interface refraction surface which concentrates all beams into one point. The geometry of the arrangement to be considered is represented in Fig. 4.9. We can imagine that

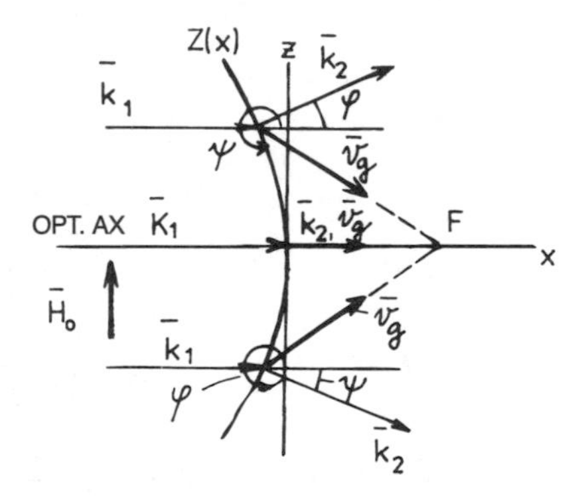

Fig. 4.9. Geometry used for solution of the refracting surface profile.

on an interface, e.g. on a step in the thickness of the layer, there is an incident surface MSW whose vectors $\vec{k}_1$ and $\vec{v}_{g1}$ are oriented in the direction of the optical axis (the direction of the x-axis in Fig. 4.9). From what was said above we know that, proportionally with deviating the interface from the normal to the optical axis, the refracted (reflected) magnetostatic beam deviates from the optical axis. This is the basic idea of the refraction interface construction, which would enable the concentration of parallel beams into one point. In other words, a profile $z(x)$ of a step is to be found which focuses all beams into F. Let $z(x)$ be approximated by a broken line and for each part boundary conditions (4.7) are given. From them we will find

$$\tan\Theta = \frac{k_1 \sin\varphi_1 - k_2 \sin\varphi_2}{k_1 \cos\varphi_1 - k \cos\varphi_2}$$

Since in our case $\varphi_1 = 0$ we can write

$$\tan\Theta = -\frac{k_2 \sin\varphi_2}{k_1 - k_2 \cos\varphi_2}$$

Furthermore $\tan\Theta = \mathrm{d}a/\mathrm{d}z$ and then

$$\frac{\mathrm{d}z}{\mathrm{d}x} = \frac{k_2 \sin\varphi_2}{k_2 \cos\varphi_2 - k_1} \tag{4.10}$$

The beam axis orientation of the ith interval of the broken line interface in the point F is given by

$$\frac{z_i}{x_i - F} = \tan\psi \tag{4.11}$$

where x_i and z_i are co-ordinates of the centre of the ith broken line interface interval. If the length of individual intervals tends to zero, we will obtain a continuous curve, and we will suppose that x, z are the co-ordinates of the curve we search for. The relationship between angle ψ and φ is given by equation (4.4), variables k_1 and k_2 are given by dispersion equations for the media 1 and 2. The task to compute curve $z(x)$ is then solved for concrete parameters of the medium. The computation shows that the profile of the focusing step changes substantially depending on the focal length and frequency. Changing the frequency, the focal length (chromatic aberration) changes. The experimental results of the investigation of the magnetostatic beam passing through the focusing step, obtained by the given method, are shown in Fig. 4.10. The focal length of the lens was 2 mm. As can be seen there is satisfactory agreement between theoretical and experimental results.

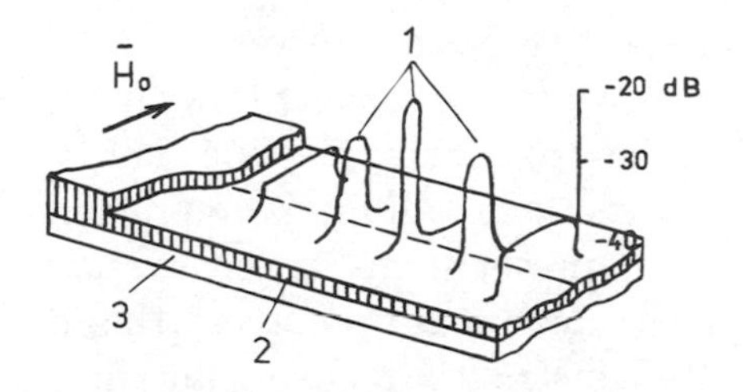

Fig. 4.10. Experimental profiles (1) of MSW beam propagating through focusing step, (2) YIG layer, (3) GGG substrate.

Convex mirror

In an analogous way the profile of the reflecting boundary (of a mirror) focusing beams into single point (focus) can be computed. In contrast to common optical focusing mirrors, which are concave, the MSW focusing mirror has to be convex. This fact is understandable from Fig. 4.9 and it follows from the anisotropic character of surface MSW propagation. Also in this case strong chromatic aberration exists, i.e. dependences of the focal length on the frequency.

Dispersive prism

The dependence of the refraction angle on the frequency of magnetostatic waves enables us to implement the MSW analogue of an optical dispersion prism. Such a prism can be implemented in two variants, either by changing the layer width or by metal-coating of the layer. The calculation of the dependence of the deviation of beam transmitted on the frequency does not differ from the computation of the beam refracted on the interface, but the computation complexity is twice as big (two interfaces). To obtain the maximal dispersion angle in dependence on the frequency it is necessary to choose the prism angles in an optimal way. Computation shows that the optimal dispersion angle is about $0.15\,\mathrm{grad\,MHz^{-1}}$. The schematic representation of experimental results for a beam passing through such a prism at frequencies $\approx 4\,\mathrm{GHz}$ which are close to the computed frequencies is given in Fig. 4.11.

MSW Focusing transducer

One of the interesting peculiarities of MSW quasioptics is the construction of an MSW focusing antenna. The antenna forming (in a geometrical approximation) a surface MSW in a layer was investigated above. If the antenna is of finite length, then it forms an MSW bundle with complex internal structure which is connected with the diffraction peculiarities of MSW propagation (this question will be discussed in more details in the next section). What is important is the fact that such a beam does not preserve its structure, and gradually, as it propagates in the layer, broadens. To avoid this it is necessary to focus this beam, i.e. to construct a focusing antenna. The methods of geometrical optics given above enable us to compute the slope of this antenna. The computation of the shape of the

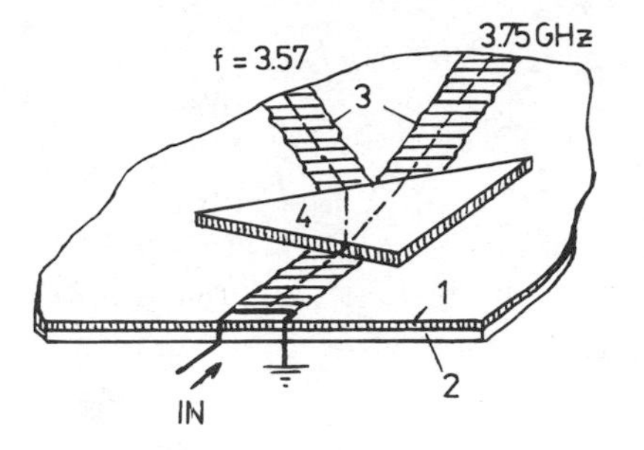

Fig. 4.11. Schematic representation of experimental results of beam transition through the disperse prism for two frequencies, 1 — YIG, 2 — substrate, 3 — MSW beam, 4 — metallic prism.

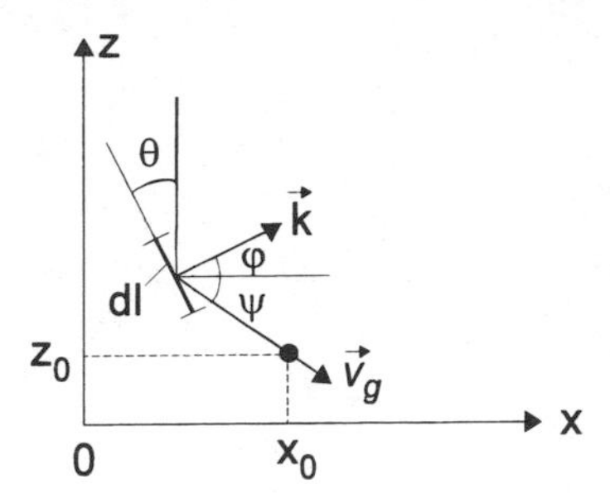

Fig. 4.12. Element of a radiation antenna.

focusing transducer is analogous to the computation of a collecting MSW mirror. In a similar way as above, first of all the curvilinear transducer is approximated by a broken line, each interval of which is analogous to the surface antenna investigated in Section 2.6, and all given assumptions in Section 2.6 are valid. The direction of the wave vector $\vec{k}$ is in each interval perpendicular to the antenna. Let us investigate an element dl of the transducer (Fig. 4.12) with the co-ordinates of the centre (x, z). Let the focus be (x_0, z_0). The line connecting points (x, z) and (x_0, z_0) represents the direction of the group velocity vector. The directions of the group velocity vector ψ and wave vector φ are related by the formula (4.4). In our case: for ψ

$$\tan\psi = \frac{z - z_0}{x - x_0} \tag{4.12}$$

Similarly, in the case of a focusing mirror it is quite easy to write the differential equation for the curve describing the profile of the excitation antenna we are looking for:

$$\frac{\mathrm{d}z}{\mathrm{d}x} = -\frac{1}{\tan\varphi} \tag{4.13}$$

Solving equations (4.12) and (4.13) together with equation (4.4) and with the equation for the surface wave (4.2), the profile of the curve $z(x)$ will be computed. In Fig. 4.13 the computed profile of the focusing microstrip antenna with focal distance $500t$ (where t is the layer

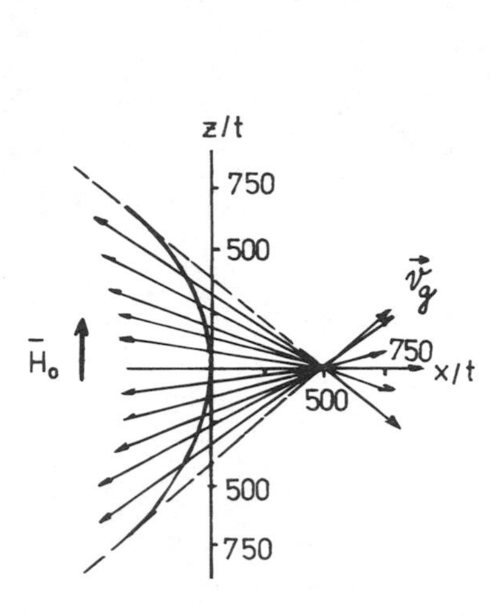

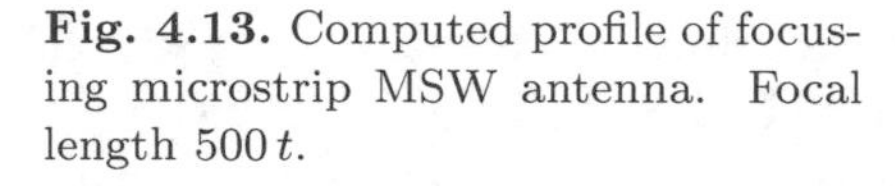

Fig. 4.13. Computed profile of focusing microstrip MSW antenna. Focal length $500\,t$.

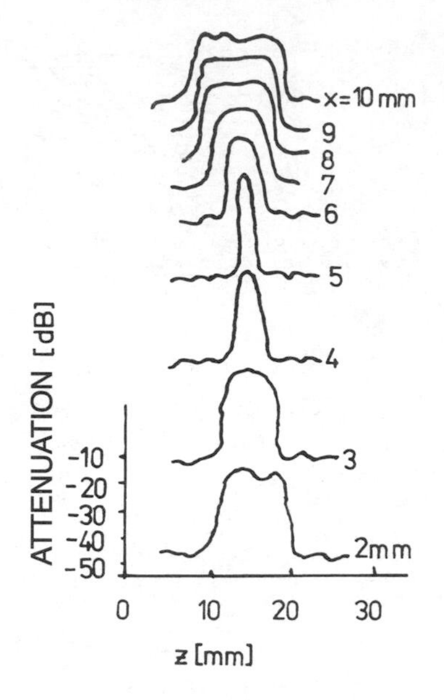

Fig. 4.14. Experimental results obtained by a focusing antenna. $M_0 = 140\,\mathrm{kA\,m^{-1}}$, $H_0 = 68\,\mathrm{kA\,m^{-1}}$, $f = 4340$ MHz, $t = 18\,\mu$m.

thickness) for parameters $\omega_H = 0.5\,\omega_M$, $\omega = 0.875\,\omega_M$ is shown. The direction of the group velocity is indicated by arrows. As follows from Fig. 4.13 the geometry of the focusing MSW antenna is entirely different from the geometry of similar "common" radiophysical antennas or "common" focusing mirrors (by "common" we mean an isotropic medium). An anisotropic MSW antenna really focuses all beams going out from its convex side into the focus and scatters beams going out from its concave side. This is an usual property of these antennas. Like other quasioptical MSW elements for focusing the transducer shows a strong dependence of its properties on frequency, i.e. it has strong chromatic aberration. The antenna depicted in Fig. 4.13 focuses only the waves of a given frequency. For waves with other frequencies the focus dissolves and shifts in the x-axis direction. With increasing frequency the focus comes near to the excitation antenna and vice versa. This frequency dependence can be useful for the construction of some MSW elements.

In Fig. 4.14 the experimental results obtained with a focusing antenna are presented. In the figure the results of a surface MSW beam profile measurement by means of an induction probe for various distances from the focusing antenna (the distance changed in steps of 1 mm from 2 to 10 mm) are given. The aperture (size) of the excitation antenna was 8 mm. With increasing distance the beam narrows (focuses) and then begins to widen. If frequency changes, the focus shifts in a good agreement with the theory. Overall accordance of theory and experiment was also very good.

The methodology of quasioptical measurements of MSW is described in detail in the monograph [1] and is similar to that described in the previous chapter (see also [31]).

4.3 MSW beam propagation considering diffraction phenomena

In previous sections magnetostatic beam propagation was investigated and it was supposed that the beam already existed and that it was approximately uniform in cross-section. While propagating, the beam diffuses and its boundaries are blurred, but in spite of this it is possible to implement its focusing, reflection and refraction. The system of assumptions is called, as already given above, the approximation of geometrical optics. In contrast to the task of isotropic geometrical optics, in the case of magnetostatic waves we have to deal with anisotropic (in the sense of microwaves) media, which substantially complicates the investigated geometrical picture. This is given by the complex dependence of the directions of MSW phase and group velocities on one another and on the location of the excitation (refraction, reflection) plane. In practical applications the situation is even more complicated.

MSW beam formation by an antenna with finite aperture (length) will be investigated. This task (strictly speaking) cannot be solved by the methods of geometrical optics because it represents the diffraction problem of wave optics. Let a flat MSW transducer with length 2δ lie on an infinite ferrite layer with thickness t. The layer plane is identical with the co-ordinate plane x, z and let the transducer be oriented in the z-axis direction (field H_0). Let the co-ordinate system centre be identical with the centre of the excitation antenna. For the sake of simplicity we will suppose that high-frequency current is uniform in the whole transducer i.e. the length of the magnetostatic wave in the strip $\lambda \gg 2\delta$; and that only surface MSW are excited by the transducer. From a strict (theoretical) point of view, the profile of the wave beam excited by a transducer of finite length oriented in the z-axis direction will be described by a two-dimensional magnetostatic potential $\psi(x,z)$, which must fulfil Helmholtz's equation and the boundary conditions on the transducer surface. Helmholtz's equation can be obtained from Walker's equation in the following way. Let the magnetostatic potential have the form

$$\psi(x,y,z) = \psi_1(y)\psi_2(x,z)$$

where $\psi_1(y)$ describes the distribution of the microwave field (and of potential) through the thickness of the sample and $\psi_2(x,z)$ in the layer plane (this is the distribution we are looking for). Then from Walker's equation it follows that [1]

$$\frac{1}{\psi_1}\frac{\partial^2\psi_1}{\partial y^2} + \frac{1}{\psi_2}\left(\frac{\partial^2\psi_2}{\partial x^2} + \frac{1}{\mu}\frac{\partial^2\psi_2}{\partial z^2}\right) = 0 \tag{4.14}$$

Using Fourier's method for separation of variables we obtain

$$\frac{1}{\psi_1}\cdot\frac{\partial^2\psi_1}{\partial y^2} = k_y^2$$

or

$$\frac{1}{\psi_2}\left(\frac{\partial^2\psi_2}{\partial x^2} + \frac{1}{\mu}\frac{\partial^2\psi_2}{\partial z^2}\right) = -k_y^2 \tag{4.15}$$

Using the substitution $z' = z\sqrt{\mu}$ we will obtain Helmholtz's equation for ψ_2 in the co-ordinates x, z' (the index 2 at $\psi(x,z')$ will be omitted hereinafter) in this form

$$\frac{\partial^2\psi}{\partial x^2} + \frac{\partial^2\psi}{\partial z'^2} + k_y^2\psi = 0 \tag{4.16}$$

Having solved equation (4.16) we go back into the co-ordinate system (x,y,z). Neglecting edge effects on the transducer edges the boundary conditions for magnetostatic potential can be written as follows

$$\psi(0,z) = \begin{cases} \eta(z) & \text{for} \quad -\delta \le z \le \delta \\ 0 & \text{for} \quad z > |\delta| \end{cases} \tag{4.17}$$

Thus the space distribution of high-frequency magnetostatic potential has the form of the function truncated in the z-axis direction (in the interval 2δ) and it can be expressed as the sum of Fourier components with various wave numbers k_x, k_z. Only waves propagating from the transducer to the half-plane $x > 0$ will be investigated. If the method of the potential decomposition into surface waves is used for $\psi(x, z)$ we have

$$\psi(x, z) = \int_{-\infty}^{\infty} \Phi(k_z) \mathrm{e}^{j(k_x x + k_z z)} \mathrm{d}k_z \qquad (4.18)$$

where $\Phi(k_z)$ is the spectrum of plane waves created by the given distribution of potential on the transducer (the Fourier image of the transducer):

$$\Phi(k_z) = \frac{1}{2\pi} \int_{-\delta}^{\delta} \eta(z) \mathrm{e}^{-jk_z z} \mathrm{d}z \qquad (4.19)$$

Since the assumption $\lambda \gg 2\delta$ was used, i.e. current and potential are constant along the length of the transducer, $\eta(z) = 1$ and then

$$\Phi(k_z) = \frac{\sin(k_z \delta)}{\pi k_z} \qquad (4.20)$$

Substituting (4.20) into (4.18) the values of the magnetostatic potential of surface MSW in an arbitrary point of the half-plane of the thin layer ($y > 0$) can be computed. If it is supposed that the wave intensity is proportional to $|\nabla\psi(x, z)|^2$, it is possible to determine the profile of a wave bundle for various distances from the excitation antenna. For example, in Fig. 4.15 are given the computed profiles of the distribution $|\nabla\psi(x, z)|^2$ for various distances from the antenna for two frequencies, one close to the lower and one close to the upper band frequency of the existence of surface MSW at $H_0 = 70\,\mathrm{kA\,m^{-1}}$. The dashed curves starting from the edge points of the transducers depict the sectors of admissible orientations of energy propagation for the given frequency. As follows from Fig. 3.12 (or Fig. 4.1) for surface MSW the sector of admissible orientations of group velocity with increasing frequency, i.e. limiting to the upper limiting frequency, widens. For frequencies close to the lower band frequency of surface MSW, where k_x and φ_c are small, the shape of

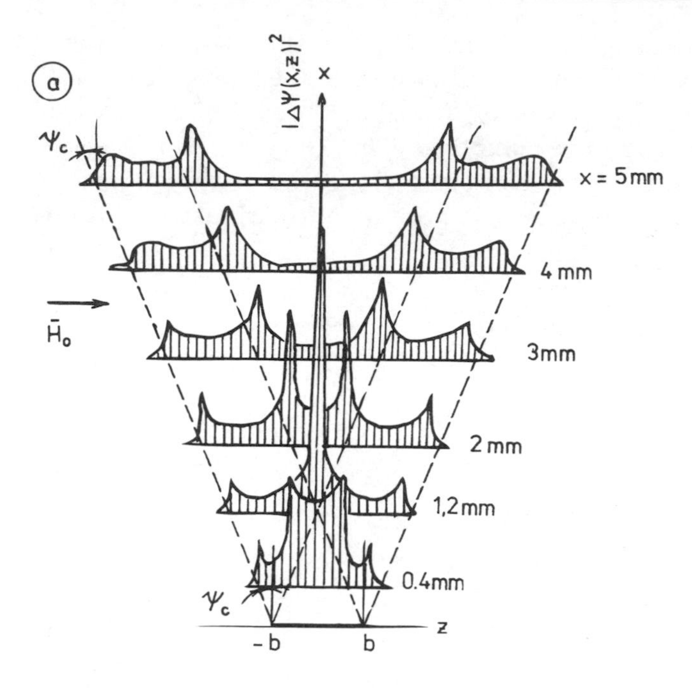

Fig. 4.15a. Computed profiles of distribution $|\Delta\psi(x,z)|^2$. $M_0 = 140\,\mathrm{kA\,m^{-1}}$, $H_0 = 70\,\mathrm{kA\,m^{-1}}, b = 1\,\mathrm{mm}, t = 1\,\mu\mathrm{m}$, critical angle φ_c, $f = 4263\,\mathrm{MHz}, \varphi_c = 40°$.

the wave beam has the following peculiarities (Fig. 4.15a). Firstly, wave intensity rapidly decreases on the boundaries of the sector of admissible energy transmission directions. Secondly, in the intersection of sector borders from different transducer edges, the rapid increase of wave intensity in the centre of the beam can be observed, i.e. as if the beam is focused. Thirdly, at further beam propagation (behind the point of "focus") it is as if the beam has split into two beams, where the width of each of them is close to the length of the excitation antenna. Such behaviour of the wave beam of surface MSW close to the lower band frequency can be explained by the fact that in this area the dependence $\psi(\varphi)$ (Fig. 4.1) has a flat section. The corresponding value of the angle ψ is almost identical with the critical angle φ_c for the given frequency. This, in principle, means that most of the surface waves forming the MSW beam have the same orientation of propagation of energy (group velocity), identical with the value of ψ on the flat section of the dependence $\psi(\varphi)$. Thus in a wave bundle, this direction and the direction symmetrical to it, with regard to the x-axis, represent directions of prevailing propaga-

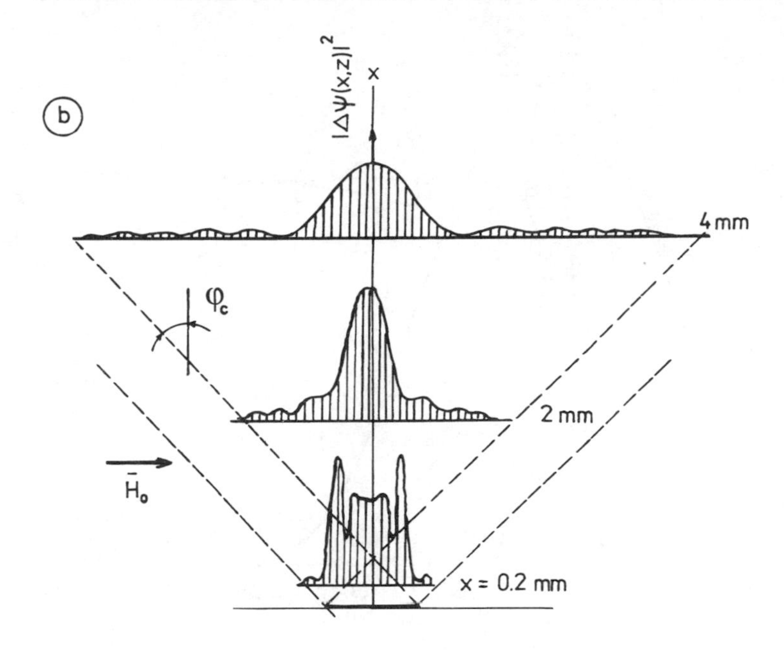

Fig. 4.15b. Computed profiles of distribution $|\Delta\psi(x,z)|^2$. $M_0 = 140$ kA m^{-1}, $H_0 = 70\,\text{kA m}^{-1}, b = 1\,\text{mm}, t = 1\,\mu\text{m}$, critical angle φ_c, $f = 4655\,\text{MHz}, \varphi_c = 65°$.

tion of surface wave energy. This causes the splitting of the surface MSW beam near the lower band frequency.

With increasing of frequency the sector of admissible directions of energy transmission by surface MSW widens, k_x and φ_c increase and the flat section on the curves $\psi(\varphi)$ disappears (Fig. 4.1). As a result, the behaviour of the beam becomes similar to the wave for which constraints, with regard to the directions of energy transmission, exist. In this case the profile of the wave bundle is formed inside the wide spectrum of admissible directions of group velocities, it does not have sharp boundaries, the bundle seems to dissolve, and its profile is analogous to the beam profile in an isotropic medium. The maximum of the wave beam intensity at large distances from the antenna at the upper band frequencies lies always on the x-axis (Fig. 4.15b).

The behaviour of an MSW wave bundle can be more exactly described by the formula (4.19). In accordance with (4.20) the dependence of the spectrum density of surface waves $|\Phi(k_z)|^2$ radiated by the antenna of a given length as a function of k_z (the Fourier image of the transducer) will be constructed. The dependence is

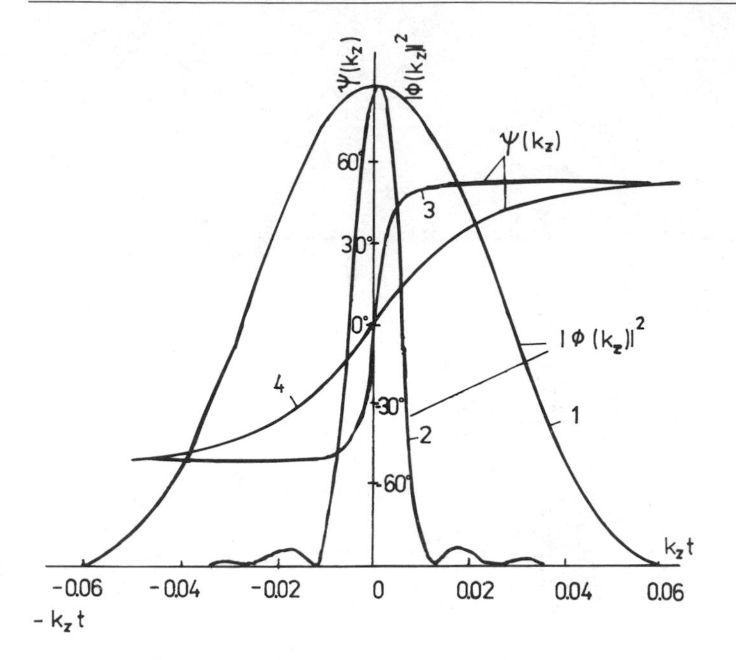

Fig. 4.16. The dependence of the spectrum density of planar waves, 1 — short antenna, 2 — long antenna, 3 — close to the lower band frequency, 4 — close to the upper band frequency.

presented in Fig. 4.16. In this figure the dependences $\psi(k_z)$ for two values of frequency are depicted. The comparison of curves 1, 2 and 3, 4 enables one to assess the behaviour of the beam. For example, let us investigate the situation described by curves 1 and 3.

A short antenna (emitter) excites a wide spectrum of wave numbers; the width of the basic band is π/δ (curve 1). Moreover, the basic band $|\Phi(k_z)|^2$ includes two flat sections of the dependence $\psi(k_z)$ (curve 3) for $\psi = \pm 45°$. This determines the character of the wave beam. Most of the surfaces with various k_z transmit energy in these two directions. Thus the beam splits. The long antenna (curve 2) has a narrow spectral diagram which does not include the flat sections $\psi(k_z)$ of curve 3. In this case the beam is analogous to a beam in an isotropic medium. For the upper limiting frequency (curve 4) the dependence $\psi(k_z)$ does not have flat sections and the beam does not split under any conditions. The analysis shows also that beams in metal-coated layers have analogous characteristics. The investigation of $\psi(k_z)$ dependences and $|\Phi(k_z)|^2$ enables us to predict the behaviour of a wave bundle at the rotation of the excitation antenna

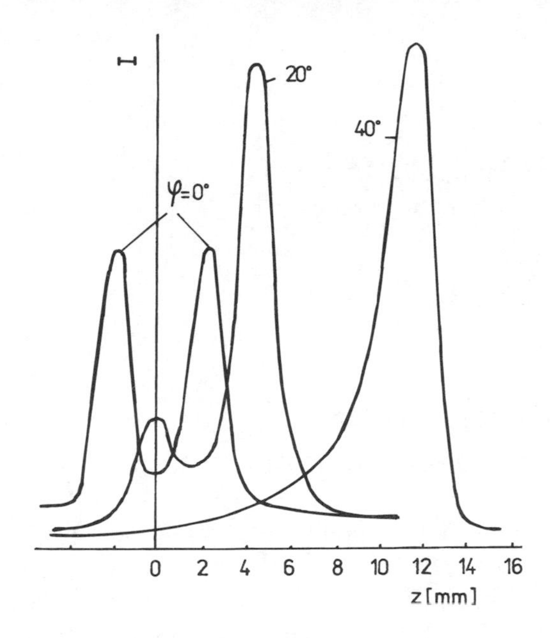

Fig. 4.17. Experimental profiles of wave beams for various rotations of the transducer with regard to magnetic field. I — intensity of the beam in arbitrary units.

with regard to the magnetic field $\vec{H}_0$. In this case the dependence $|\Phi(k_z)|^2$ is the same and the curve $\psi(k_z)$ shifts up (down) and to the left (right) with regard to the point 0°. Let us suppose now that we have a short antenna (curve 1) and that we are operating near to the lower band frequency (curve 3). The shift of curve 3 down and to the left with regard to curve 1 results in the fact that the lower flat section of curve 3 goes out of the zone of curve 1 and the prevailing part of the MSW energy radiates according to the upper flat section of curve 3. The formerly symmetrical split beam is "transposed", and the positions of its maxima becomes unsymmetrical with regard to the x-axis, one maximum getting smaller and the other bigger. After further rotation only one maximum remains. In Fig. 4.17 the experimentally measured profiles of wave beams for various emitter rotations with regard to the magnetic field H_0 (changing the angle φ) are depicted. The amplitude and direction of magnetic field remain unchanged. From the dependence it can be seen that approximately for $\varphi \cong 20°$ the "transposition" of maxima occurs and for $\varphi \cong 40°$ only one maximum remains [1].

Chapter 5

NONLINEAR PROPERTIES OF MAGNETOSTATIC WAVES

5.1 Behaviour of magnetic media in a strong high-frequency magnetic field

Up to now linear processes in ferromagnetic media have been investigated, i.e. processes at very small amplitudes of external high-frequency field $\vec{h}$ and at time-dependent magnetization $\vec{m}$. The Landau–Lifshitz equation for the vector of magnetization (1.21) is in the general case nonlinear. It was linearized using the assumptions

$$|\vec{h}| \ll |\vec{H}_0| \qquad \text{and} \qquad |\vec{m}| \ll |\vec{M}_0| \tag{5.1}$$

The latter condition enables us to neglect in the equation of motion (1.21) the product of time dependent variables and to obtain the solution in the form of linear formulae relating amplitudes of components of high-frequency magnetization and magnetic field. In this approximation the vector $\vec{M}$ precesses with infinitely small amplitude, i.e. its position almost does not differ from an equilibrium position. As was shown in Chapter 1, the length of the vector $|\vec{M}| = \text{const.}$, i.e. at precession the end point of the vector $\vec{M}$ moves on the surface of a sphere. Nevertheless, in the linear approximation this deviation is so small that this area of the sphere can be approximated by a plane. This was done when we used the method of complex amplitudes (formula (1.30) etc.).

The condition (5.1) does not have to be always fulfilled. In ferromagnetic materials there exist a number of interesting and important phenomena which are caused by a nonlinear relationship between the time dependent magnetization and high-frequency field. This relationship usually shows at sufficiently large amplitudes of the high-frequency magnetic field, when conditions (5.1) do not hold. Among these phenomena can be included processes connected with nonlinear ferromagnetic resonance, parametric amplification of high-frequency signals, processes of detection and transformation of frequency, propagation of nonlinear waves, transmission of a signal through a nonlinear medium, etc. Most of the references dedicated to nonlinear phenomena are from the 1950 s – 1960 s (e.g. [19, 20] and bibliographies in them). Only recently have papers dealing with nonlinear phenomena in the propagation of magnetostatic waves appeared. Before we approach their study, let us recall the nonlinear phenomena in ferromagnetic materials connected with uniform nonlinear precession of a magnetization vector in a strong high-frequency field and corresponding effects. Later we will discuss nonlinear phenomena in the propagation of magnetostatic waves.

Above all the question arises of which excitations should be considered as strong ones and when it is necessary to use nonlinear formulae. It is supposed in the literature that nonlinearity arises in the case when the amplitude of excitation in a magnetic medium is comparable with any of its parameters, e.g. with the resonance line width [20]. We will use this definition hereinafter.

Let us consider a uniformly magnetized ferromagnetic material with saturation magnetization $\vec{M}_0$. An external magnetic field $\vec{H}_0$ is applied in the direction of the z-axis while static, state $\vec{M}_0 \| \vec{H}_0$. If this system leaves the equilibrium state, the magnetization vector $\vec{M}_0$ begins the precession around the direction of the magnetic field, whereby its amplitude remains constant during this motion (Chapter 1), i.e.

$$|\vec{M}| = |\vec{M}_0| = \text{const.} \tag{5.2}$$

The end point of vector $\vec{M}_0$ at any moment of its motion is located on the sphere S (Fig. 5.1). As we will see, this motion can be rather complicated, as for example at forced precession under the influence of a high-frequency field. However, under any conditions the vector end point cannot leave the sphere S. This property of $\vec{M}_0$ is one of the most important for the creation of nonlinear parametric phenomena. As follows from Fig. 5.1 the formula (5.2) can be written in the form

$$M_0^2 = m_x^2 + m_y^2 + M_z^2 = \text{const.}$$

or

$$M_z^2 = M_0^2 - (m_x^2 + m_y^2) \tag{5.3}$$

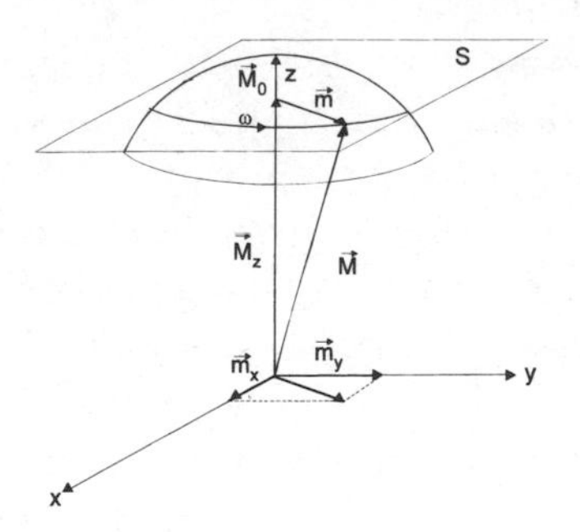

Fig. 5.1. Motion of the end point of the magnetization vector on a sphere.

where m_x and m_y are the transverse components of the high-frequency vector of magnetization and M_z is the longitudinal component (Fig. 5.1) in the direction of constant magnetic field $\vec{H}_0$ (Fig. 5.1). In a linear theory a small amplitude of all excitations was supposed, and considering (5.3) m_x^2; $m_y^2 \ll 1$ (small value of precession angle) we obtained $M_z = M_0 =$ const. and the end point of vector $\vec{M}$ at enforced oscillations rotated in the plane Σ parallel to the xy-plane (Fig. 5.1). As already stated, at large amplitudes this does not hold and the vector $\vec{M}$ end point at enforced precession will move on the sphere S. This must cause a reduction of the longitudinal component of magnetization M_z.

The change of longitudinal component of the magnetization vector has a fundamental importance in nonlinear and parametric processes which can be observed in an isotropic ferromagnetic material magnetized to saturation. Let us investigate the basic peculiarities of a nonlinear processes on the basis of the solution of a nonlinear equation of motion in the form

$$\frac{\mathrm{d}\vec{M}}{\mathrm{d}t} = -\mu_0\gamma(\vec{M} \times \vec{H}) \tag{5.4}$$

where $\vec{H} = \vec{H}_0 + \vec{h}$ ($\vec{h}$ is the variable magnetic field). The individual components are given by

$$\begin{aligned} \frac{\mathrm{d}m_x}{\mathrm{d}t} &= \mu_0\gamma[-m_yH_0 + M_zh_y - m_yh_z] \\ \frac{\mathrm{d}m_y}{\mathrm{d}t} &= \mu_0\gamma[m_xH_0 - M_zh_x + m_xh_z] \\ \frac{\mathrm{d}M_z}{\mathrm{d}t} &= -\mu_0\gamma[m_xh_y - m_yh_x] \end{aligned} \tag{5.5}$$

where $H_0 = H_z$ is the z component of the constant magnetic field; h_x, h_y, h_z are components of the alternating magnetic field; and m_x, m_y, M_z are components of the magnetization vector $\vec{M}$.

Instead of the third equation in (5.5) it is better to use the equivalent condition (5.3) taking into account that in the general case variables m_x and m_y are small, although in this case finite. Then we can write

$$M_z \cong M_0 \left(1 - \frac{1}{2M_0^2}(m_x^2 + m_y^2)\right) \tag{5.6}$$

Let us investigate the internal precession of the magnetization vector under the influence of a given external high-frequency magnetic field by solving the system of equations (5.5) together with (5.6). We will use the method of successive approximations [19]. The investigation of nonlinear phenomena in electronics, already taking into account small terms of second order, results in useful conclusions and is a good basis for understanding possible nonlinear effects. We will also proceed in this way. We will restrict the method of successive approximations to second-order terms.

The principle of the method of successive approximations will be shown, for the sake of simplicity, in the case of a pure transverse high-frequency magnetic field with arbitrary polarization, varying harmonically with time, i.e.

$$\vec{h} = \mathrm{Re}(\vec{h}_0 e^{j\omega t})$$

with components

$$\begin{aligned} h_x &= h_{0x} \cos \omega t \\ h_y &= h_{0y} \cos(\omega t + \varphi_y) \end{aligned} \tag{5.7}$$

As follows from linear theory (Chapter 1) the components of time-dependent magnetization in a linear approximation are

$$\begin{aligned} m_x &= \chi h_x + j\chi_a h_y \\ m_y &= -j\chi_a h_x + \chi h_y \\ m_z &= 0 \end{aligned} \tag{5.8}$$

where

$$\chi = \frac{\mu_0 \gamma M_0 \omega_H}{\omega_H^2 - \omega^2}; \qquad \chi_a = \frac{\mu_0 \gamma M_0 \omega}{\omega_H^2 - \omega^2},$$

and m_z is the time-dependent component of the magnetization vector in a linear approximation. The formulae (5.8) will be used as the first approximation of the solution of equations (5.5). First of all, let us substitute (5.7) into (5.8) and then find formulae for real components of m_x and m_y. In the first approximation we will obtain (neglecting quadratic terms)

$$\begin{aligned} m_x &= \mathrm{Re}\left[\chi h_{0x} \mathrm{e}^{j\omega t} + j\chi_a h_{oy} \mathrm{e}^{j(\omega t + \varphi_y)}\right] \\ &= \chi h_{0x} \cos\omega t - \chi_a h_{0y} \sin(\omega t + \varphi_y) \qquad (5.9.1) \\ m_y &= \mathrm{Re}\left[-j\chi_a h_{0x} \mathrm{e}^{j\omega t} + \chi h_{0y} \mathrm{e}^{j(\omega t + \varphi_y)}\right] \\ &= \chi_a h_{0x} \sin\omega t + \chi h_{0y} \cos(\omega t + \varphi_y) \qquad (5.9.2) \end{aligned}$$

Now, we will substitute (5.9) into (5.6) and determine the component M_z in the second approximation (with quadratic terms) in the form

$$\begin{aligned} M_z = M_0 - \frac{1}{2M_0} &\left[[[\chi h_{0x} \cos\omega t - \chi_a h_{0y} \sin(\omega t + \varphi_y)]^2\right. \\ &\left. + [\chi_a h_{0y} \sin\omega t + \chi h_{0y} \cos(\omega t + \varphi_y)]^2\right] \qquad (5.10) \end{aligned}$$

After relevant modifications we will obtain for the time-dependent component of longitudinal magnetization, $m_z = M_z - M_0$, the following formula [19]

$$\begin{aligned} m_z = \frac{1}{M_0} &\left[(\chi^2 + \chi_a^2)(h_{0x}^2 + h_{0y}^2) - 4\chi\chi_a h_{0x} h_{0y} \sin\varphi_y\right. \\ &\left. + (\chi^2 - \chi_a^2)[h_{0x}^2 \cos(2\omega t) + h_{0y}^2 \cos(2\omega t + \varphi_y)]\right] \qquad (5.11) \end{aligned}$$

Now we substitute M_z (formula (5.10)), taking into account (5.11), into the formulae for χ and χ_a instead of M_0. The modified susceptibilities and values obtained for χ and χ_a will be then substituted into (5.9.1) and (5.9.2). The results are the values of m_x and m_y in the second approximation (with quadratic terms). In this way all three components of the time-dependent magnetization vector m_x, m_y and m_z can be obtained; they contain nonlinear terms. Repeating this process the third approximation can be obtained, etc. However, this process is used very seldom because in practical applications in most cases we deal with small values of high-frequency magnetic fields ($h_0 \ll H_0$) and for the qualitative treatment of nonlinear phenomena it is sufficient to use the linear equations (5.9) for the components m_x and m_y and nonlinear formulae (5.11) for the component m_z. In such an approximation the relationship between m_x and h_x and m_y and h_y, respectively, remain linear and terms of higher order can be neglected. The longitudinal component of the magnetization coupling with alternating magnetic field is nonlinear and plays a decisive role at finite amplitudes of high-frequency magnetic field. As follows from (5.10) and (5.11), taking into account nonlinearity results in two effects. The first is connected with the reduction of the longitudinal component because of the presence of terms in m_z which do not depend on time (the first and second terms in m_z). The reduction is proportional to the second power of the amplitude of the high-frequency magnetic field. The reduction of the longitudinal component of magnetization (in the direction of H_0) indicates that ferrite at nonlinear excitation seems to be "demagnetized". The second effect is connected with the fact that due to motion of the magnetization vector a component with frequency 2ω appears. This component of motion of vector $\vec{M}$ is also connected with the change of the longitudinal component of time-dependent magnetization.

We have given above a simple consideration of nonlinearity in the form of quadratic terms in formulae for the longitudinal component of magnetization that results in the emergence of components varying with a frequency which is different from the frequency of an external excitation force. This again results in rather complicated forced motion of the vector $\vec{M}$ and in the emergence of a number of interesting effects. As an example we will investigate nonlinear motion of the magnetization vector in a transverse high-frequency magnetic field of various polarizations.

We will express the longitudinal component of magnetization m_z (5.11) in the form of two components, constant ΔM_0 (which does not depend on time) and time-dependent Δm_z. We get

$$\Delta M_0 = \frac{1}{4M_0}\left[(\chi^2 + \chi_a^2)(h_{0x}^2 + h_{0y}^2) - 4\chi\chi_a h_{0x} h_{0y} \sin\varphi_y\right] \tag{5.12}$$

$$\Delta m_z = \frac{1}{4M_0}(\chi^2 - \chi_a^2)[h_{0x}^2 \cos(2\omega t) + h_{0y}^2 \cos(2\omega t + \varphi_y)] \tag{5.13}$$

We will investigate the behaviour of ΔM_0 and Δm_z in transverse high-frequency fields of various polarizations.

1. Circular polarization of an external high-frequency field

In this case $h_{0x} = h_{0y} = h_0$, $\varphi_y = \pm\pi/2$, where "+" corresponds to a counterclockwise polarized field and sign "−" to a clockwise polarized field. From (5.9.1), (5.9.2), (5.12) and (5.13) for the time-dependent components m_x, m_y and Δm_z, and also for the constant component ΔM_0, it follows that

$$\begin{aligned} m_x &= (\chi \pm \chi_a) h_0 \cos(\omega t) \\ m_y &= (\chi \pm \chi_a) h_0 \sin(\omega t) \\ \Delta m_z &= 0 \\ \Delta M_0 &= \frac{1}{2M_0} (\chi \pm \chi_a)^2 h_0^2 \end{aligned} \tag{5.14}$$

In (5.14) the upper sign corresponds to clockwise polarization and the lower to counterclockwise polarization. From (5.14) it follows that transverse components m_x, m_y either have a resonant character or have not. With clockwise polarization the variable

$$\chi_+ = \chi + \chi_a = \gamma\mu_0 M_0 / (\omega_H - \omega)$$

exhibits resonance at $\omega = \omega_H$. For counterclockwise polarization

$$\chi_- = \chi - \chi_a = \gamma\mu_0 M_0 / (\omega_H + \omega)$$

does not have a resonant character.

As for the longitudinal component, this at circular polarization does not contain variable components. Only a reduction of the longitudinal component of magnetization M_0 by a value ΔM_0 proportional to the second power of the amplitude of the high-frequency field occurs.

2. Elliptical polarization of an external high-frequency magnetic field

In this case $h_{0x} \neq h_{0y}$, $\varphi_y = \pm\pi/2$ and from formulae (5.12) and (5.13) for longitudinal components we have

$$\Delta M_0 = \frac{1}{4M_0} [(\chi^2 + \chi_a^2)(h_{0x}^2 + h_{0y}^2) \pm 4\chi\chi_a h_{0x} h_{0y}] \tag{5.15}$$

$$\Delta m_z = \frac{1}{4M_0} (\chi^2 - \chi_a^2)(h_{0x}^2 - h_{0y}^2) \cos(2\omega t) \tag{5.16}$$

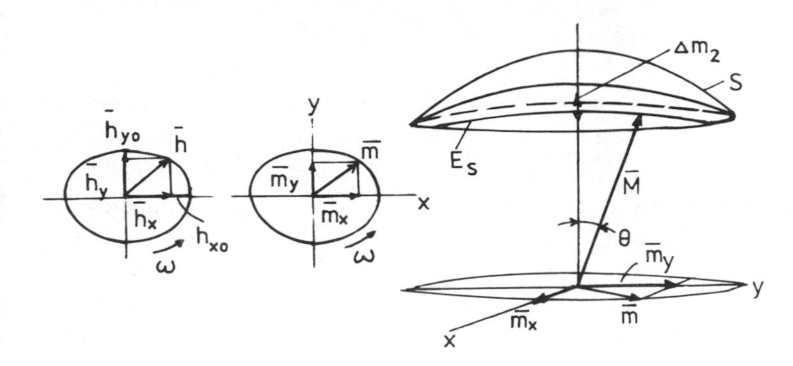

Fig. 5.2. Motion of the magnetization vector on a sphere in the case of elliptical polarization. $\Delta m_z \cong \cos 2\omega t$, S — sphere surface, E_S — ellipse on the sphere surface.

and for transverse components of the magnetization vector from (5.9) we obtain

$$
\begin{aligned}
m_x &= (\chi h_{0x} \pm \chi_a h_{0y}) \cos(\omega t) \\
m_y &= (\chi_a h_{0x} + \chi h_{0y}) \sin(\omega t)
\end{aligned}
\tag{5.17}
$$

In (5.15) up to (5.17) the upper sign corresponds to clockwise polarization and the lower one to counterclockwise polarization. As follows from the formulae obtained the motion of the vector $\vec{M}$ in this case has a complex character. The vector $\vec{M}$ moves on an ellipse, where this ellipse is located on the surface of sphere S, as is shown in Fig. 5.2. During the $\vec{M}$ vector motion the angle Θ between $\vec{M}$ and the z-axis changes continuously with frequency 2ω. This leads to the result that the longitudinal component of the vector $\vec{M}$ also oscillates with frequency 2ω. Thus in this case two kinds of motion of the vector M are taking place simultaneously: transverse motion with frequency ω and longitudinal motion with frequency 2ω. Moreover, as in the case of circular polarization, the constant component of magnetization decreases by the value given in (5.15). The cases of forced motion of vector $\vec{M}$ investigated can be used for detection and to double the frequency [19]. In the latter case, as follows from (5.16), the most effective is a linear polarization of a high-frequency field at which either $h_{0x} = 0$ or $h_{0y} = 0$. In this case the amplitude of the doubled-frequency component is a maximum. In spite of this these MSW devices are not used in practical applications.

3. Nonlinear ferromagnetic resonance

Nonlinearity of ferrite changes the characteristics of ferromagnetic resonance. It should be noted that the amplitude of ferromagnetic resonance is given by the formula (1.52) and depends on losses (loss parameter α) as well as on the amplitude of magnetization M_{0z}. It was shown above that in a nonlinear regime M_{0z} begins to depend on the amplitude of the high-frequency magnetic field and decreases with its increase. This should result in a decrease of the magnitude of maximum magnetic susceptibility, which actually happens. In a linear approximation the resonance line width and magnetic susceptibility in resonance are given by

$$\Delta\omega = \alpha\omega_H, \qquad \chi_{\max}(0) = \gamma\mu_0 M_0/(2\alpha\omega_H) \tag{5.18}$$

where $\chi_{\max}(0)$ is the maximum value of susceptibility given by the formula (1.52) for infinitely small amplitudes. As follows from the formulae (5.10) and (5.11), with increasing amplitude of the external high-frequency magnetic field the longitudinal component of magnetization M_z begins to decrease, i.e. $M_z < M_{0z}$. Considering a nonlinear change of M_z the magnitude for the maximum value of magnetic susceptibility the following formula can be obtained:

$$\chi_{\max} = \frac{\gamma\mu_0 M}{2\alpha\omega_H} \qquad m'_z = \chi_{\max}(0) m_z, \tag{5.19}$$

where $m'_z = M_z/M_0$, and M_z is given by the formulae (5.10), (5.11) or (5.12).

As follows from (5.19) the resonance value of susceptibility decreases with increasing amplitude of the high-frequency field in the same way as the longitudinal component of magnetization decreases. In the quadratic approximation the decrease of M_z is proportional to the second power of the high-frequency magnetic field amplitude, i.e. $m'_z \simeq 1 - Gh^2$, where G is constant which can be computed (formula (5.12)). The calculated dependence $m'_z(h^2)$ is represented in Fig. 5.3. However, experiment gives quite different results. In the same figure the experimental points of the dependence $\chi_{\max}/\chi_{\max}(0)$ [20] are depicted. As can be seen, the nonlinear descent of susceptibility begins much earlier than follows from the simple theory and is much steeper. The theoretical analysis used does not include all nonlinear phenomena and quadratic terms only do not explain the experimental results

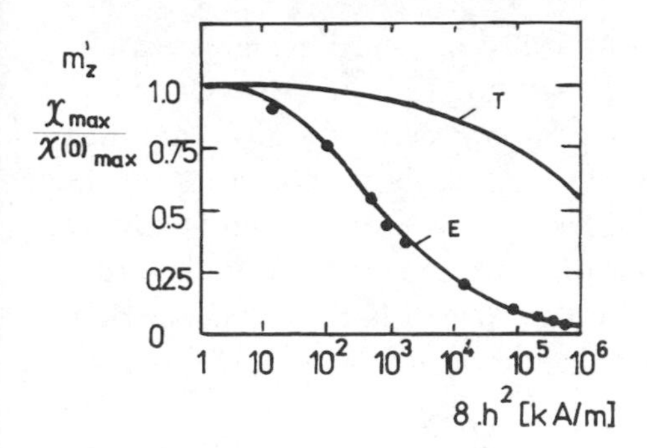

Fig. 5.3. The comparison of computed and experimental dependences $m_z'(h^2)$. T — theory m_z', E — experiment $\chi_{\max}/\chi(0)_{\max}$.

obtained for the ratio $\chi_{\max}/\chi_{\max}(0)$. The heart of the matter is in the fact that the decrease of the ratio $\chi_{\max}/\chi_{\max}(0)$ happens much earlier than the incorporation of quadratic terms into the equation of motion justifies. The correct explanation of this phenomenon was first given by Suhl [2]. He showed that reaching some "threshold" value of signal amplitude causes in ferrite a parametric amplification of the thermal waves at the frequency which is equal to one half of the signal frequency applied. It gives so-called parametric excitation of the first order. Excitation of spin waves is at the expense of the energy of uniform precession oscillations at ferromagnetic resonance, which again results in supplementary attenuation of this precession. This fact causes anterior saturation of ferromagnetic resonance and the resonance curve extension.

Let us proceed now to the investigation of a simple mechanism of parametric amplification of magnetostatic waves in magnetic materials.

5.2 Parametric excitation (amplification) of magnetostatic waves in magnetic materials

The principle of parametric excitation of oscillations consists in the following. Time change of one of the oscillating system parameters results in a frequency shift; in the case of nonlinear elements in the shift of two different frequencies. This enables us to transform the input energy with frequency ω (excitation frequency) into the energy of oscillations with frequency ω_1 (own frequency). The simplest example of a mechanical parametric oscillating system is a see-saw if we are sitting on it and swinging on our own. During the oscillation period given by frequency ω_1 we twice stand up and bend down with excitation frequency ω, i.e. in this case $\omega = 2\omega_1$. The amplitude of see-saw oscillations increases at this. An analogous experiment can be done with a bob on a thread of variable length. If the bob is swinging and we change the thread length twice during a period in the same phase, the amplitude of oscillations will begin to increase.

For a quantitative description of parametric amplification we will investigate a simple LC circuit [21]. Let a circuit parameter, which is time-dependent, be the capacitance which changes with frequency ω, i.e.

$$C(t) = C_0 + \Delta C \cos(\omega t + \phi),$$

and let frequency ω be double that of the circuit's own frequency $\omega_1 = 1/\sqrt{(L_0 C_0)}$. The change of capacitance may occur e.g. by the shift of capacitor plates. Let the shift of excitation phase with regard to the phase of charge change on the capacitor be chosen in such a way that if the charge is maximum, the capacitor plates are moving away (work is performed) and if the charge is equal to zero, capacitor plates are coming nearer to each other. In such a way twice during the period the circuit is supplied with energy from an external source and the amplitude of oscillations begins to increase. In the case that the phase of the excitation signal is opposite, the oscillations will be damped. We will describe this process in a quantitative way. The equations for current and voltage in the circuit are

$$\frac{\mathrm{d}i}{\mathrm{d}t} = -\frac{1}{L_0} u_c(t) \tag{5.20.1}$$

$$\frac{\mathrm{d}}{\mathrm{d}t}[(C_0 + C_1(t))u_c] = i \tag{5.20.2}$$

where $C_1(t) = \Delta C \cos(\omega t + \phi)$. We will introduce the amplitudes of the circuit's own oscillations "a" and "a^*" [21]:

$$a = \frac{1}{2}\sqrt{L_0(J + j\omega_1 C_0 U_c)} \tag{5.21.1}$$

$$a^* = \frac{1}{2}\sqrt{L_0(J - j\omega_1 C_0 U_c)} \tag{5.21.2}$$

where U_c and J represent complex amplitudes of voltage and current for harmonic signal with frequency ω_1. Considering (5.21.1) and (5.21.2) the equation system (5.20) may be written as

$$\frac{\mathrm{d}a}{\mathrm{d}t} = j\omega_1 a - \frac{\mathrm{d}}{\mathrm{d}t}\left[\frac{C_1(t)}{2C_0}(a - a^*)\right] \tag{5.22.1}$$

$$\frac{\mathrm{d}a^*}{\mathrm{d}t} = -j\omega_1 a^* + \frac{\mathrm{d}}{\mathrm{d}t}\left[\frac{C_1(t)}{2C_0}(a - a^*)\right] \tag{5.22.2}$$

For $C_1(t) = 0$ the system (5.22) transforms into equations of an elementary oscillator in the form of normal oscillations. These equations describe two independent circuit oscillations which are represented by rotating phasors which rotate in a complex plane in opposite directions. If $C_1 \neq 0$, these oscillations are bound parametrically (by means of a changing parameter) and their amplitudes may change. Let $C_1 \ll C_0$ (approximation of a weak binding) and let us suppose

$$a(t) = A(t)\mathrm{e}^{j\omega_1 t} \tag{5.23}$$

where $A(t)$ is slowly changing amplitude of oscillations, i.e.

$$\frac{\mathrm{d}A(t)}{\mathrm{d}t} \ll j\omega_1 A_1$$

If we express the time dependence of variable capacitance by means of complex functions in the form

$$C_1 = \frac{\Delta C}{2}\left[\mathrm{e}^{j(\omega t + \phi)} + \mathrm{e}^{-j(\omega t - \phi)}\right] \tag{5.24}$$

and substitute (5.32) and (5.24) into (5.22), neglecting the terms with frequency $\pm 3\omega_1$, we can write the equation system (5.22) in the form

$$\begin{aligned} \frac{\mathrm{d}a}{\mathrm{d}t} &= j\omega_1 a + C_{12}\mathrm{e}^{2j\omega_1 t}a^* \\ \frac{\mathrm{d}a^*}{\mathrm{d}t} &= -j\omega_1 a^* + C_{21}\mathrm{e}^{-2j\omega_1}a \end{aligned} \tag{5.25}$$

where the coupling coefficient $C_{12} = j\omega_1(\Delta C/4C_0)\mathrm{e}^{j\phi} = C_{21}^*$. The equations (5.25) represent a system of related equations with variable coefficients. Using (5.23) these equations can be transformed into a system with constant coefficients:

$$\begin{aligned}\frac{\mathrm{d}A}{\mathrm{d}t} &= C_{12}A^* \\ \frac{\mathrm{d}A^*}{\mathrm{d}t} &= C_{21}A\end{aligned} \tag{5.26}$$

the solution of which, having substituted the exponential function $\mathrm{e}^{\delta t}$, is

$$\delta_{1,2} = \pm\delta = \pm|C_{12}| = \omega_1 \frac{\Delta C}{4C_0} \tag{5.27}$$

Equation (5.27) shows that in the circuit there may exist increasing as well as damped oscillations. If we use the initial conditions

$$a(0) = |a(0)|\mathrm{e}^{j\vartheta} = \frac{1}{2}\sqrt{L_0[i(0) + j\omega_1 C_0 u_c(0)]}$$

we can write the solution of the system (5.25) considering (5.27) in the form [21]

$$a(t) = \frac{|a(0)|}{2}\mathrm{e}^{j(\omega_1 t+\vartheta)}\left[[1 + j\mathrm{e}^{j(\phi-2\vartheta)}]\mathrm{e}^{\delta t} + [1 - j\mathrm{e}^{j(\phi-2\vartheta)}]\mathrm{e}^{\delta t}\right] \tag{5.28}$$

An analogous formula can be written also for $a^*(t)$. To obtain increasing oscillations it is necessary to choose in an appropriate way the phase of excitation (the capacitor plates are moving away if the charge is maximum and coming nearer if it is equal to zero). From (5.28) it follows that if

$$(\phi - 2\vartheta) = (3/2)\pi + 2n\pi \;\; (n = 0,\; \pm1;\; \pm2; \ldots)$$

i.e.

$$\mathrm{e}^{j(\phi-2\vartheta)} = -j$$

then in the circuit there exist only increasing oscillations (signal amplification). For

$$(\phi - 2\vartheta) = (\pi/2) + 2n\pi \;\; (n = 0;\; \pm1;\; \pm2; \ldots)$$

only damped solutions exist in the circuit.

This is an elementary theory of parametric signal amplification in a single oscillation circuit. In general, when in an oscillation circuit frequencies ω_1 and ω_2 exist, equations of parametrically coupled oscillations are in the form [21]

$$\begin{aligned}\frac{\mathrm{d}a_1}{\mathrm{d}t} &= j\omega_1 a_1 + C_{12}\mathrm{e}^{j\omega t}a_2^*,\\ \frac{\mathrm{d}a_2}{\mathrm{d}t} &= -j\omega_2 a_2^* + C_{21}\mathrm{e}^{j\omega t}a_1\end{aligned} \tag{5.29}$$

Oscillations a_1 and a_2 will be coupled in an active way if the following condition is fulfilled for excitation (pumping) frequency

$$\omega = \omega_1 + \omega_2 \tag{5.30.1}$$

where ω_1 is the signal frequency and ω_2 is the so-called free frequency. In the example with the oscillation circuit $\omega_1 = \omega_2 = \omega/2$, i.e. the signal frequency was equal to the free frequency. It is noteworthy that for systems with waves propagating with frequencies ω_1 and ω_2 it is necessary to write another condition:

$$\vec{k} = \vec{k}_1 + \vec{k}_2 \tag{5.30.2}$$

where $\vec{k}$ is the wave vector of an excitation wave and $\vec{k}_1$, $\vec{k}_2$ are wave vectors of eigenwaves of the system without the presence of an excitation wave. It can be shown that the formula (5.30.1) represents the law of conservation of energy in a parametric process and the formula (5.30.2) the law of conservation of momentum. It should be also noted that the conditions (5.30.1) and (5.30.2) describe a parametric resonance of the first order.

Let us investigate now briefly the possible mechanisms of a parametric coupling in magnetic medium, which is described by the equations (5.5). We will introduce complex variables

$$m_\pm = m_x \pm jm_y \tag{5.31}$$

where $m_+ = m_-^*$, which characterize the motion of magnetization vector in a transverse plane (x, y). Considering (5.31) and (5.6) the first two equations of (5.5) for transverse components can be written in the form

$$\frac{\mathrm{d}m_+}{\mathrm{d}t} = +j\omega_H m_+ + j\mu_0\gamma h_z m_+ - j\mu_0\gamma M_0\left(1 - \frac{1}{2}\frac{m_+m_-}{M_0^2}\right)h_+ \tag{5.32.1}$$

$$\frac{\mathrm{d}m_-}{\mathrm{d}t} = -j\omega_H M_- + j\mu_0\gamma h_z m_- + j\mu_0\gamma M_0\left(1 - \frac{1}{2}\frac{m_+m_-}{M_0^2}\right)h_- \tag{5.32.2}$$

Neglecting the nonlinear term in parentheses, i.e. we put $M_z \simeq M_0$ and the system of equations (5.32) separates into two equations for m_+ and m_- in which parametric coupling is implemented due to the term $h_z m_\pm$. It should be noted that parametric excitation is not possible if the end point of the magnetization vector describes a circular trajectory, i.e. oscillations of $m_\pm$ are circularly polarized. On the contrary, parametric excitation is possible if the oscillations of $m_\pm$ are elliptically polarized. In fact this motion can be decomposed into two circular motions of vectors with different lengths rotating in opposition to each other. For example, oscillations of m_+ can be in the form

$$m_+ = a\mathrm{e}^{j\omega_1 t} + b\mathrm{e}^{-j\omega_1 t} \tag{5.33}$$

where $a \neq b, \omega_1$ is a signal frequency. Excitation with frequency $\omega = 2\omega_1$ will couple these oscillations and parametric coupling will appear. If the phase of the excitation field is properly chosen, energy with frequency $\omega \simeq 2\omega_1$ is supplied to the system and the precession of the magnetic moment with frequency ω_1 becomes larger. In other words, if we begin to "rock" the component m_z (Fig. 5.3 and formula (5.11)) with frequency 2ω, M_z begins to change and thus also χ and χ_a, which again, through equation (5.9), begin to "rock" m_x and m_y with frequency ω_1. This is roughly the picture of parametric amplification of oscillations in a ferrite medium. In this case the mathematical description is rather simple. For parallel pumping, $h_z = h_0\cos(\omega t + \phi)$ and from the reduced form of the equation (5.32.1) for m_+, for example, we have

$$\frac{\mathrm{d}m_+}{\mathrm{d}t} = j\omega_H m_+ + j\mu_0\gamma h_0\cos(\omega t + \phi)m_+$$

From this, taking into account (5.33), the system of parametrically coupled equations, analogous to the system (5.25) for the oscillation circuit, can be obtained. As was shown above this has an increasing amplitude solution. For transverse pumping when both oscillating systems (with frequency ω_1 and also $\omega \simeq 2\omega_1$) are polarized in a transverse plane, the physics of a parametric interaction is much more complicated. This process is described in detail in [1].

Let us investigate now the parametric instability of magnetostatic spin waves. The model to be investigated will be the following. Let us assume that a spin lattice is at room temperature at which spins (magnetic moments) are in chaotic thermal oscillations and propagating in ferrite in the form of chaotic thermal spin waves (an experimental investigation of such thermal oscillations will be described in the next chapter). Thus an arbitrarily ordered spin system at room temperature is characterized by an infinite system of chaotic spin waves, frequencies of which lie in a thermal frequency band. These waves are random: they arc not coupled to each other and represent thermal noise. However, if to such a system an external signal with frequency ω and wave vector $\vec{k}$ is applied, two spin waves from the given infinite spectrum with frequencies ω_1 and ω_2 and wave vectors $\vec{k}_1$ and $\vec{k}_2$ fulfilling (5.30.1) and (5.30.2) have the possibility of the coupling between themselves. In this case parametric coupling appears between thermal waves ω_1 and ω_2 in accordance with (5.25). The corresponding signal with frequency ω represents an excitation signal, and ω_1 and ω_2 represent signal and free frequencies. Parametric instability of (amplification) of spin waves arises. In the very same way two regular magnetostatic waves can be coupled together. In this case, usually the so-called degenerated case, $\omega_1 = \omega_2 = \omega/2$ is implemented. Let us give a necessary mathematical description of this process and investigate existing constraints (see e.g. [20 – 24] and corresponding bibliography). For the sake of simplicity we will consider the simplest model of parametric excitation (amplification) of spin waves in an infinite medium [20]. We will suppose that the high-frequency excitation field is parallel to $\vec{H}_0$ and that it is uniform, i.e.

$$h_z = h_0 \cos(\omega t) \tag{5.34}$$

We will use the equation of motion (5.4) and take into account that the magnetic field H contains in the general case constant and also time-dependent external fields and an exchange field. We will write complex variables for transverse components of the vector $\vec{m}$ in the form (5.31) and using the equation of motion we will obtain for complex conjugate amplitudes an equation of the same type as (5.32). Then we will expand the transverse components $m_{\pm}$ into the space Fourier series and after some modifications we will obtain the system of coupled differential equations for the kth space harmonic magnetization (kth spin wave) $b_{\pm k}$:

$$\frac{db_k}{dt} = j(\omega_k + j\alpha_k)b_k - G_k b^*_{-k}\cos(\omega t) \quad (5.35.1)$$

$$\frac{db^*_{-k}}{dt} = -j(\omega_k - j\alpha_k)b^*_{-k} + G^*_k b_k \cos(\omega t) \quad (5.35.2)$$

where $\omega_k = [\omega'_H(\omega'_H + \omega_M \sin^2\theta_k)]^{1/2}$, $\omega'_H = \omega_H + \beta\mu_0\gamma M_0(ak)^2$ $\omega_H = \mu_0\gamma H_0, \beta$ is an exchange constant, a is a lattice constant, $\omega_M = \gamma\mu_0 M_0$, θ_k is the polar angle of the kth spin wave, α_k is a loss parameter in the sense of (1.22) for the kth spin wave, and

$$G_k = j\frac{\mu_0\gamma h_0\omega_M \sin^2\theta_k}{2\omega_k} \times \mathrm{e}^{2j\phi_k}$$

is the so-called coupling parameter, where ϕ_k is the azimutal angle of the kth spin wave.

Equations (5.35) will have a harmonic solution if the conditions of parametric resonance (5.30.1) and (5.30.2) are fulfilled. For a space uniform excitation (pumping) given by the equation (5.34) $\vec{k}_1 = -\vec{k}_2$ ($\vec{k} = 0$) and $\omega_{k1} = \omega_{k2} = \omega/2$ (this means that the degenerate regime is taken into account). From (5.35) it can be seen that parallel excitation cannot excite spin waves propagating in the direction of magnetic field H_0 ($\theta_k = 0, \quad G_k = 0$). This follows from the fact that such spin waves have a circular polarization of magnetic moment and for them $\Delta m_z(2\omega_H) = 0$ [formula (5.16)]. We are looking for the solution of the system of equations (5.35) in the form

$$\begin{aligned} b_k &= b_{0k}\mathrm{e}^{j(\omega t/2)+\delta t} \\ b^*_{-k} &= b^*_{0-k}\mathrm{e}^{-j(\omega t/2)+\delta t} \end{aligned} \quad (5.36)$$

where δ is the constant of the amplitude rise of the kth spin wave. After substitution of (5.36) into (5.35) we have a system of algebraic equations, the determinant of which has to be equal to zero. Thus

$$(\omega_H - \frac{\omega}{2})^2 + (\alpha_k + \delta)^2 - \frac{1}{4}|G_k|^2 = 0$$

and for $\delta_{1,2}$ we have

$$\delta_{1,2} = -\alpha_k \pm \left[\frac{1}{4}|G_k|^2 - \left(\omega_H - \frac{\omega}{2}\right)^2\right]^{1/2} \tag{5.37}$$

From (5.37) it follows that the condition of instability of spin waves is

$$\left[\frac{1}{4}|G_k|^2 - \left(\omega_H - \frac{\omega}{2}\right)^2\right]^{1/2} \geq \alpha_k \tag{5.38}$$

As can be seen, δ has a maximum positive value for $\omega_H = \omega/2$, i.e. when the frequency of the amplified spin wave is equal to one half of the excitation frequency. From equation (5.37) the threshold value of the high-frequency magnetic field amplitude corresponding to the condition $\delta = 0$ can be determined in the form

$$h_{0c} = \frac{4\omega_k[\alpha_k^2 + (\omega_H - \omega/2)^2]^{1/2}}{\mu_0 \gamma \omega_H \sin^2\theta_k} \tag{5.39}$$

For $h_0 > h_{0c}$ parametric excitation (amplification) of spin waves is possible in a ferrite. The minimum value of the threshold field, as follows from (5.39), corresponds to $\omega_H = \omega/2$, i.e.

$$(h_{0c})_{\min} = \frac{2\omega\alpha_k}{\mu_0 \gamma \omega_M \sin^2\theta_k} \tag{5.40}$$

From (5.40) it follows that the threshold value of the high-frequency magnetic field amplitude is proportional to the loss parameter for the kth spin wave and the frequency ω. This value depends substantially also on the angle θ_k and has its minimum for $\theta_k = \pi/2$.

As it can be seen from (5.37) δ is proportional to the value h_0, i.e. instability of spin waves is possible (if losses are small) for sufficiently low amplitudes of the excitation high-frequency magnetic field. As follows from Chapter 2, for each type of magnetostatic wave a lower limiting frequency (for $k \to 0$) exists, below which propagation of magnetostatic waves is not possible. Obviously the excitation frequency has to fulfil the condition

$$\omega \geq 2\omega_{k\,\min}(k \to 0) \tag{5.41}$$

The condition (5.41) shows that in parametric processes of the first order the excitation frequency ω must not be lower than twice the lower limiting frequency of the magnetostatic spin wave spectrum. If the excitation frequency is given, then this condition determines the lower limit of magnetostatic spin wave instability. To sum up, for parametric processes the threshold power of excitation field and the lower band frequency exist for a possible instability of spin waves. The first condition is in principle given by losses in the given material and the second one by the excitation frequency used.

It should be noted at the end of this section that in an analogous way parametric processes at perpendicular pumping of parametric instability can be investigated. Resonance phenomena connected with the precession of the magnetization vector are fundamental for them; unfortunately a mathematical treatment is in this case very complicated.

5.3 Nonlinear losses in MSW propagation

The origin of MSW losses in ferrites is rather complicated (see e.g. [2] and bibliography). Here should be included mainly losses caused by relaxation processes at low levels of power, and losses caused by MSW scattering on various inhomogeneities (chemical inhomogeneities, dislocations, etc.), losses caused by the coupling of MSW with other types of wave (e.g. acoustic) and transmission of energy into these types of wave. Also Joule losses are possible in metallic plates surrounding the ferrite, etc. All these losses were characterized by the generalized loss parameter α, which is present in the Landau–Lifshitz equation of motion, where the essence of these losses was not interesting for us. We can say that all the given losses did not depend on signal amplitude, i.e. they were "linear" losses characterizing a medium and they were not connected with the power transmitted by magnetostatic wave.

There exists, however, one more source of losses, which behaves in an absolutely different way and thus also the parameters of the medium will depend on the amplitude of the high-frequency signal.

These are parametric losses described in the previous section. A regular signal, playing the role of an excitation signal, excites two thermal spin waves which begin to take over the energy of the regular input signal, which causes its attenuation. What is important is the fact that this parametric amplification of spin waves is proportional to input power (formula (5.37)), i.e. parametric losses (power "taken away" by thermal spin waves) will increase with the rise of input signal. The medium becomes nonlinear in the magnitude of losses. If, in a similar way as above, we characterize losses by the parameter α, then for parametric losses α is a function of input signal, i.e. $\alpha(P_{in})$, and increases with it. This is very important for the understanding of many phenomena connected with nonlinear phenomena of MSW [1]. First of all, it holds for nonlinear ferromagnetic resonance ((5.1) and Fig. 5.3).

For nonlinear parametric amplification of MSW the limits of the frequency band in which MSW can propagate and the fulfilment of conditions (5.41) are very important. These facts determine the frequency limits for parametric effects of the first order for MSW.

The spectrum of spin waves is given by the following formula [1]:

$$\omega_k = [\omega'_H(\omega_H + \omega_M \sin^2 \theta_k)]^{1/2} \tag{5.42}$$

where $\omega'_H = \omega_H + \beta\mu_0\gamma M_0(ak)^2$. From this it follows that the minimum frequency of spin waves is

$$\omega_{k\,\mathrm{min}} = \omega_H \tag{5.43}$$

It should be noted that for spin waves with large k it is not essential to consider boundary conditions because the size of a real ferrite specimen is much bigger than the length of a spin wave. We should note also that for parametric excitation of magnetostatic spin waves, the magnetostatic wave has to fulfil both condition (5.30.1) and condition (5.30.2) ($\vec{k}_{\mathrm{MSW}} = \vec{k}_{1\mathrm{sw}} + \vec{k}_{2\mathrm{sw}}$). For spin waves with large values of k this condition can be fulfilled because thermal spin waves can be excited for an arbitrary value of the angle θ_k (Fig. 5.4a). In Fig. 5.4a these diverge from each other. In principle it is possible to excite magnetostatic waves also at small values of $\vec{k}$ if they lie within the magnetostatic area. In this case both excitation and amplified waves will propagate in approximately the same direction (Fig. 5.4b). For these waves we also have $\omega_{\mathrm{min}} = \omega_H$. And, at last, in all cases the frequency of the wave has to fulfil the condition

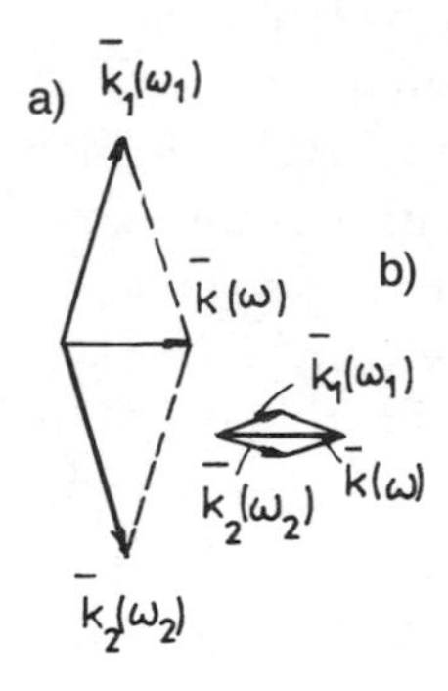

Fig. 5.4. The dependence of wave vectors of MSW for two cases of the condition (5.30.2)

$$\omega > 2\omega_H \tag{5.44}$$

As was given in Chapter 2, volume MSW propagate in a frequency band the maximum width of which is given by

$$\omega_H \leq \omega \leq \sqrt{[\omega_H(\omega_H + \omega_M)]}$$

Surface MSW in a free ferrite plate has a maximum band for

$$\sqrt{[\omega_H(\omega_H + \omega_M)]} \leq \omega \leq \omega_H + (1/2)\omega_M$$

and in a metal-coated layer

$$\sqrt{[\omega_H(\omega_H + \omega_M)]} \leq \omega \leq \omega_H + \omega_M$$

Thus the maximum width of the total spectrum of MSW is

$$\omega_H \leq \omega \leq \omega_H + \omega_M$$

We will divide these inequalities by $\omega_M = \mu_0\gamma M_0$. For volume MSW we have

$$(\omega_H/\omega_M) \leq (\omega/\omega_M) \leq \sqrt{[(\omega_H/\omega_M)(1 + \omega_H/\omega_M)]} \tag{5.45.1}$$

for surface MSW in a free ferrite layer

$$\sqrt{[(\omega_H/\omega_M)(1 + \omega_H/\omega_H)]} \leq \omega/\omega_M \leq (1/2) + \omega_H/\omega_M \tag{5.45.2}$$

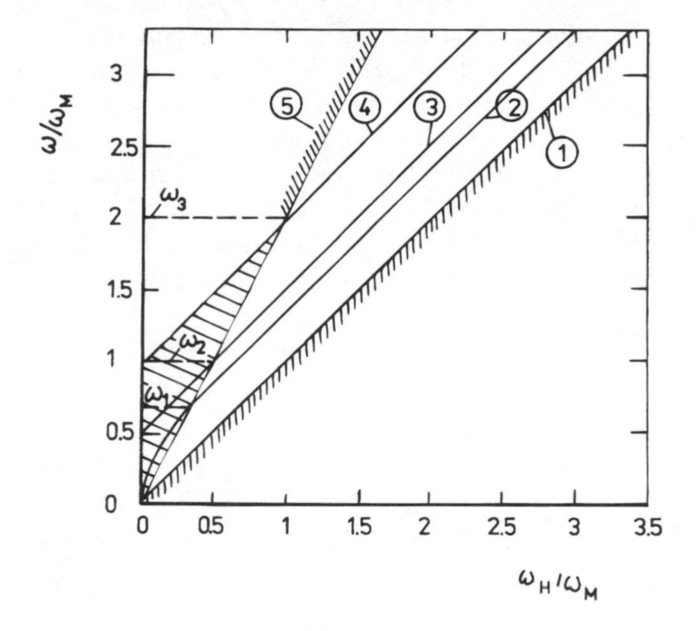

Fig. 5.5. Lines of frequency limits for volume and surface MSW as functions of ω_H/ω_M, 1 — $\omega = \omega_H$, 2 — $\omega = \sqrt{\omega_H(\omega_M + \omega_H)}$, 3 — $\omega = \omega_H + \omega_M/2$, 4 — $\omega_H + \omega_M$, 5 — $\omega = 2\omega_H$. The area $\omega < \omega_H$ is a restricted area for spin waves.

and for surface MSW in a metal-coated layer

$$\sqrt{[(\omega_H/\omega_M)(1 + \omega_H/\omega_M)]} \leq \omega/\omega_M \leq 1 + \omega_H/\omega_M \tag{5.45.3}$$

In Fig. 5.5 the curves of frequency limits for volume and surface MSW as a function of $\omega_H/\omega_M = H_0/M_0$, and also curves $\omega = \omega_H$ and $\omega = 2\omega_H$ are shown. The hatched area in Fig. 5.5 corresponds to the area of MSW existence in which $\omega \geq 2\omega_H$. In this area the parametric amplification of waves at frequency $\omega/2$ is possible, which results in nonlinear attenuation of the MSW excitation wave with frequency ω. As has already been given, resulting from the existence of the parametric mechanism, the amplification and thus also the loss parameter α are proportional to the input signal, i.e. $\alpha(P_{\text{in}})$. For all other frequency zones of the existence of MSW (Fig. 5.5) first-order parametric processes resulting in the instability of MSW are not possible. From Fig. 5.5 we can see that the curve corresponding to the upper band frequency for volume MSW intersects the curve $\omega = 2\omega_H$ at the point $\omega_1 = 0.65\omega_M$, for surface MSW in a free slab in the point $\omega_2 = \omega_M$, and for surface MSW in a metal-coated slab at the point $\omega_3 = 2\omega_M$.

For a thin YIG layer $M_0 = 140\,\mathrm{kA\,m^{-1}}$ and corresponding frequencies are 3.2 GHz, 4.9 GHz and 9.8 GHz. Nonlinear losses caused by parametric excitation of spin waves for magnetostatic waves in YIG structures can be observed only in the long-wave area of the high-frequency spectrum. In the short-wave area of the high-frequency band (for $f > 9.8$ GHz) parametric losses cannot exist and the loss parameter, with changing input signal power, remains constant until observable decrease of the longitudinal component of magnetization M_z begins (5.1).

For this there exists a convincing experimental proof [1]. In Fig. 5.6 the experimental characteristics $P_{\text{out}} = f(P_{\text{in}})$ of a trans-

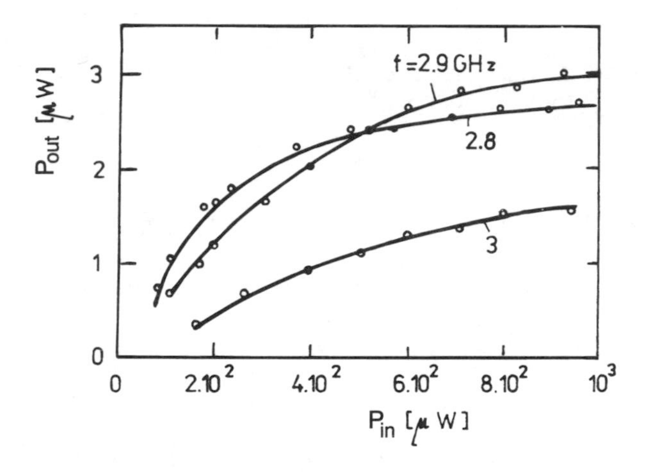

Fig. 5.6. Experimental dependence of the transmission line output power on the level of input power for surface MSW.

mission line on a thin YIG layer at the excitation of surface MSW in the band 2.8 – 3.0 GHz are presented. Common "linear" losses in the transmission line for normal power levels (dispersion on nonhomogeneities, adaptation losses, etc.) were about 20 dB. On increasing the input power the output power at first increases in a linear way and then for $P_{\text{in}} \simeq 400$–$500\,\mu$W begins to saturate. This can be explained only by supplementary losses in YIG as a result of spin wave excitation at a half-frequency. For these waves the condition for parametric resonance $f > 2f_h$ is fulfilled because in this case $f_H = 1.12$ GHz.

These facts play an important role in the short-wave band, i.e. at frequencies which are higher than the parametric frequency limit

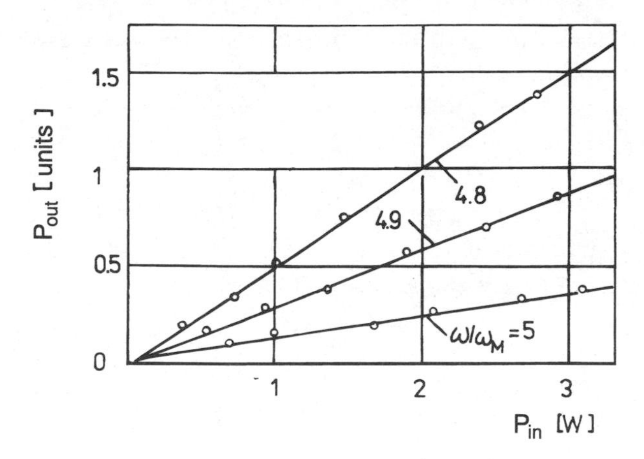

Fig. 5.7. The dependence of transmission line output power for surface MSW on the level of input power in millimetre wave frequency band. $M_0 = 140\,\mathrm{kA\,m^{-1}}$, $H_0 = 624\,\mathrm{kA\,m^{-1}}$, $\omega_H/\omega_m = 4.45$.

($f > 9.8$ GHz for YIG). In these bands the losses are determined only by linear processes and do not depend on the input signal amplitude. In contrast to all high-frequency devices in which losses always increase with frequency, in high-frequency MSW elements losses are almost independent of the frequency (we mean the basic, not "parametric" part of losses). Moreover, as a result of the absence of the parametric mechanism, the losses do not even depend on the input signal power. The amplitude characteristics of an MSW transmission line in the millimetre wave frequency band are shown in Fig. 5.7. As can be seen, in accordance with the theory, the characteristics have a linear character at these frequencies [1].

5.4 Lines loaded by ferrite with an inverse dynamic nonlinearity

A strip line loaded by a ferrite medium in which MSW can be excited has an interesting characteristic connected with the transmission of signals with various power level. This characteristic can be experimentally observed only in a long-wave frequency domain when the condition $\omega > 2\omega_H$ is fulfilled and it is connected with the parametric mechanism of nonlinear losses of MSW which was investigated above. The essence of this phenomenon is based on the fact that a signal with a low power level (below some threshold value) propagating in such a structure will be damped more than a signal with a higher power level.

A typical example of such a structure is given in Fig. 5.8. In

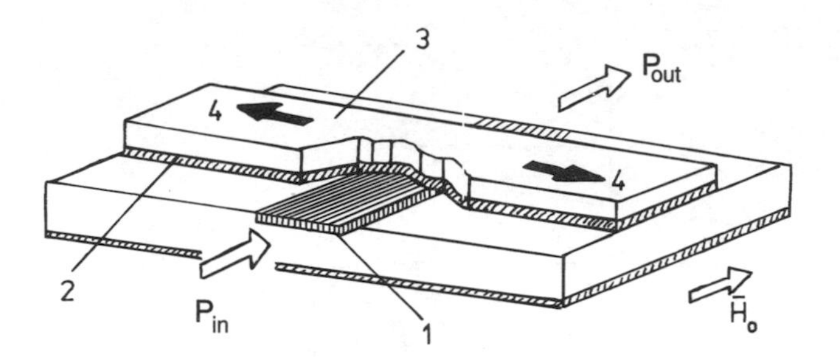

Fig. 5.8. A typical case of a structure used in a nonlinear regime. 1 — microstrip, 2 — ferrite, 3 — substrate, 4 — surface MSW.

principle it is a microstrip line loaded by a ferrite layer (the layer is placed on a microstrip line). A constant magnetic field is oriented parallel to the strip and its amplitude is chosen in order to excite effectively surface MSW in a thin layer. These will propagate perpendicularly to the microstrip [25]. The special behaviour of such a microstrip line can be explained by means of the curves in Fig. 5.9. Line 1 in Fig. 5.9 corresponds to the linear amplitude characteristic of the line $P_{\text{out}} = f(P_{\text{in}})$ in the case when the magnetic field in Fig. 5.8 $\vec{H}_0 = 0$. For a defined value of $\vec{H}_0$ the high-frequency current

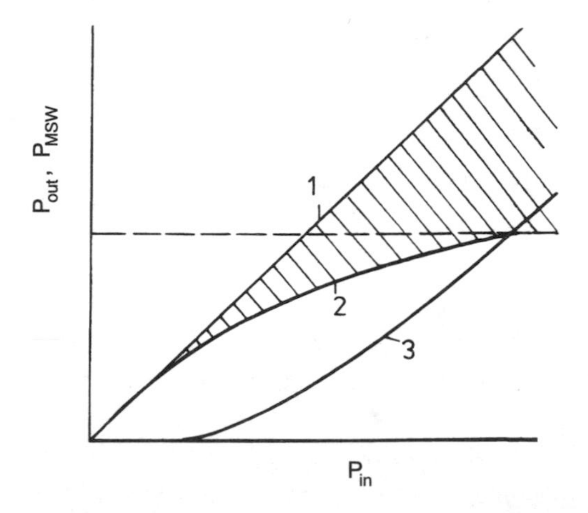

Fig. 5.9. Power dependences for the line in Fig. 5.8. 1 — $P_0(H_0 = 0)$, 2 — P_{MSW}, 3 — $P_{\text{out}} = P_{\text{in}} - P_{\text{MSW}}$.

excites a magnetostatic wave in a microstrip. As is demonstrated by experiments for low levels of input power, almost the whole electromagnetic energy in the microstrip is transformed into MSW energy. This causes a large attenuation of electromagnetic waves at the output of the microstrip line. Thus for a weak signal the line will have a large attenuation. With increase of the input power, after the threshold level of input signal, with frequencies fulfilling the conditions of parametric resonance, was reached, the gradual saturation of MSW amplitude occurs (curve 2 in Fig. 5.9). In this case a great part of the electromagnetic energy will be transformed from the input to the output of the microstrip line. Signal attenuation begins to decrease, which corresponds to the hatched area in Fig. 5.9. The amplitude characteristic of such a line is represented by curve 3 in Fig. 5.9. As can be seen from this, the signal with a low level of noise will be attenuated more than the signal with a higher noise level. In principle, such a line has an inverse nonlinear dynamic characteristic. Inversion here expresses the relationship with nonlinear characteristics of common high-frequency elements. Here it is essential that "threshold" power corresponds to sufficiently low values of input signal amplitude (Section 5.2) and for this reason in such a structure a nonlinearity is seen already at small power levels of input signal. Such a line can be used for example for attenuation of noisy signals which have a power level less than some threshold, in contrast to normal signals, the power level of which is almost always higher than "threshold". For this reason devices designed on this basis are called signal-to-noise enhancers.

In the monograph [1] an original way of explaining this phenomenon is proposed. It is based on the dependence of the efficiency of MSW excitation by a microstrip line on losses in a ferrite. Intuitively we can suppose that the increase of losses in ferrite lowers the efficiency of excitation of MSW. On the other hand, the less efficiently the MSW are excited, the bigger the part of the electromagnetic energy is transferred without losses from the input to the output of a microstrip line. For this reason, if losses in ferrite are small, the excitation of MSW is efficient, electromagnetic energy is transformed into MSW and the signal in a microstrip line is strongly attenuated. If losses in ferrite are large, then excitation of MSW is weak and thus electromagnetic energy is not transformed into the ferrite (by means of MSW) and the signal is transferred from the input to the output of the microstrip line without attenuation. This is a complete analogy with the excitation of a resonator with losses, which is connected to a straight waveguide in such a way that one part of the energy is "taken away" by the resonator. It is obvious that the bigger the quality factor of the resonator, i.e. the less lossy it is, the more efficiently it is excited and the bigger the part of the power taken away from the waveguide. On the output of the waveguide the power will decrease and the signal will be strongly attenuated. The efficiency of resonator excitation with increasing losses decreases, the power from the waveguide is in less measure taken away into the resonator and the power is transferred through the waveguide freely. Attenuation in the waveguide decreases. In the case of large losses it is not possible to excite the resonator at all.

The question is what can cause the change of losses in the ferrite. One possible mechanism was described in the previous section. It was the mechanism of nonlinear losses caused by parametric excitation of thermal spin waves in a ferrite. The higher the signal in a microstrip line, the more efficient is the excitation of thermal spin waves at half-frequency, the higher the loss parameter $\alpha(P_{\text{in}})$ and the less efficiently the surface MSW are excited with a frequency equal to the signal frequency. Attenuation of MSW in the microstrip decreases and the signal is transformed with smaller losses through the microstrip. As a proof of this, in experiments the mechanism of nonlinear suppresion of a weak signal is observed only at frequencies lying within the frequency limits of parametric instability (for YIG it is $f > 9.8\,\text{GHz}$) [26].

The exact analysis of nonlinear MSW excitation connected with a parametric instability of spin waves is very difficult because the theory of nonlinear losses is only under development. For this reason, hereinafter a qualitative model of MSW excitation in a ferrite structure for various levels of input signal will be given [27, 28], which sufficiently describes the characteristic features of energy transmission in structures with an inverse dynamic nonlinearity.

Let us consider a piece of microstrip line loaded by a ferrite (Fig. 5.10a). It represents a transmission line with losses and wave impedance Z_0. In the general case Z_0 is a complex variable. We will suppose that the microstrip line is matched from both ends and losses in it are caused only by the MSW excitation.

The time average (during one period) of high-frequency power

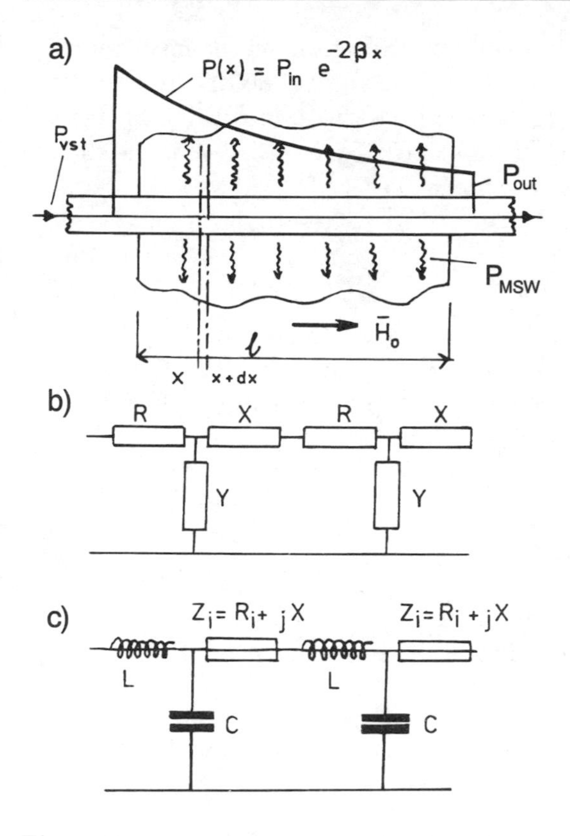

Fig. 5.10. Transmission line model of a microstrip line loaded by ferrite.

transmitted through the cross-section x is

$$P(x) = \frac{1}{2} J^2(x) \mathrm{Re}(Z_0) \tag{5.46}$$

where $J(x)$ is the surface line current density in the microstrip and $\mathrm{Re}(Z_0)$ is the real component of wave impedance. The change of average power in the microstrip line per length $\mathrm{d}x$ (Fig. 5.10a) can be described in the form

$$\mathrm{d}P(x) = -\mathrm{d}P_{\mathrm{MSW}}$$

where P_{MSW} is the power connected with MSW excitation (power "transmitted into ferrite"). In accordance with the formula (2.71) the MSW power radiated into the ferrite can be expressed in the form

$$\mathrm{d}P_{\mathrm{MSW}} = \frac{1}{2} J_0^2 R_i \,\mathrm{d}x \tag{5.47}$$

where R_i is the real component of radiation impedance per unit length of the microstrip and J_0 is the amplitude of the line high-frequency current density in the microstrip line per length $\mathrm{d}x$. If $J(x) = J_0$, then considering formulae (5.46) and (5.47) for the change of power in the line $\mathrm{d}P(x)$ we can write

$$\mathrm{d}P(x) = -P(x)\frac{R_i}{\mathrm{Re}(Z_0)}\mathrm{d}x \tag{5.48}$$

Integrating equation (5.48) under the assumption $R_i = \text{const.}$, for power $P(x)$ propagating in the microstrip line we obtain the following formula:

$$P(x) = P_{\mathrm{in}} \exp\left(-\frac{R_i}{\mathrm{Re}(Z_0)}\right) x = P_{\mathrm{in}}\mathrm{e}^{-2\varkappa x} \tag{5.49}$$

where P_{in} is the power at the input of the microstrip line and $\varkappa = (1/2)(R_i/\mathrm{Re}(Z_0))$ is the coefficient of attenuation in the microstrip line. In this case the power transformed into MSW (power of MSW excitation) is

$$P_{\mathrm{MSW}} = P_{\mathrm{in}}(1 - \mathrm{e}^{2\varkappa l}) \tag{5.50}$$

where l is the total length of the microstrip line (Fig. 5.10a). From (5.50) it also follows that the efficiency of MSW excitation is given by $\varkappa l$, and for $\varkappa l \to \infty$ the total input of electromagnetic power is transformed into power of MSW. In the frequency band lying below the limit of parametric instability, the variable $\varkappa$ with the rise of input power above the "threshold" has to decrease, which results in the decrease of efficiency of MSW excitation. How will it manifest itself in our model? For explanation let us consider again a piece of a microstrip line with ferrite as a transmission line with the distributed parameters (Fig. 5.10b), where R is the real component of the impedance per unit length and X and Y are reactance and admittance of the line per unit length respectively. The wave impedance of such a line is [29]

$$Z_0 = \sqrt{\frac{R + jX}{jY}} \tag{5.51}$$

and the attenuation constant is

$$\varkappa = \sqrt{\left[\frac{1}{2}Y[(R^2 + X^2)^{1/2} - X]\right]} \tag{5.52}$$

Comparing the formula (5.52) with the attenuation constant in (5.49) we have

$$R_i/(2\mathrm{Re}(Z_0)) = \sqrt{\left[\frac{1}{2}Y[(R^2 + X^2)^{1/2} - X]\right]}$$

It can be shown that the radiation resistance of MSW is equivalent to the resistance of microstrip line per unit length (if the line does not have any other losses). From the last formula and from (5.51) it follows that

$$R_i = R$$

As was shown in Chapter 2, the excitation power of MSW considering losses in ferrite has both real and imaginary components, which means that it is a complex variable. For this reason an equivalent circuit of a transmission line with distributed parameters for the analyzed structure with ferrite can be depicted as in Fig. 5.10c, where L and C represent inductance and capacitance per unit length of the line without losses and $Z_i = R_i + jX_i$ is a complex impedance of the line after MSW excitation. We will suppose that the real component of this impedance R_i is equal to the real component of radiation impedance of MSW and is connected with the power transmitted by MSW, and the imaginary component X_i is determined by the imaginary component of the MSW radiation impedance and represents an excited power in ferrite (Chapter 2). In this case the power in the line is determined by the formula (5.40) where the attenuation coefficient, in accordance with (5.52), can be written in the form

$$\varkappa = \frac{\pi\sqrt{2}}{\lambda_0}\left[[(1 + X_i/\omega L)^2 + (R_i/\omega L)^2]^{1/2} - (1 + X_i/\omega L)\right]^{1/2} \tag{5.53}$$

where λ_0 is the length of an electromagnetic wave in the microstrip line. The model of MSW excitation in the line investigated is based on the fact that starting from some "threshold" value of the input

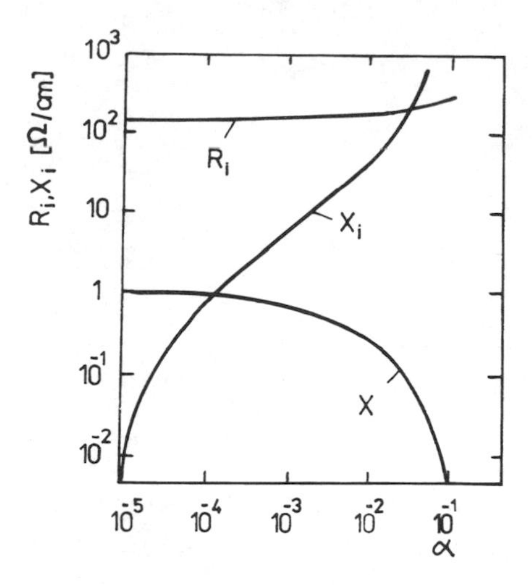

Fig. 5.11. Dependences of the radiation resistance R_i and reactance X_i on the loss parameter α. $\omega_M/\omega_H = 4.9$, $\omega/\omega_H = 2.94$, $w/d = 3$, $t/d = 5$.

power, the parameter of internal losses α begins to increase due to parametric excitation of thermal spin waves. Increase of the parameter α results in a change of the radiation resistance of MSW, i.e. a change of R_i and X_i values in (5.53). This again results in a change of the parameter $\varkappa$ in (5.51), i.e. the power transmitted through the microstrip line changes. In this way the nonlinear treatment of MSW excitation can be transformed to a linear treatment of wave excitation in a medium with losses. However, the fact that losses are a function of the input signal has to be taken into account, i.e. the parameter α depends on the input power. In Fig. 5.11 the dependences of R_i and X_i on the loss parameter α for a special case of surface MSW excitation are given (formula (2.80)). As can be seen, with an increase of the parameter α the reactive component of the radiation impedance X_i rapidly grows and for $\alpha \simeq 10^{-2}$ the value of X_i is comparable with the real component of the radiation impedance R_i. The rise of X_i, as follows from (5.53), results in a decrease of the value of $\varkappa$ in (5.50) and attenuation in the microstrip line decreases (Fig. 5.11). The line transmits a high-level signal well and a low-level one badly. The influence of inverse dynamic nonlinearity appears. It should be noted that the proposed model is only qualitative, i.e. it does not allow computation of the dependence of losses in ferrite $\alpha(P)$ on the input power. Currently such dependence can be obtained only on the basis of experimental data. In

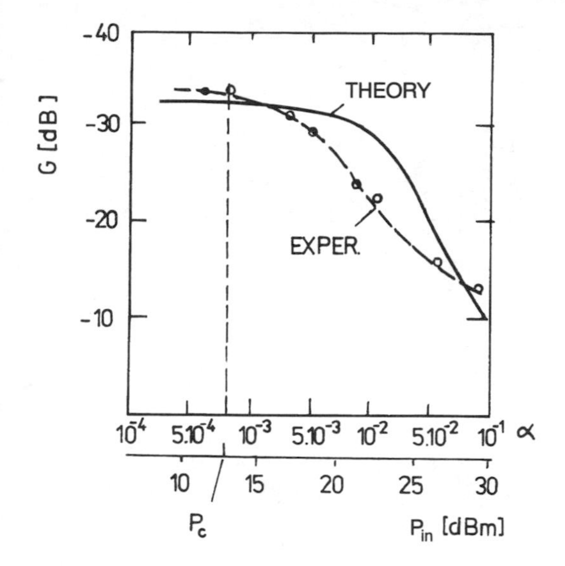

Fig. 5.12. Theoretical dependence of the transmission coefficient G on the loss parameter α. $f = 3\,\mathrm{GHz}$, $M_0 = 140\,\mathrm{kA\,m^{-1}}$, $H_0 = 140\,\mathrm{kA\,m^{-1}}$, $l = 0.53\,\mathrm{cm}$, $l_1/t = 5$, $w/t = 5$. P_c — threshold power.

the end it should be noted that an analogous effect connected with the rise of the input resistance of a ferrite delay line (particularly with a steep increase of the reactive component of this impedance) with an increase of the input power was experimentally observed in [30].

On the basis of the results obtained the dynamic characteristic of the investigated structure as a function of the loss parameter α can be computed. In Fig. 5.12 the theoretical dependence of the transmission coefficient $G = 10\log(P_{\text{out}}/P_{\text{in}})$ of the microstrip line with a thin YIG layer on the parameter α (full line) is depicted. In the same figure (dotted line) an analogous experimental dependence on the magnitude of the input power is given (in dBm), i.e.

$$P_{in}(\text{dBm}) = 10\log(P_{\text{in}}/\text{mW})$$

Starting from some level of the input power ($P_{\text{in}} > P_c$, $\alpha > 10^{-3}$) attenuation of the signal in this line decreases, the theoretical and experimental dependences are in qualitative agreement.

In Fig. 5.13 the theoretical dependences of the transmission coefficient G in the band of surface MSW excitation for various losses α are presented. As can be seen, for small α ($\alpha \simeq 10^{-5} \div 10^{-3}$) G

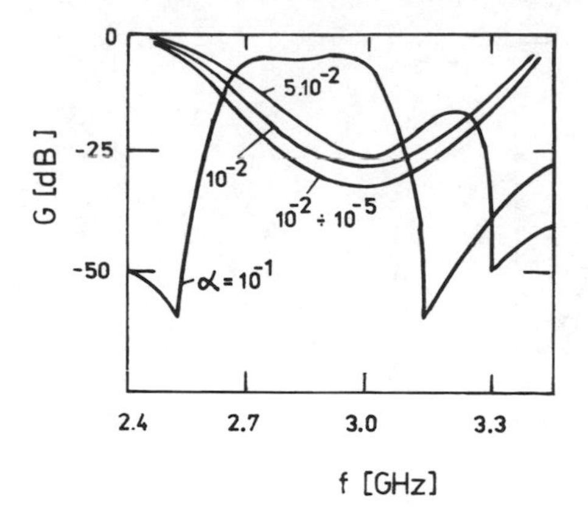

Fig. 5.13. Theoretical dependence of the transmission coefficient G in the band of surface MSW excitation for various values of the loss parameter α. $\lambda_0/l = 2$, $\omega_M/\omega_H = 4.9$, $l_1/t = 5$, $w/t = 3$.

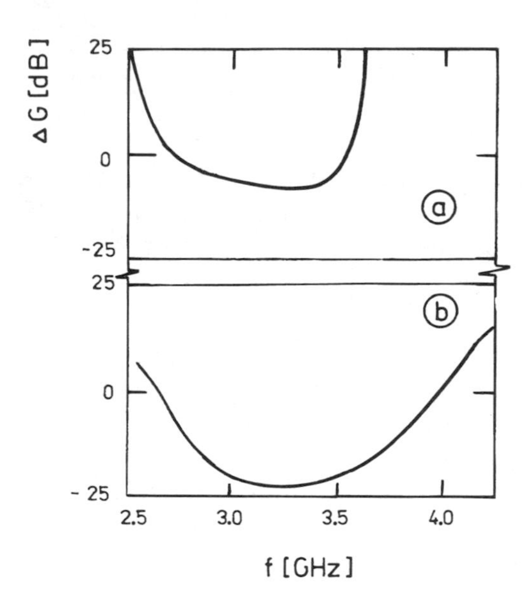

Fig. 5.14. The comparison of theoretical and experimental frequency dependence of ΔG. a) theory — $\Omega_M = 4.9$, $l_1/t = 5$, $w/t = 3$, $\lambda_0/l = 2$, b) experiment $\Omega_M = 4.2$, $w/t = 1$, $l_1/t = 2$.

depends weakly on α and is determined by the dependence R_i in the pass band; G is a minimum in the middle of the band and is equal to zero on its boundaries. With the rise of α, the attenuation coefficient in the middle of the band decreases and on its boundaries increases.

In Fig. 5.14a the calculated dependence of the variable $\Delta G = G_0 - G_2$ (where G_0 are values of G for $\alpha < 10^{-3}$ and G_2 are values of G for finite α) is given, and in Fig. 5.14b an analogous dependence (where G_0 are values of G for $P_{\text{in}} > P_c$ and G_2 are values of G for $P_{\text{in}} < P_c$) obtained from an experiment is presented. As can be seen these dependences are in approximate qualitative agreement.

Chapter 6

MAGNETOSTATIC WAVE SPECTROSCOPY

6.1 Physical methods of investigation of magnetostatic waves

In previous chapters we dealt with basic questions of the theory of the MSW excitation and propagation, in the case of their linear as well as nonlinear excitation. In this and the following chapters we will concentrate on experimental issues with regard to the practical application of the theoretical knowledge obtained, partly from the point of view of the investigation of magnetic material properties and partly from the point of view of the construction of specific microwave components and devices using the properties of MSW. For efficient and usable results of the theory from the point of view of practical application it is necessary to verify basic theoretical assumptions and corresponding results in an experimental way. Also the construction of models of individual components of magnetostatic structures which can serve as initial elements for the computer design of microwave devices using, as substructures, MSW elements requires an experimental verification of these models and, as the case may be, an experimental determination of the free parameters which are inputs into these models. From this follows the need for suitable experimental methods for the investigation of the properties of excited magnetostatic waves in corresponding geometrical configurations. This chapter is therefore dedicated to some selected basic experimental methods.

As follows from the theoretical analysis given above the dispersion relations (i.e. the dependence of the frequency of propagating MSW on the applied stationary magnetic field and wave vector) and losses of propagating MSW as a function of the physical parameters of the magnetic material used, depending on the type of excited

MSW and the geometrical arrangement, belong to the basic characteristics of magnetostatic waves. In principle such experimental techniques should be found, which could provide the direct information about the excited spin wave spectrum. It should be noted that this terminology is not absolutely correct. This is a consequence of the fact that the waves investigated are slow electromagnetic waves, which follows from the solution of Maxwell's equations, whilst spin waves are of purely quantum mechanical character and arise as the consequence of an exchange interaction between individual adjacent spins of the magnetic system. In spite of this physical distinction, we will not differentiate between these waves. It is obvious that one of the basic possibilities for this direct information is the use of electromagnetic waves in the microwave and millimetre frequency bands, which correspond to a frequency domain of elementary excitations of a considerable number of commonly used magnetic materials. In this context it should be noted that most of our knowledge of excitations in magnetic systems has been obtained just through the use of microwave experiments. It was shown, however, that microwave experiments were not able to give an answer to some questions which arose either in connection with the advance of theory in individual areas of the investigation of magnetic excitations or as a consequence of the development of new magnetic materials. This is true especially where the properties of a selected group of excited spin waves had to be investigated and where the implementation and interpretation of microwave experiments is difficult or impossible. For this reason also other experimental methods were investigated which would overcome the deficiencies of microwave experiments mentioned. One possibility is the interaction of an excited magnetic system with light, which can be included in the domain of optical methods of the investigation of magnetic excitations (here belong the Brillouin and Raman scattering of light, absorption of light in magnetic materials and the method of the investigation of total reflection, the so-called ATR method).

Until recently inelastic scattering of neutrons was the most useful experimental technique for the investigation of dispersion properties of magnetic excitations. The details of this experimental technique and the mechanism of neutron scattering are described e.g. in the monographs [36] and [37]. Inelastic scattering of neutrons gives information about the energy of magnetic excitations — magnons — in the whole range of the Brillouin zone. Such experimental results can be used for the design of models of spin configuration and dispersion. The scattering of neutrons on magnetic excitations is, however, limited to the domain of small energies, that is of small amplitudes of investigated wave vectors, and by the resolution of this experimental method which is bound up with the existence of Bragg's maximum of scattering in this domain.

Electron spectroscopy is another possibility which is widely used for the investigation of excitations of a different kind on a crystal surface. There is some hope that surface magnetic excitations could be detected in this way (see e.g. [38]).

Also, infrared spectroscopy provides some possibilities for magnetic excitation investigation. In this case infrared radiation interacts by its magnetic component oriented perpendicularly to the orientation of magnetic moments with magnetically ordered matter, and in the case that the frequency matches the frequency of existing magnetic excitations the resonance absorption of this radiation occurs. The external magnetic field (which is necessary) is usually homogeneous throughout the sample, which means that the total wave vector of the system has to be equal to zero. The application of infrared spectroscopy, however, also enables the investigation of more complicated effects, e.g. the excitation of two (or more) magnons by means of absorption of a single photon etc. In this case individual magnons can have non-zero wave vectors of magnetic excitations, because the requirement mentioned prescribes that the total sum of wave vectors of excitations included into this interaction has to be equal to zero. From the experimental data obtained information can again be deduced about dispersion properties of the magnetic system investigated. By this technique mainly anti-ferromagnetic materials were investigated (details can be found e.g. in the publication [39] which has the nature of an overview).

With regard to the orientation and experience of the authors we will hereinafter deal in more detail with optical and microwave methods of the investigation of properties of magnetic excitations.

6.2 Optical methods of investigation of magnetic excitations.

As has already been mentioned, magnetic excitations in solid material can be investigated also by means of their interaction with light, i.e. by means of optical methods. From them was set up, as one of the standard methods, the technique of light scattering. In this experiment, the sample is iluminated by monochromatic light and the scattered light is investigated. The light scattering is caused by nonhomogeneities in the sample. In the case that the nonhomogeneities are static, scattering is elastic, i.e. a shift of frequency in scattered light does not occur. If the nonhomogeneities are dynamic and light with them interacts in the area of its penetration into the matter, the scattering is inelastic and a frequency shift appears in the spectrum. If this frequency shift is up to a frequency of 300 GHz ($10\,\mathrm{cm}^{-1}$), then we speak about Brillouin and sometimes also about Brillouin–Mandelstam light scattering, named after the authors who independently described this phenomenon [40, 41]. If the shift of frequency is bigger, then we speak about Raman light scattering. The first experimental observations of light scattering on magnetic excitations were made by Fleury and his coworkers [41], who in 1966 observed temperature-dependent Raman light scattering from one and two magnon excitations in antiferromagnetic FeF_2.

The first experimental studies in the field of Brillouin light scattering were published by Sandercock and Wettling [43] in 1973. The increasing importance of the latter experimental technique for the investigation of magnetic excitations testifies to the fact that recently three works [44, 45, 46] and the monograph [47] were dedicated to this topic.

The theory of scattering of magnetic excitations, in principle, differs only a little from the corresponding theories of light scattering on other particles, which have been known for a longer time, and an interested reader can refer to them e.g. in the book [48]. From a classical point of view light scattering can be viewed as light reflection on wavefronts of collective plane wave excitations. Taking into account that wavefronts propagate through a medium with phase velocity v_p, then this motion results in a Doppler shift in the reflected light, which can be either in the direction of lower or higher frequencies according to whether the wave vector $\vec{k}$ of excitation has a component in the same or in the opposite direction to the direction of the incident light. Moreover, if we take into account the fact that in Brillouin scattering the excitations investigated have frequencies of several orders smaller than the frequency of the incident light ω_0, i.e. $\omega_0 \gg \omega_m$, we can arrive at two basic conditions for frequency and wave vector in the process of light scattering on a magnetic excitation in the form

$$\begin{aligned} \omega_s &= \omega_0 \pm \omega_m \\ \vec{k}_s &= \vec{k}_0 \pm \vec{k}_m \end{aligned} \tag{6.1}$$

where the index s corresponds to scattered light, the index 0 to the incident light and the index m to magnetic excitation. The same formula can be derived from a quantum mechanics point of view on inelastic scattering, while formulae (6.1) represent the law of conservation of energy and the law of conservation of momentum of a quasiparticle. The $+$ sign corresponds to decay of a quasiparticle in the scattering process (so-called anti-Stokes' frequency shift) and the $-$ sign corresponds to the creation of a quasiparticle (so-called Stokes' frequency shift). From the formulae (6.1) it follows that in the case of thermal excitations (i.e. excitations excited only by thermal motion of a lattice), which can move in an arbitrary direction with a wide spectrum of possible frequencies and wave vectors, by a suitable geometrical arrangement of an experiment we can select one special wave vector $\vec{k}_m$, for which the corresponding frequency, fulfilling the formula (6.1), has to appear in the observed spectrum.

It may seem that the quantitative description of the interaction of light with magnetic excitations should be based on the interaction between a magnetic dipole moment of spin fluctuation and the vector of magnetic field intensity of incident light. However, Elliot and London [49] showed that this interaction is too weak and that a much more efficient mechanism is an electric interaction. Magnetic excitations in a magnetic material cause a spatial modulation of medium permittivity and the light scattering is a consequence of the fluctuation of permittivity. For fluctuations of the dipole moment $\vec{P}(r,t)$ of a medium, induced by incident light with vector of electric field intensity $\vec{E}(\vec{r},t)$ then [50]

$$\Delta P_l \cong \sum_n \Delta\varepsilon_{ln}(\vec{r},t) E_{0n}(\vec{r},t) \tag{6.2}$$

These fluctuations are thus caused by fluctuations of components of the permittivity tensor ε_{ln}. In the case of magnetic excitations, $\Delta\varepsilon_{ln}$ is to be considered in relationship to the excited states in a magnetically ordered system. On the basis of experience we can suppose that the influence of the magnetic state on the tensor of permittivity will be small and therefore this dependence can be written in the form of the series [47]

$$\Delta\varepsilon_{ln} = \sum_i f_{lni} M_i + \sum_i \sum_j g_{lnij} M_i M_j \tag{6.3}$$

The coefficients of the series f_{lni} are components of the so-called linear magneto-optical tensor and g_{lnij} of the so-called quadratic magneto-optical tensor. The solution of Maxwell's equations considering (6.2) and (6.3) enables us to find formulae for vectors of electric and magnetic field intensities of scattered light associated with the fluctuations of polarization and in this way acquire all the necessary information about the magnetic system. Published papers, based on these solutions, dealing with the aspects of scattering on magnetic excitations have concentrated mainly on the following issues:

— the analysis of the mechanisms of light scattering
— the relationship of scattered light to magneto-optical phenomena
— the relationship of intensities of the Stokes' and anti-Stokes' lines in a spectrum of scattered light for given properties and shape of the investigated material
— polarization phenomena

One important item of knowledge for the preparation and evaluation of an experiment of light scattering on magnetic excitations is the consequence of the study of selection rules for the polarization of scattered light, according to which the rotation of the polarization vector of scattered light towards the incident light, usually about 90°, occurs at the scattering of light on magnetic excitations. The theoretical analysis of the light scattering is extensive and because of the limited extent of this monograph we will not go into details. The interested reader is referred to the corresponding literature (see e.g. [54 – 56, 210]).

As has already been mentioned, the geometrical arrangement of the experiment is of extraordinary importance because according to (6.1) it determines the selection of the wave vector and in this way also the frequency of the excitation investigated. The geometrical arrangement is of course conditioned on the transparency of the material and also on the selected magnitude of the wave vector, or the other way round: the magnitude of the wave vector is determined by the geometrical arrangement. In the experiment light is incident on the sample and scattered light is registered at a given angle to the incident light. In principle there exist three basic geometrical arrangements of the experiment for the investigation of light scattering, namely the configuration of forward scattering, 90° scattering, and backscattering, from which various constraints follow for the magnitude of the wave vector.

The configuration of forward scattering is schematically depicted in Fig. 6.1a. Light from a laser is incident on the sample and scattered light is intercepted by a collecting lens located behind the sample; the axis of the lens is parallel to the direction of the incident light. As follows from the vector diagram, for wave vectors of incident and scattered light and the wave vector of excitation, the size of the collecting lens, or as the case may be the diaphragm behind it, determines the magnitude of the wave vector of scattered light and in this way also the magnitude and orientation of the wave vector of the excitation k_m investigated. The use of the diaphragm behind the lens enables the selection of the direction and amplitude of the wave vector of the investigated excitation in the range $(0 - k_{max})$, in which the practical value of $k_{\max}$ given by the distance of the lens from the sample is usually $k_{\max} \cong 0.1\, k_0$. However, it is obvious that this geometrical arrangement can be used only for transparent materials.

The arrangement for 90° scattering is schematically depicted in Fig. 6.1b. One part of the incident light is scattered and the other part either passes through the sample or is reflected by its surface. Direct reflected light is separated by a diaphragm and scattered light is again intercepted by a collecting lens. As follows from the vector diagram the amplitude of the wave vector of scattered light and also of the investigated excitation is again determined by the geometrical arrangement. However, since $k_s \cong k_0$, the change of k_m as a consequence of the finite size of the collecting lens and its location with regard to the sample is small, the value of the wave vector of the investigated excitation for this geometrical arrangement can be only $k_m \cong k_0\sqrt{2}$.

The situation for backscattering is schematically depicted in Fig. 6.1c. In this case the collecting lens for scattered light is usually also a focusing lens for incident light. In this case it is necessary to consider carefully also the properties of the investigated material. Let us consider first a transparent material. Incident light with wave vector k_0 and with angle ψ to an interface, refracts into the other

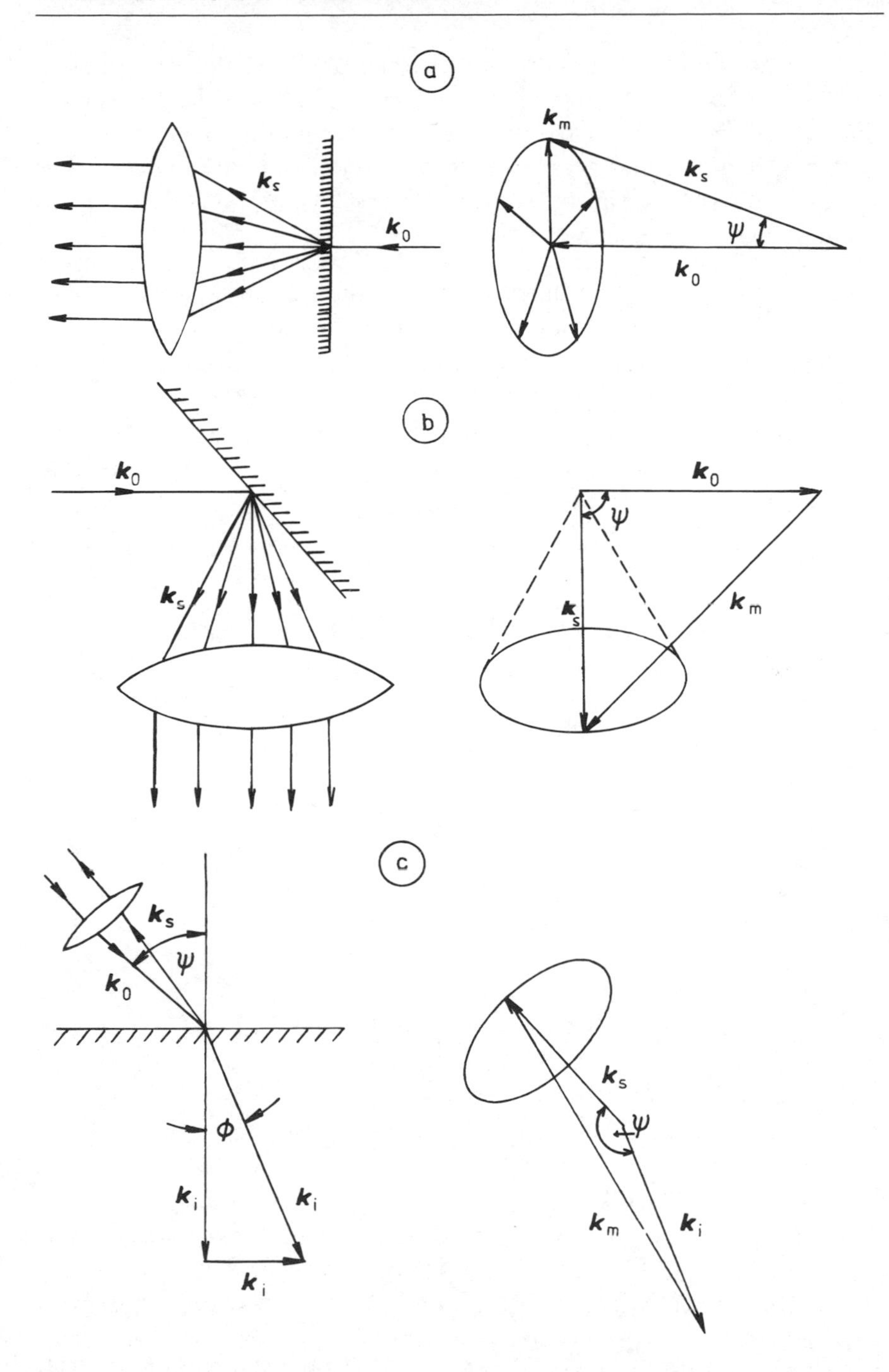

Fig. 6.1. Schematic representations of scattering geometry: a) forward, b) 90°, c) back.

medium at the angle φ. By means of the Snellius–Descartes law of refraction and the geometry it can be shown in a simple way that in the case of backward scattering, when the angle between the wave vector of incident and scattered light $\simeq 180°$, the value of the wave vector k_m of the excitation is given by the formula [53]

$$k_m = 2nk_0 \tag{6.4}$$

where n is the refractive index of the medium being investigated.

The situation is considerably different in the case of a partially transparent and partially opaque material. As a consequence of absorption the light only partially penetrates into the material, which results in the fact that the perpendicular component of the wave vector of light in the material $k_i^\perp$ does not have to be conserved and its value can be in the range of $\Delta k_i^\perp$ given by the formula [53]

$$\frac{\Delta k_i^\perp}{k_i^\perp} \simeq \frac{2n_2}{n_1} \tag{6.5}$$

where n_1 and n_2 are the real and imaginary components of the refractive index of the material. The influence of uncertainty in $k_i^\perp$ on the observed wave number of magnetic excitation is indicated in Fig. 6.2. The uncertainty of $k_i^\perp$ also causes uncertainty in the perpendicular

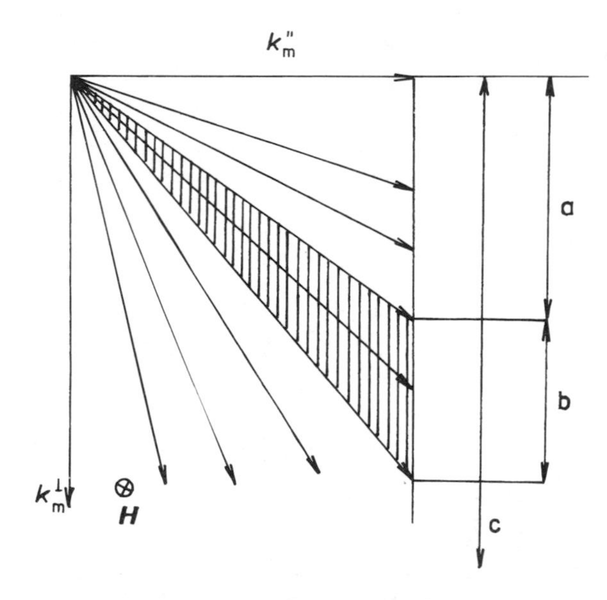

Fig. 6.2. The wave vector in the case of backscattering for a) transparent, b) semi-transparent, c) opaque material. The magnetic field lies in the surface plane.

component of $\Delta k_m^\perp$ of the wave number of the investigated excitation which will depend on the coefficient of absorption of the investigated material. Thus in the case of a transparent material, to the angle of incidence these corresponds a single value of the wave vector k_m; however, in the case of a partially transparent material the wave vector can lie in the hatched area in Fig. 6.2, and in the case of opaque material within the interval c (where δ is the penetration depth). Fig. 6.2 corresponds to a magnetic material magnetized in the surface plane. Since according to (6.1) to each k_m these corresponds ω_m, to the uncertainty of $\Delta k_m^\perp$ these will correspond the frequency band $\Delta\omega_m$, which will influence the shape of the observed spectrum. In the case of metals the optical absorption of which is high, the uncertainty of $k_m^\perp$ can be very large (e.g. for Fe it is $\Delta k_m^\perp \simeq 10^6\,\mathrm{cm}^{-1}$). On the other hand, the spectrum of a surface wave, the wave vector of which is given by the component $k_i^\parallel$, remains sharp because of the parallel component of the vector $k_i^\parallel$ and it means also that $k_m^\parallel$ remains the same.

The second important variable which has to be determined from an experiment is the frequency shift given by (6.1). Characteristic values of frequencies of magnetic excitations (of the acoustic branch) are usually in the range up to 100 GHz. However, the resolution of lattice monochromators in this frequency band is not sufficient. At present there exists only one interferometer, the Fabry–Perot one (FPI), which has sufficient resolution, and for this reason it is almost exclusively used for the observation of Brillouin scattering of light on magnetic excitations. The FPI is comprised of two plane mirrors located parallel to one another. The resonant frequency of light passing through this arrangement depends on the distance between the mirrors and the following expression holds:

$$\lambda = 2L/p \tag{6.6}$$

where p is a integer number corresponding to the order of diffraction, λ is the wavelength of light and L is the distance between the mirrors. For a fixed value of L and a given order of diffraction p there is a frequency shift from one maximum of intensity to the other (Fig. 6.3), also called the spectral range of an interferometer, equal to $c/(2L)$, where c is the velocity of light in a vacuum. If we want to observe a light signal the frequency of which differs from the basic frequency of light by less than the given spectral range, the distance between

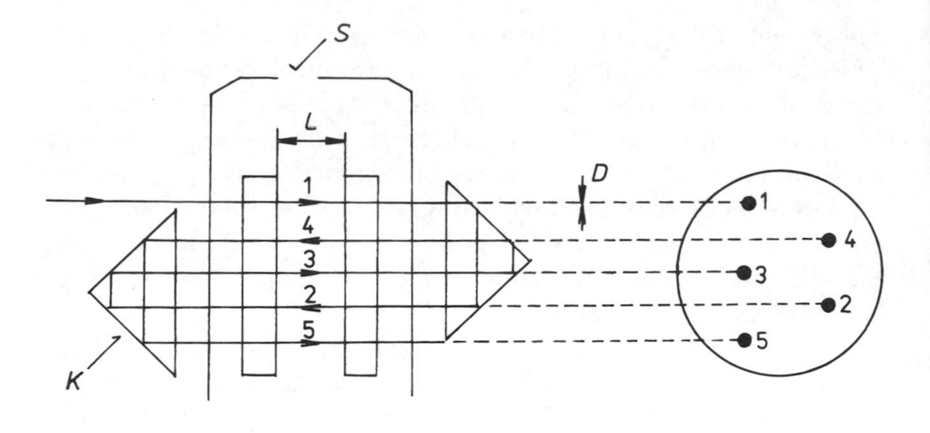

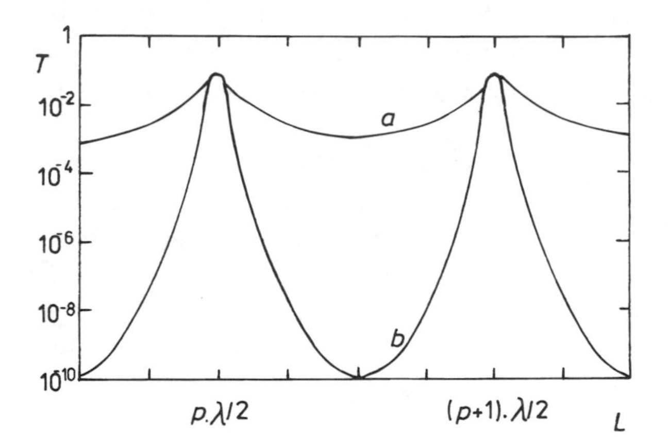

Fig. 6.3. Five-pass configuration of FPI and theoretical transition curves for a) one, b) five passes. K — corner cube, S — shades, D — beam diameter, L — the distance between mirrors.

the mirrors has to change in order to fulfil equation (6.6). Thus by continuous change of the location of mirrors we can observe a spectrum within the spectral range of the interferometer. This change of location has to be done while fulfilling the plane-parallelism of mirrors with high accuracy. The other problem is the low intensity of light scattered on magnetic excitations. Especially in the geometry of backward scattering, the intensity, as a result of elastic scattering, is higher by 5–6 orders than the intensity of light scattered on dynamic excitations. If we take into account that the contrast of a standard FPI, defined as the ratio of maximum and minimum intensity of light passing through an interferometer (Fig. 6.3) is of the order 10^3, then with regard to the ratio of useful and useless light intensity, it is obvious that a standard FPI cannot be used for the observation of Brillouin light scattering on magnetic excitations. These were limiting factors for the observation of light scattering on opaque materials for a long time. Substantial progress was made by J. R. Sandercock [54–56] who constructed a scanned electronically stabilized multi-transitional FPI in which a light beam passes through the same interferometer several times. If we introduce a contrast of single pass FPI C_1 and the coefficient of transmission of the interferometer T_1 as the ratio of the maximum intensity of light going through the interferometer to the light intensity at the input of the interferometer, then the resultant contrast C_j and the coefficient of transmission T_j for j passes of light through the interferometer is given by

$$\begin{aligned} C_j &= C_1^j \\ T_j &= T_1^j \end{aligned} \tag{6.7}$$

The contrast and the coefficient of transmission of an interferometer are functions of the parameters of mirrors, e.g. reflectivity, absorption, roughness, etc. Necessary contrast while preserving sufficient intensity can be achieved by suitable selection of the number of transitions through the interferometer and the parameters of the mirrors. A schematic depiction of the change of contrast for one and five passes through an interferometer is shown in Fig. 6.3. In practical applications the optimum number of passes is 4 to 6. The essence of Sandercock's improvement is in the use of a sophisticated system for multiple direction of a beam through an interferometer combined with electronic scanning and with long-term control of the mirror-parallelism. Electronic scanning and control of the plane-parallelism of mirrors is implemented by means of piezoelectric transducers and a feedback circuit, which reacts to the change of the intensity of light passing through the interferometer. The construction of a mechanical system is also important [56]. More details about the system can be found in references to Sandercock's works. Another improvement was achieved by the location of two multi-pass interferometers in a tandem arrangement on one common mechanical system [57], which also solved the problem of long-term synchronization of more interferometers. Nowadays, there exist interferometers of this construction which have contrast higher than 10^{12}. A block diagram of a typical arrangement for backscattering is shown in Fig. 6.4.

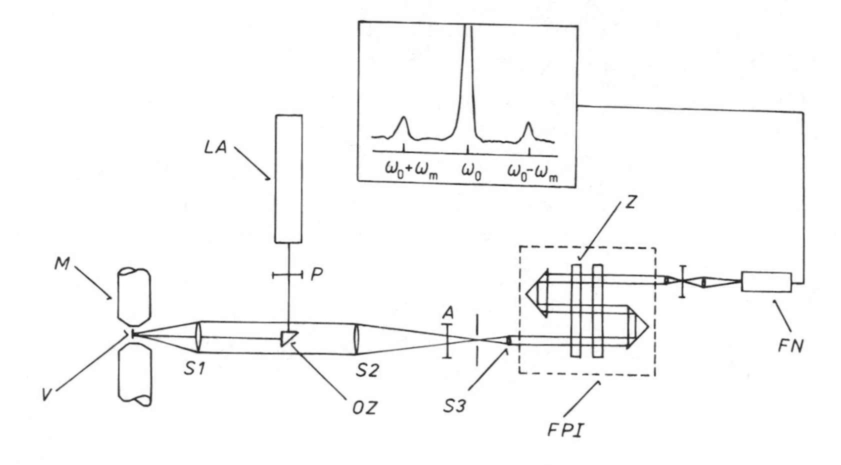

Fig. 6.4. Block diagram of experiment arrangement in the case of backscattering, LA — laser, M — magnet, V — sample, S1, S2, S3 — lenses, OZ — reflecting mirror, Z — interferometer mirrors, P — polarizer, A — analyzer, FN — photomultiplier.

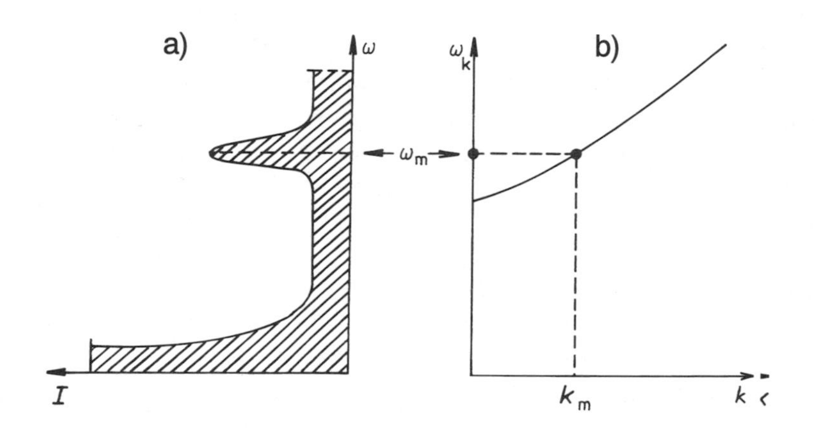

Fig. 6.5. Schematic representation of the spectrum of light scattering on spin wave: a) spectrum, b) dispersion relation.

Light from a laser is focused on a sample by the lens S1 which concurrently serves as a collecting lens for the scattered light. This

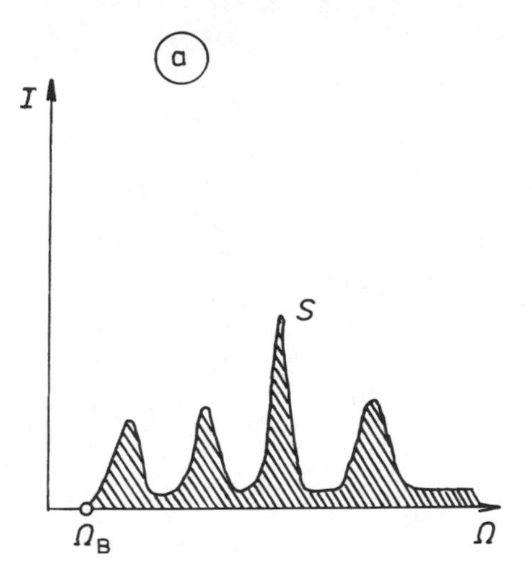

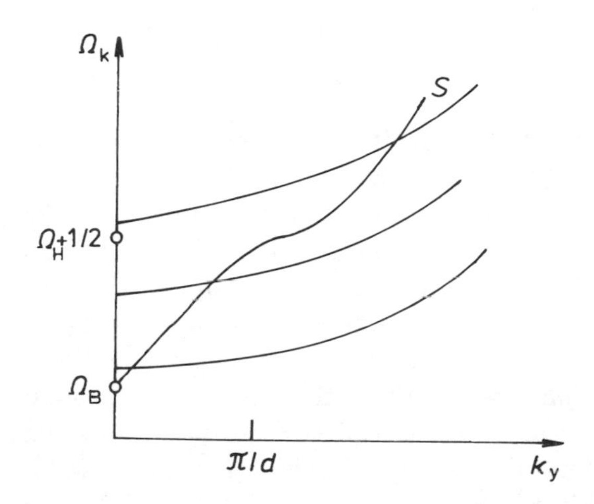

Fig. 6.6. Planar geometry. Schematic representation of the influence of the thickness on the spectrum of scattered light for spin waves with $k = k_y$. a) thin slab, b) thick slab (p. 176), c) half-space (p. 177). t — slab thickness, Ω — normalized frequency, S — mark of surface wave.

refracted light is focused into the aperture of an input collimator by the lens S2 and from the collimator light enters a multi-pass FPI. The polarizer P and the analyzer A are for the case of magnetic excitations always turned by 90°. Outgoing light from the FPI is applied to a sensitive photomultiplier with low noise level (low dark

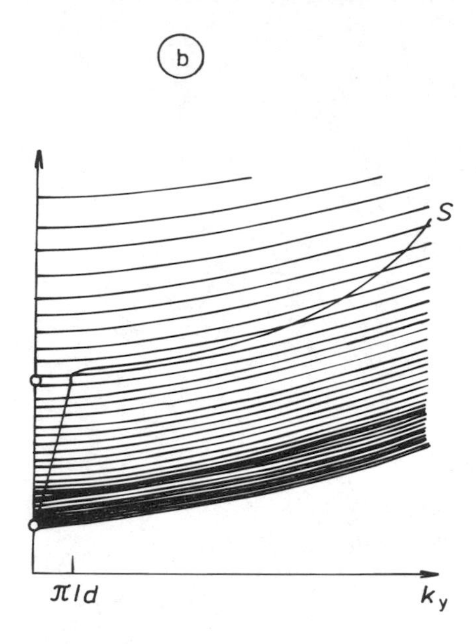

Fig. 6.6b.

counts), which is connected to the circuit for the registration of the number of photons and to the multichannel analyzer.

Up to now we have dealt with a geometrical arrangement for the experiment and with the possibilities of measuring a relatively small signal frequency shift, and furthermore, with small intensity. Let us consider now what changes in the spectrum of light can be expected in the case of light scattering on excited magnetic excitations. In the case of a transparent material, the equations (6.1) have to hold strictly. By determination of the geometry of experiment is also given the wave vector of excitation. The dispersion relation of spin waves is schematically represented in Fig. 6.5b. After the excitation of a spin wave, with the wave vector k_m and corresponding frequency ω_m we can observe the rise of intensity of scattered light (Fig. 6.5a). The advantage of this experiment is that the corresponding frequency of the spin wave can be changed in a simple way by variation of the orientation and magnitude of a static magnetic field H. This enables experimental determination of the dependence $\omega_k(H, k_m)$. Light scattering on the planar slabs magnetized in various directions enables the observation of various types of magnetostatic waves, as was described in previous sections, i.e. surface, forward and backward volume magnetostatic waves. For a planar sample also the strong in-

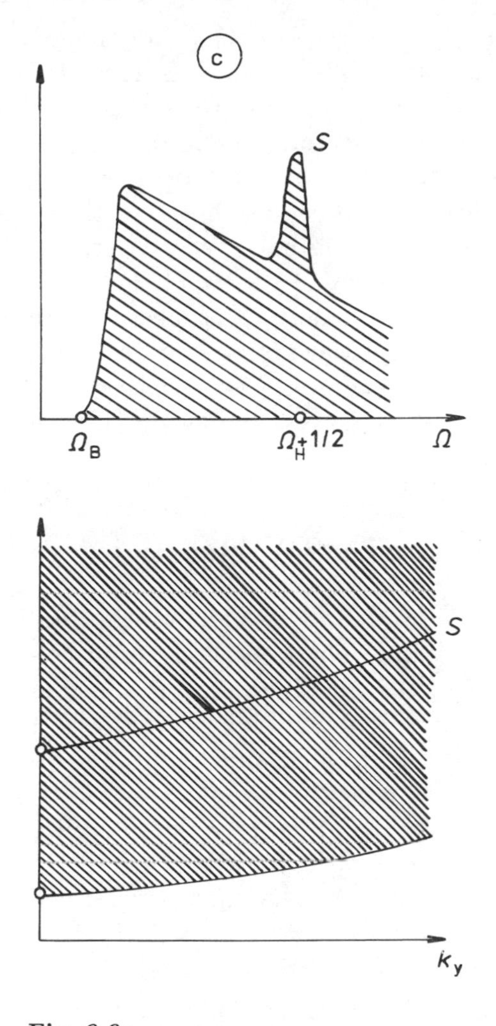

Fig. 6.6c.

fluence of the sample thickness on the character of spin waves in the case of a sample magnetized in the surface plane is expected and corresponding spectra of the Brillouin light scattering are schematically represented in Fig. 6.6a, b, c. Fig. 6.6a corresponds to a thin slab where corresponding volume branches of the spectrum are separated from each other as a result of an exchange shift and discrete values of the normal component of a wave vector. The branch of the surface wave intersects individual volume branches, where the effects of repulsion are neglected. The corresponding Brillouin spectrum will contain well separated maxima given by the conditions of the experiment. Fig. 6.6b represents the case of a slab which is about ten times thicker. The rise of frequency of a surface wave is sharpened and is shifted to the lower values of wave vector. Simultaneously the thickening of exchange branches in the direction of the low frequency of a spin band occurs. Similar thickening occurs also in the spectrum of scattered light. Fig. 6.6c corresponds to a half-space. A surface wave is present in this limit from frequency Ω_B for $k \to 0$ and the condensation of volume states results in a continuous spectrum. In the spectrum of scattered light a surface wave is present. Camley, Rahman and Mills in [58] dealt with the computation of spectra dependent on the sample thickness; other corresponding details can also be found in [58].

6.3 High-frequency methods of MSW investigation

By high-frequency methods we will understand experimental methods in a microwave frequency band. These methods, in principle, can be divided into four categories. The first category includes a modified classical resonator method, by which information about a magnetic system for one frequency is obtained practically, and simultaneously the phenomenon of ferromagnetic resonance is used. The second category includes the methods based on a direct excitation of magnetostatic waves by the use of microstrip transducers. In the third category excitation of MSW by means of microstrip transducers is also used; however, for the registration of excited MSW an inductive sensor probe is used. The fourth category includes methods using electromagnetic wave propagation in waveguide structures with a ferrite slab.

We shall now deal with the above mentioned experimental methods.

6.3.1 Ferromagnetic resonance

The first ferromagnetic experiments made by Griffiths [59] were theoretically explained by Kittel [60] as early as 1948; Kittel also predicted the spin wave resonance [61], which was subsequently experimentally verified by Seavey and Tannenwald [62] in 1958. The spin wave resonance proved itself to be a useful method for the investigation of dispersion characteristics at small values of the wave number k. An review of spin wave resonance and relevant issues can be found in Puszkarski [63].

In the beginning ferromagnetic resonance was used mainly for the investigation of losses in ferromagnetic materials by measuring the so-called resonance line width ΔH. Some information about dispersion properties at one frequency could be obtained also from the measurement of the shift of the resonance field as a function of material parameters. In principle, it was the determination of the loss parameter of these materials, measured mainly at a constant frequency (given by the resonant frequency of the used material), by measurement of the dependence of the imaginary part of the diagonal component of the permeability tensor as a function of the external magnetizing field. Rőschmann and Winkler [64] developed a method which enables the separation of the influence of porosity and anisotropy in polycrystalline materials with a narrow resonance line. This method is based on the measurement of frequency and temperature dependence of ΔH in a non-resonant (and for this reason broad-band) structure. In this particular case the fact that an anisotropic field decreases with temperature much more than magnetization was used. At present this method seems to be one of the most useful for the investigation of properties of polycrystalline materials, also for the reason that it enables full automation with on-line processing by computers.

Ferromagnetic resonance (FMR), however, has also proved itself as an efficient tool for the investigation of MSW. In the dependence on a geometrical arrangement (the orientation of the applied magnetic field, the location of a sample, etc.) a high-frequency field couples with magnetostatic waves in a ferrite sample which results in selective excitation of a series of magnetostatic modes. In the case that individual magnetostatic modes are identified and their wave vectors $\vec{k}$ are known, basic forecasts of the theory can be testified by means of measured dispersion relations.

The classical approach to FMR experiments used microwave spectrometers, in which the sample was located usually in a cavity resonator with high quality factor Q. Corresponding spectra were then obtained by the change of external magnetic field. A new simple system is presented in [65] in which the cavity resonator is replaced by a short-circuit waveguide that enables the construction of a relatively broad-band system in which the frequency is swept at constant external magnetic field. The FMR system is schematically depicted

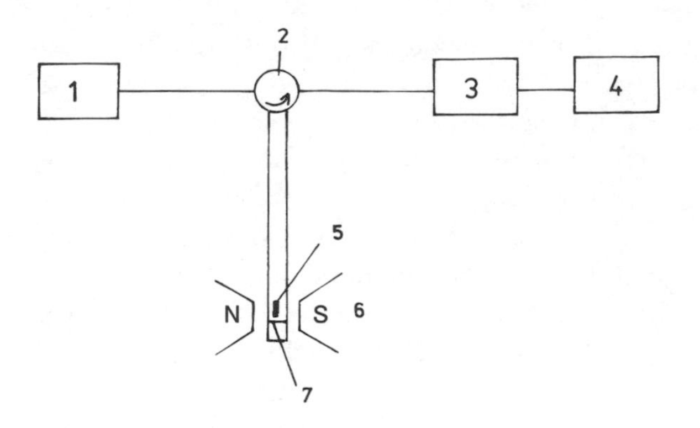

Fig. 6.7. Schematic representation of a broad-band FMR system. 1 — generator, 2 — circulator, 3 — network analyzer, 4 — data processing, 5 — sample, 6 — magnet, 7 — adjustable short-circuit.

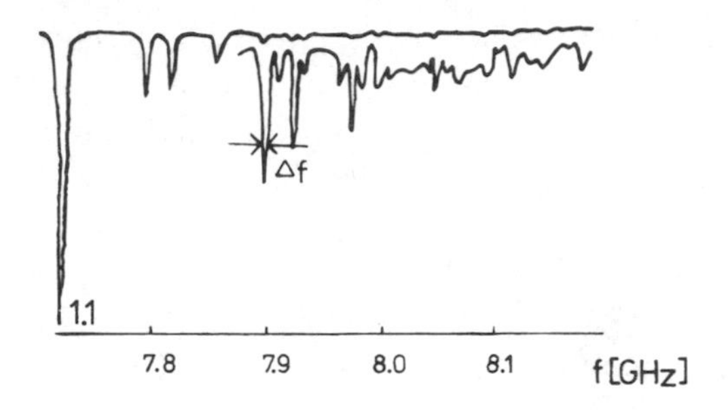

Fig. 6.8. Typical experimental spectrum for a 45 μm thick YIG layer.

in Fig. 6.7. Electromagnetic waves from a generator (sweeped) are directed through a circulator to a waveguide with an adjustable short at its end, which is located in a magnet. The reflected wave is processed. Either a diode detector or, for greater sensitivity and better signal processing, a network analyzer can be used for registration. A typical spectrum for a thin YIG layer 45 μm thick and with dimensions $3.9 \times 4.84\,\mathrm{mm}^2$ is shown in Fig. 6.8. Introducing the x and y components of the wave vector by the formulae

$$k_x = \frac{n_x \cdot \pi}{a_{\mathrm{eff}}}; \qquad k_y = \frac{n_y \cdot \pi}{b_{\mathrm{eff}}},$$

where n_x, n_y are integer numbers representing the half-period of a standing wave in the corresponding direction, it is possible, in a relatively simple way to identify individual excited magnetostatic modes and also experimentally either to confirm or reconstruct the dispersion dependences $\omega(k)$, where $k = (k_x^2 + k_y^2)^{1/2}$. The authors in [66] showed that by a suitable location of the sample at points of various symmetry of a high-frequency magnetic field standing magnetostatic waves can be selectively excited with even or odd values of n_x or n_y. Another approach is presented in [67], where the combination of a special shape of sample (circular discs) with a suitable location inside a cavity resonator was used. This combination enables investigation of simultaneous excitation of magnetostatic and exchange spin modes. FMR experiments enable the observation of standing magnetostatic waves as a consequence of refractions from the edges of the sample. The characteristic wave vector $\vec{k}$ is given by the size of the sample in this case. From the above it follows that for the excitation of waves with a high wave number k, the edges of the sample have to be given with high accuracy, which is not always a simple task, because the static magnetic field inside the sample is decisive. Furthermore, the magnetostatic waves mentioned can be observed only in materials with very low high-frequency losses. In principle therefore experiments of this type are carried out only on high quality epitaxially grown films. However, it was shown that the above mentioned approach results in good agreement between experimental results and theoretical predictions as calculated for corresponding types of excited magnetostatic waves and corresponding geometrical configurations of the sample in Chapter 3.

Up to now the methods of magnetostatic wave excitation mentioned have required the fulfilment of special conditions on the shape and accuracy of the size of the sample. An interesting modification of this approach is presented in [68], where the authors took advantage of a periodic system of induction transducers. Again, the measurement is made in a short-circuited waveguide; however, in this case the sample, in principle of arbitrary shape, is placed on a periodic lattice of passive conductors (created e.g. by evaporation on an alumina substrate) with period Δ. The sample, together with the plate, is then located in a short-circuited waveguide at the position of minimum high-frequency electric field in such a way that induced currents in the conductors create a non-uniform magnetic field in the sample. At in-phase excitation the non-uniform magnetic field will have the period Δ. Excited spin waves propagate perpendicularly to the system of conductors with the wave vector $\vec{k}$, projection of which to the plane of the film is

$$k_{\perp} = \frac{2n\pi}{\Delta}, \quad n = 1, 2, 3, \ldots \tag{6.8}$$

For the wavelength $\lambda \ll a, b$ (a, b the dimensions of the sample) non-uniform oscillations with wave number k are excited in accordance with (6.8) in the sample plane. The observed spectrum enables, by means of identification of individual excited modes, the measurement of the dispersion characteristics. Furthermore, rotation of the plate with conductors allows the study of the angular dependence of the excited magnetostatic waves with regard to the orientation of the external magnetic field.

An experimental approach developed independently by Kalinikos et al. [69] and Adam et al. [70] is based on resonant excitation of magnetostatic waves in a small area of the material investigated (usually thin YIG films). The area is limited by a non-uniform magnetic field in the form of magnetic "pit", whose boundaries fulfil the conditions of total reflection for the excited types of magnetostatic waves. Thus in this case the interface for MSW resonance is not limited by the physical edges of the sample, but by the area of a large change of demagnetizing fields which simulates real changes of the material properties at the edges of the sample. In the case described the sample is located in a hole of a given diameter (which depends on the waveguide size used) in a side wall of the waveguide for the corresponding frequency band (e.g. X-frequency band) with a short-circuit at its end. A thin sheet from a soft ferromagnetic material with a larger hole than the hole in the wall of the waveguide is located between the waveguide and the layer under investigation. The configuration is schematically depicted in Fig. 6.9a and the corresponding measured spectra without the sheet and with the sheet are shown in Figs. 6.9b, c (for a YIG layer). From the measured dependences it is obvious that the complicated spectrum of excited MSW without the presence of the sheet (excitation is in the whole sample) is transformed to the simple localized curves shown in Fig. 6.9c. The hole in the sheet of soft ferromagnetic material creates a magnetic "pit", the depth of which, for a given size and the external magnetic field used, is approximately $16\,\mathrm{kA\,m^{-1}}$, and in this way it creates the conditions for resonance of low order magnetostatic waves, whilst the area out of the pit is for given types of waves below the cut-off frequency. In [70] is presented also the condition for the capture of the magnetostatic mode in a magnetic pit with diameter D and depth δH for magnetostatic waves with wave number k near to the bottom of MSW band in the form

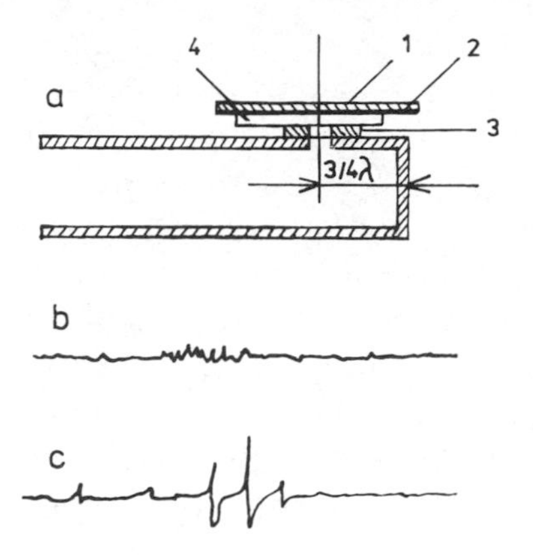

Fig. 6.9. FMR determination of thin layer properties (after [70]). a) schematic representation of configuration, b) recorded spectrum without layer 3, c) with layer 3. 1 — GGG substrate, 2 — YIG layer, 3 — soft ferromagnetic layer (125 μm), 4 — glass separation layer (160 μm).

$$D \cdot \delta H > 1.2 \cdot M_0 \cdot t \tag{6.9}$$

where M_0 is the saturation magnetization of the layer and t is its thickness. This approach is obviously suitable for nondestructive testing of local properties of YIG layers.

6.3.2 MSW excited by means of microstrip transducers

In microwave devices magnetostatic waves are usually excited by means of microstrip lines [71]. Therefore it is natural to use microstrip transducers also for the investigation of material properties by means of MSW. In principle there are two possible approaches: (a) the use of one microstrip transducer in the configuration usually used in nonlinear MSW devices (a schematic representation of such an element is shown in Fig. 5.8), and (b) the other possibility is the configuration used in delay lines. The system consists of two microstrip transducers, one of which excites magnetostatic waves which propagate in a given ferrite layer and the second serves as a receiving antenna. We will describe both possibilities of MSW generation as well as the possibilities of their use for diagnostic purposes.

(a) *Single microstrip transducer*

The structure of a microstrip line is in principle determined by two factors. To secure matching to conventional devices (to prevent reflections as a consequence of mismatch) the characteristic impedance of the line should be for example 50Ω. On the other hand, to excite MSW with sufficiently large wave numbers the microstrip should be sufficiently narrow. The sample to be investigated is located either directly on the microstrip line or above the microstrip in the distance d_h by means of a separating dielectric layer. The location of the layer or its orientation with regard to an external magnetic field determines the selection rules for the generation of individual types of magnetostatic waves. The detection of the signal passing from the input to the output of a line can be carried out either by means of a crystal detector or by means of a scalar or vector network analyzer. The excitation of individual types of waves is characterized by strong attenuation and by the change of phase of the high-frequency signal passing from the input to the output of an element when the frequency (or magnetic field) is swept through the band of MSW excitation. The dependence of the width of the MSW excitation band on the given configuration was experimentally investigated in [72] on thin YIG layers of various thickness for the case of surface magnetostatic waves. In the given case the external magnetic field was swept at a constant frequency f. The results obtained are shown in Fig. 6.10a, b, c. The results in Fig. 6.10a show the width of the MSW excitation band as a function of the layer thickness. The theoretical width of the band was computed for k in the range 0–20 mm^{-1}. The upper limit of k corresponds to the first zero value of the radiation resistance computed for the width of a transducer $a = 300\,\mu\mathrm{m}$. Fig. 6.10b illustrates the theoretical and experimental dependences of the excitation band width as a function of the transducer width a and Fig. 6.10c compares the theoretical and experimental results which could be expected for medium wave numbers k. From the comparison of results in Figs. 6.10a, b, c it follows that there is very good agreement between theory and experiment for the structure investigated. The excited high-frequency field around the microstrip in the given configuration has two components: one (h_z) is perpendicular to the dielectric substrate and the second one (h_x) is parallel to the substrate and perpendicular to the microstrip line. These components are then even and odd functions of the shift from the

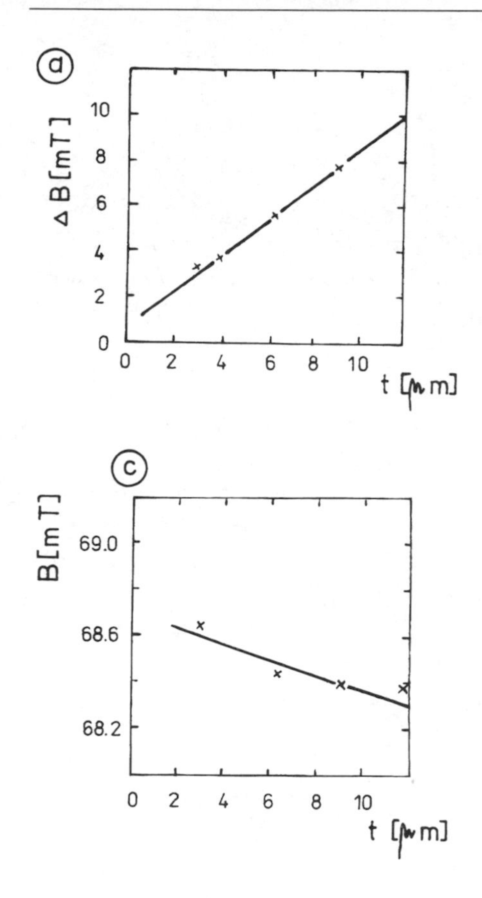

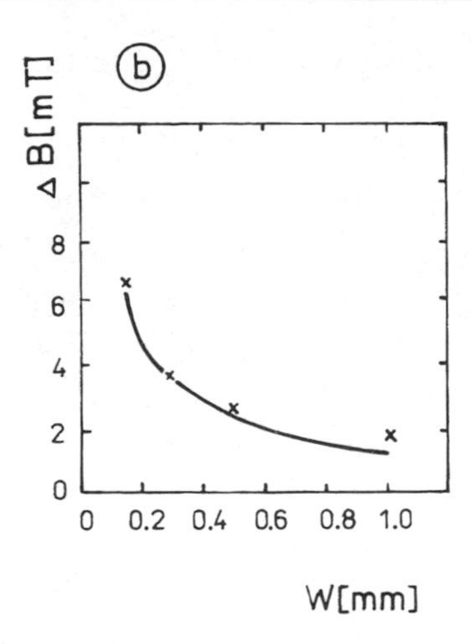

Fig. 6.10. The dependence of the excitation band width as the function of a) layer thickness, b) transducer width, c) external magnetic field as the function of layer thickness for a average wave number k. $f = 3.5\,\mathrm{GHz}$, $w = 0.3\,\mathrm{mm}$, $M_0 = 126.4\,\mathrm{kA/m}$.

microstrip in the sample plane. Along the microstrip an electromagnetic wave propagates with wavelength $\lambda_g = c/f\sqrt{(\varepsilon_{\mathrm{ef}}\mu_{\mathrm{ef}})}$, where for the microstrip line $\varepsilon_{\mathrm{ef}} \simeq 1/2$ and μ_{ef} includes the influence of susceptibility of the sample at resonance (of the real component). From analysis of the distribution of the components of the high-frequency field in the given structure as well as from the shape and symmetry of h it follows that MSW are generated.

If MSW are generated in a rectangular sample (in a thin film), the resultant field of MSW is composed of propagating MSW, depending on x, y as the function $\sin(\omega t - k_x x - k_y y)$, which remain coherent at the reflection from opposite edges of the sample. For this reason the phase change on the distance $2a$ $(2b)$ is an integer multiple of n_x (n_y) 2π. The corresponding modes therefore can be marked by a pair of integer numbers n_x, n_y with the components of the wave vector in the form

$$k_x = \frac{n_x \cdot \pi}{a}, \qquad k_y = \frac{n_y \cdot \pi}{b}.$$

It should be noted, however, that there is a basic difference between the given formulae, although they hold for both directions. The components parallel to H after the reflection from the edges form a standing wave, whilst the components perpendicular to H (representing a surface wave) propagate in one direction on a surface and they are "reflected" at the edges of the sample in such a way that they propagate in the opposite direction, but on the opposite surface of the sample. For experiments with an external magnetic field perpendicular to the sample plane, reflection in both directions creates a standing wave.

Measured spectra for the first and also for the second configuration of an external magnetic field have a similar character, as shown in Fig. 6.8 obtained from an FMR experiment. The process of their evaluation is also analogous, when it is necessary to identify results from an experiment with individual excited standing waves represented by the pair of integers n_x, n_y with theoretical dependences. Experiment enables the study of the basic properties of excited MSW in a given material, as well as the determination of material parameters and the study of the coupling between a microstrip line and excited MSW.

The second alternative of the use of a microstrip transducer for non-destructive investigation of material properties is presented in [73]. A microstrip line, with wave impedance again e.g. $50\,\Omega$, is terminated by a short- or open circuit, and represents an excitation structure as depicted in Fig. 6.11. The layer under investigation is located in this case on the surface of the transducer (marked by

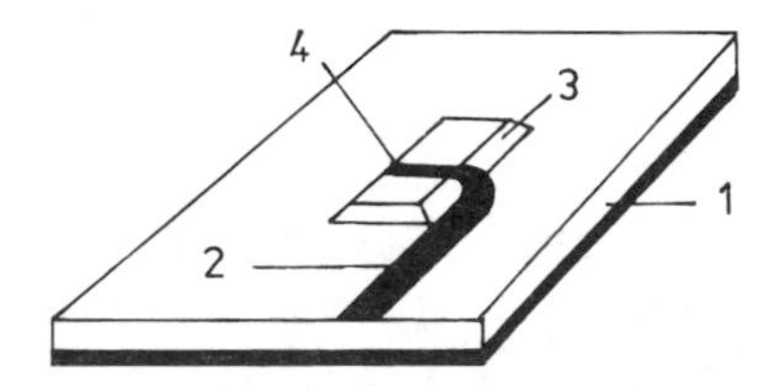

Fig. 6.11. Microstrip structure for non-destructive investigation of material properties. 1 — alumina substrate, 2 — feeding microstrip, 3 — separation layer, 4 — microstrip antenna.

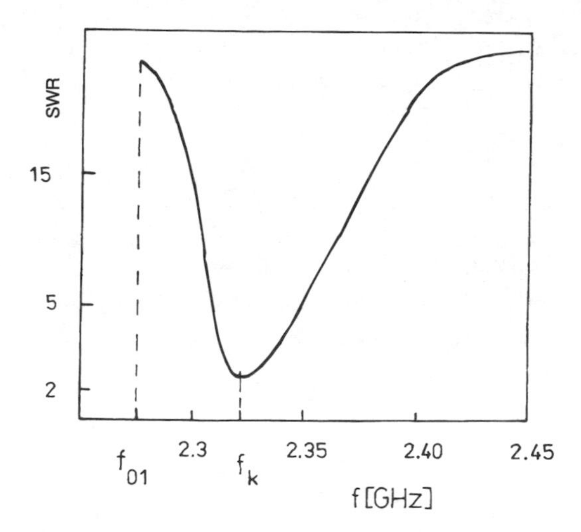

Fig. 6.12. Typical experimental dependence of SWR of structure in Fig. 6.11 on frequency.

number 4 in Fig. 6.11) which excites MSW in the layer. The mutual orthogonal arrangement of the supply line of the high-frequency signal (2 in Fig. 6.11) and the microstrip antenna eliminates the influence of line 2 on the characteristics of the transducer 4. The method of measurement of layer parameters is based on the measurement of the dependence of the standing waves ratio (SWR) at the input of line 2 as a function of frequency, e.g. for various orientations of the transducer with respect to the crystallographic axis of the measured layer and at the constant magnitude and orientation of an external magnetic field. The typical shape of such measured dependence is illustrated in Fig. 6.12. A clear minimum can be seen on it for frequency f_k, corresponding to the wave number of excited MSW k_m. From the theoretical analysis of radiation resistance and SWR for a given geometrical size of the microstrip transducer and the ferrite layer investigated, k_m can be determined. By means of the frequency f_{01} (Fig. 6.12) at which, for a given external magnetic field, MSW begin to be excited, for example, the required parameters of the magnetic layer investigated, e.g. saturation, magnetization M_s, the anisotropy constants K_1, etc., can be determined. In the case that the samples investigated are large, the parameters obtained can be considered, under some conditions, local and in this way we can test the homogeneity of the distribution of parameters of the layers investigated. This information is important especially in the case of industrial mass preparation of thin layers from the point of view of the scatter of parameters which is admissible in the construction of MSW devices.

(b) *System of delay lines*

For the measurement of losses and other properties the system of two or more microstrip transducers with an analogous arrangement to that in Fig. 6.13 representing delay lines, is usually used. The

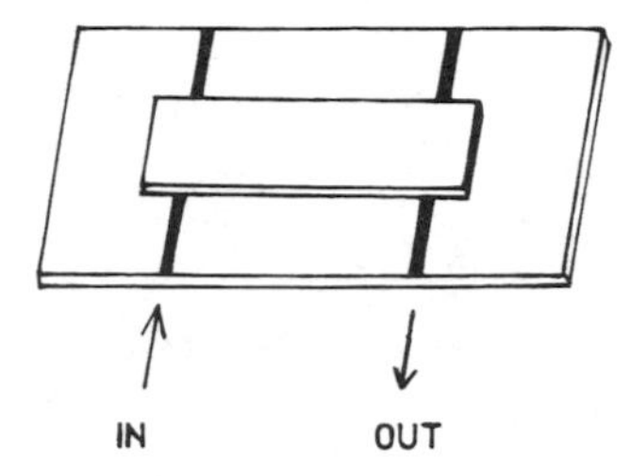

Fig. 6.13. System of microstrip lines in delay line configuration.

design of well operating MSW elements requires an understanding of the process of relaxation which is in progress with the propagation of magnetostatic waves with small (but non-zero) wave numbers k (usually in the interval $10 - 10^3\,\mathrm{cm}^{-1}$). In most papers dedicated to this topic the total losses are measured, i.e. losses composed of propagation losses and losses caused by the transformation of electromagnetic energy in a transducer. Webb et al. [74] separated the individual components from each other by the measurement of the delay time for various path lengths of MSW propagation. In [75] a surface MSW was used. For surface MSW losses per length l associated with the delay τ can be expressed by means of the constant of attenuation for which the following holds:

$$\varkappa_\tau = v_g \cdot \mathrm{Im}(k) = v_g \cdot k''$$

where

$$v_g = \partial\omega/\partial k' \qquad \text{and} \qquad k = k' - jk''$$

$\varkappa_\tau$ is measured using the interference of signals incident on two identical microstrip antennas placed at distances l_1 and l_2 from the

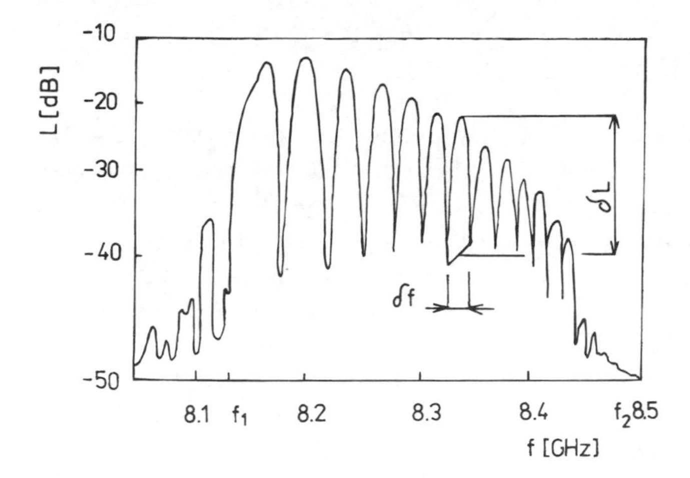

Fig. 6.14. Recorded interference signal from two microstrip antennas located at different distances from an exciting antenna.

same source antenna. The recorded interference signal is depicted in Fig. 6.14. Then for the constant of attenuation (at small losses in the sample) we have [75]

$$\varkappa_{\tau} = \delta f \cdot \ln[\tanh^{-1}(\delta L/17.3)] \tag{6.10}$$

where δf is the frequency difference of two adjacent maxima or minima and δL is the difference of losses (in dB) in adjacent maximum and minimum, as can be seen in Fig. 6.14, and 17.3 is the numerical value derived in [75]. The real component of the wave vector $k' = \mathrm{Re}(k)$ can be obtained from

$$k' = n \cdot \frac{2\pi}{\delta l} \tag{6.11}$$

where $\delta l = l_2 - l_1$ and $n = 1, 2, 3, 4, \ldots$ are the numbers of maxima, numbered from the lower limiting frequency of the MSW excitation band. It is obvious that from the measured dependence in Fig. 6.14 the dispersion of MSW investigated $f = f(k')$ can be immediately determined, comparison of which with the theoretical prediction enables the determination of other required parameters of the material, e.g. M_0, effective thickness of a layer t, etc.

6.3.3 Propagation of magnetostatic waves in a cut-off waveguide

The recently discovered transparency of a cut-off waveguide containing a ferrite slab provides another possibility for the investigation of magnetic material properties [76, 77]. To be able to use this technique it is necessary to know the basic properties of a cut-off waveguide with a ferrite slab. Theoretical analysis of such a configuration is based on work which has been known for a long time, e.g. [78, 2]. The more complicated case of propagation of electromagnetic waves in a rectangular waveguide containing an anisotropic slab was analyzed in [79]. Since a thin YIG film on GGG substrate represents in principle a multi-layer structure, we will present a modified approach for obtaining the dispersion equations for propagation of TE_{m0} modes in a rectangular waveguide "loaded" by anisotropic slabs. Restriction to TE_{m0} modes is justified because from the analysis of the experimental results obtained it follows that the investigated waveguide is cut off already for a basic mode TE_{10}. The geometry of the arrangement is illustrated in Fig. 6.15. The method used for obtaining

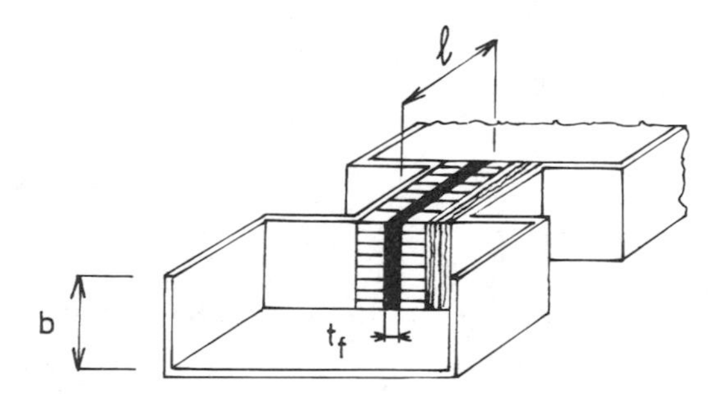

Fig. 6.15. Section of a cut-off waveguide with investigated magnetic layer.

the dispersion characteristics is analogous to the approach of transverse impedance [80] or surface permeability from Chapter 2. On the basis of a classical solution of partial waves the dispersion equation is obtained from boundary conditions, which have to be fulfilled on the interface of two different media. To obtain the characteristics of a waveguide which contains an arbitrary number of isotropic and anisotropic slabs (Fig. 6.15) it is useful for TE_{m0} modes to introduce a "longitudinal impedance" for every area by the formula

$$Z^l(x) = E_z(x)/H_y(x) \tag{6.12}$$

Using the given asumptions and Maxwell's equations for the area with a ferrite

$$Z_f^l = \frac{-j\omega\mu_p[1 + D_f.\tan(k_{pf}.x)]}{-D_f \cdot [k_{pf} - s\gamma_p(\mu_a/\mu)\tan(k_{pf}.x)] + [k_{pf}\tan(k_{pf}.x) + s\gamma_p(\mu_a/\mu)]} \tag{6.13}$$

where μ, μ_a are diagonal and non-diagonal components of the permeability tensor, $k_{pf} = (k_p^2 - \gamma_p^2)^{1/2}$, $k_p^2 = \omega^2\mu_p\varepsilon_f$, $\mu_p = (\mu^2 - \mu_a^2)/\mu$ is the so-called transverse permeability, ε_f is the permittivity of a ferrite, γ_p is the propagation constant in the y-axis direction, ω is the angular frequency, $s = \pm 1$ represents the direction of propagation (the direction of applied external field) and D_f is an integration constant, which has to be determined from the boundary conditions. An analogous equation can be obtained for the area of dielectric if we substitute in (6.13) $\mu_a \rightarrowtail 0$, $\mu = \mu_0$, $\varepsilon_f \rightarrowtail \varepsilon_d$, $k_{pf} \rightarrowtail k_d$ and $D_f \rightarrowtail D_d$. Let us suppose now that the thickness of the ith ferrite layer is t_{f_i} and let this be loaded for $x = 0$ by the impedance Z_{i-1}^l and for $x = t_{f_i}$ by the impedance Z_i^l. Boundary conditions require fulfilment of the equalities $Z_f^l(x = 0) = Z_{i-1}^l$ and $Z_f^l(x = t_{f_i}) = Z_i^l$. After substitution of x into (6.13) and elimination of the unknown constant D_f we have

$$\begin{aligned} Z_i^l = &\left\{\omega\mu_{p_i}\left[k_{pf_i}Z_{i-1}^l + \left(sk_{pf_i}\frac{\mu_{a_i}}{\mu_i}Z_{i-1}^l + j\omega\mu_{p_i}\right)\tan(k_{pf_i}t_{f_i})\right]\right\} \\ &\div\left\{\omega\mu_{p_i}k_{pf_i} + \left[jZ_{i-1}^l\left(k_{pf_i}^2 + \gamma_p\left(\frac{\mu_{a_i}}{\mu_i}\right)^2\right) - \omega\mu_{p_i}s\gamma_p\frac{\mu_{a_i}}{\mu_i}\right]\right. \\ &\left.\times\tan(k_{pf_i}t_{f_i})\right\} \end{aligned} \tag{6.14}$$

which relates impedances on both sides of a ferrite slab. By substitution of $t_{f_i} \rightarrowtail t_{d_i}$, $k_{pf} \rightarrowtail k_d$ and $\mu_a \rightarrowtail 0$, $\mu = \mu_0$ in (6.14) we obtain an equation which relates the impedances on both sides of a dielectric slab. The formula (6.14) for a ferrite and dielectric slab with the condition for Z^l on the waveguide walls enables us, in principle, to solve any dispersion problem (under given approximation) by means of computers.

For experiment a waveguide section of the cut-off waveguide with a ferrite slab is located between the poles of an electromagnet and

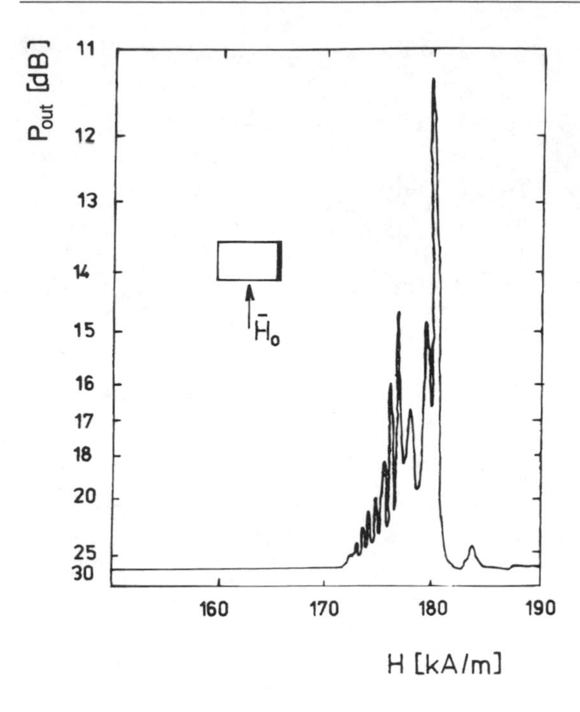

Fig. 6.16. Experimental dependences of a power transmitted through a cut-off waveguide. YIG plate very close to waveguide sidewall.

the dependence of the signal transmitted as a function of external magnetic field (at constant frequency) or as a function of frequency (at constant external magnetizing field) is recorded. The recorded dependence of the power transmitted when the plate is located close to the waveguide side wall is illustrated in Fig. 6.16. Comparison of the measured spectrum with theoretical computations enables determination of the parameters of the layer under investigation.

6.3.4 Investigation of the propagation of magnetostatic waves by means of an induction probe

Recently the most interesting experimental and theoretical results in the investigation of the physics and applications of MSW have been achieved by the study of the propagation of MSW beams. We dealt with MSW beams from a theoretical point of view in Chapter 4. Almost all experimental results in this connection have been obtained by the technique of recording the characteristics of propagating MSW by means of an induction probe. This experimental technique was introduced by N.P. Vlannes [81] and improved by A. Vashkovsky and his coworkers [33]. In approaches presented up to now the emphasis was put on the investigation of the whole system composed of the cavity resonator system, waveguide, microstrip lines, and magnetic material under test (usually thin YIG layer) and not on the investigation of the character of MSW itself. However, to better understand the physics and possible applications of MSW it is necessary to study MSW. The principle of the measurement method, which enables a high spatial resolution, is based on scanning of electromagnetic fields excited by MSW. Electromagnetic waves have an electric as well as magnetic component. Since the magnetic component of magnetostatic waves is emphasized, it is appropriate to investigate these waves by scanning of the magnetic component of the electromagnetic field. If we express the space–time dependence of high-frequency magnetic field in the form

$$\vec{h}(t) = \vec{h} \cdot \mathrm{e}^{j(\omega t - \vec{k}\cdot\vec{r})} \tag{6.15}$$

where $\vec{h}$ is the complex vector amplitude of this field, ω its angular frequency and $\vec{k}$ the wave vector of magnetostatic wave propagation (where $|k| = 2\pi/\lambda_k$ and λ_k is a wavelength of this wave), $\vec{r}$ is a position vector, and t is time. From the equality of tangential components on the interface for the components of a high-frequency magnetic field on the surface of a sample it follows that the corresponding component of the field inside the sample is scanned. A schematic representation of the induction probe used is shown in Fig. 6.17. For the induced voltage on the ends of a loop delete

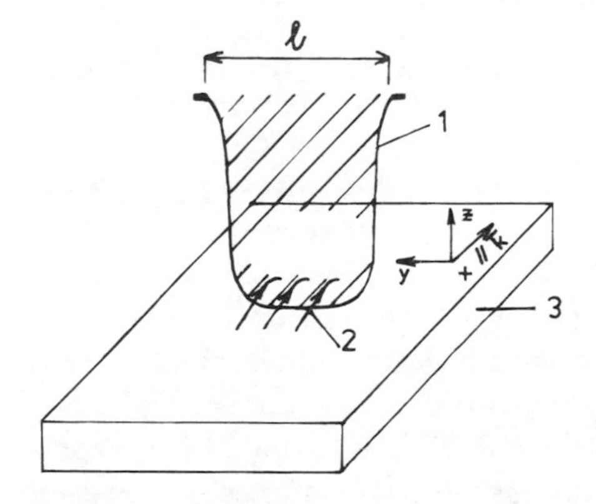

Fig. 6.17. Schematic representation of a induction probe. 1 — loop, 2 — lines of magnetic flux, 3 — ferrite layer.

$$U_i = \int_S j\omega\mu_0 \vec{h}_e \cdot \mathrm{d}\vec{S} \tag{6.16}$$

where $\vec{b}_e = \mu_0 \vec{h}_e$ and $\vec{h}_e$ represents high-frequency magnetic field on an outer surface of the sample (Fig. 6.17). This high-frequency field can be theoretically determined from the solution of MSW propagation for a given geometrical arrangement. The depicted induction probe is sensitive to the x and y components of the high-frequency magnetic field. A maximum induced voltage U_i can be found by its rotation. The normal to the plane of the loop is in this case parallel to the direction of the vector $\vec{k}$. From the measured induced voltage it is thus possible to draw conclusions about the profile and propagation direction of MSW. A schematic representation of such an experimental arrangement is in Fig. 6.18. MSW is in this case

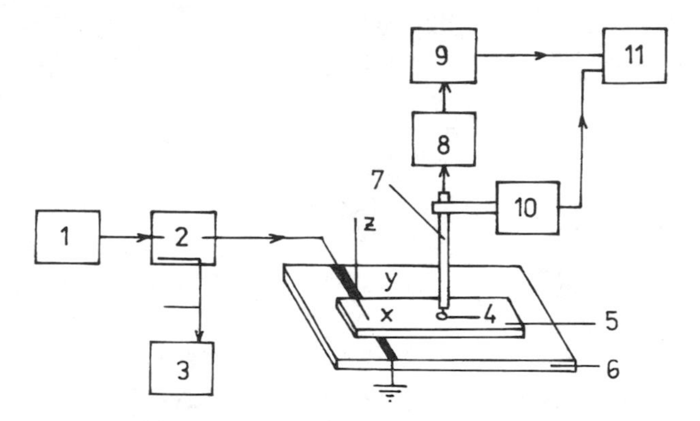

Fig. 6.18. Schematic representation of experimental arrangement with a induction probe. 1 — generator, 2 — directional coupler, 3 — power meter, 4 — induction probe, 5 — YIG sample, 6 — dielectric substrate, 7 — high-frequency cable, 8 — detector, 9 — amplifier, 10 — positioner, 11 — x, y recorder.

excited by a microstrip antenna and the excited wave propagates in the x-axis direction. To separate fast electromagnetic waves from MSW (velocities differ by 10^3) impulse measurements, which enable differentiation of the two signals, can be carried out or it is necessary to take into consideration the presence of both signals. The ratio of the intensity of fast electromagnetic waves to the intensity of MSW decreases with the distance to the excitation antenna. Typical profiles obtained by moving the probe across the sample for constant distance from the excitation antenna and for constant frequency and constant external applied magnetic field are illustrated in Figs 4.10

and 4.14. The dispersive character of MSW is obvious from these figures. Since the measured induced voltage also includes the information about the wave number of the MSW being investigated (see 6.15), the measurement enables one to experimentally determine the dispersion relation. Furthermore, the experimental arrangement also enables the direct observation of a profile of MSW in samples of larger size, which resulted in the experimental discovery of beams of magnetostatic waves as well the direct experimental investigation of the incidence, reflection and refraction of magnetostatic beams, i.e. quasioptics of magnetostatic beams, discused in Chapter 4. The details of the use of this technique for the determination of some parameters of magnetic materials will be presented in the next section.

6.4 Diagnostics of magnetic materials by means of MSW

In previous sections we dealt with selected experimental methods which enabled a direct or indirect investigation of a selected group of magnetostatic waves. Excitation and propagation of magnetostatic waves are, however, bound not only to the given geometry but also to the material properties of the medium in which the wave propagates. In other words, the investigation of propagation of a selected group of magnetostatic waves provides the possibility to study some basic properties of magnetic materials. For this reason we will call this experimental approach MSW diagnostics of magnetic materials. Thus under MSW diagnostics we will understand, in principle, the determination of material parameters through the adjustment of verified theoretical results to the obtained experimental dependences, with adjustable variables representing the parameters we are looking for. As follows from the presented description of individual experimental methods, a direct measurement of losses and sections of dispersion relations of MSW in given materials is possible. From this follows the possibility of indirect determination of all parameters which are connected with the mentioned methods. This includes the resonance line width ΔH, its dependence on the magnetic field or wave vector, saturation magnetization M_0, exchange parameter D, the constants of cubic or uniaxial anisotropy K_1, K_2, K_u, thickness or effective thickness of the layer t, difference between surface and volume parameters, etc. It should be noted, however, that with regard to indirect determination of these parameters the accuracy of measurement can be influenced by many circumstances given by the precision of measurement of other known parameters, the properties of the approximation procedure used, the adequacy of the model describing given physical phenomena, etc. An exact determination of magnetic field, which occurs in all equations, either for dispersion or for investigated losses, is also a problem: only the external magnetic field can be measured, but in equations the internal field occurs, which is computed either by introducing corresponding demagnetization factors or it can be determined from an experiment if it is viewed as one of the possible parameters. In both cases, however, the accuracy of its determination can be much smaller than the theoretical estimation of the accuracy of the parameter investigated. For these reasons the determination of magnetic parameters from indirect measurements with an accuracy of (10–15) % is considered sufficient. In spite of these complications these experimental results give important information about the properties of the magnetic materials investigated and especially in cases when the information cannot be obtained by other means, or if it can be obtained in another but more complicated way and with the same accuracy. Furthermore, such measurements usually give sufficient and fast information about the technology of their preparation, which enables corrections to be made to achieve the required properties. Hereinafter we will show with concrete examples the possibilities of obtaining necessary material information by means of, MSW diagnostics and experimental methods described in previous sections. Examples will be divided into two groups, those determined from the measurement of dispersion relations and those obtained from measured losses. The necessary theoretical foundations were either described in previous chapters or will be briefly presented if needed.

6.4.1 Determination of parameters from measured dispersion relations

All experimental approaches presented in this chapter are appropriate for the measurement of dispersion relations. It is noteworthy that sufficient information, i.e. measurement of sections of dispersion relations in a sufficiently wide frequency band and range of wave numbers by means of FMR or by excitation of standing MSW by means of microstrip transducers is effective only for materials with very low magnetic losses in the given frequency band. This approach was used e.g. in [82] and in some other references. For the high-frequency range and for wave numbers close to the centre of the Brillouin zone the most effective method for the measurement of scattering relations seems to be (at least up to now) the technique of Brillouin light scattering. This technique can be used for a wide range of materials of interest from an application point of view. Scattering can be either on thermal excitations or on pumped ones. The light scattering on thermal magnetic excitations in a thin Fe layer is investigated in [83]. Thermally excited magnetostatic waves as a function of the magnitude of the component of the wave vector $\vec{k}$ in the plane of the layer at constant external magnetic field were investigated. This enables the investigation of dispersion relations in a relatively wide range of possible wave numbers. The geometry of the experimental arrangement used is illustrated in Fig. 6.19a. The magnetic field was in the plane of a layer. The magnitude of the wave vector of the surface excitation can be simply determined from the experimental geometry as

$$k_{ms} = 2k_0 \sin \Theta_0 \tag{6.17}$$

($k_0 = \omega/c$ is a wave number of incident light) because the component of the wave vector parallel to the surface is conserved also for materials with high optical absorption, which without any doubt the Fe layer is. As a consequence of large absorption a rich spectrum arises, similar to the spectrum which is presented in Fig. 6.6a. The change of the wave vector of a surface wave due to (6.17) can be achieved by changing the angle of incidence Θ_0 at which the light is incident on the sample. To keep interpretation simple the thickness of the layer 34.8 nm was chosen in order to shift the nearest exchange mode (the frequency of which is $\simeq 1/t$ of the layer) to higher frequencies. The results obtained are presented in Fig. 6.19b. With regard to the fact that the frequency of the exchange mode is given by the component of the wave vector perpendicular to the plane of the layer, the change of the magnitude of the wave vector component parallel to the layer plane does not have any influence on the frequency, which can be clearly seen also in the experimental results. The component of the wave vector parallel to the layer plane does have, however, a strong influence on the frequency of a surface magnetostatic wave. Furthermore, the conditions of the experiment

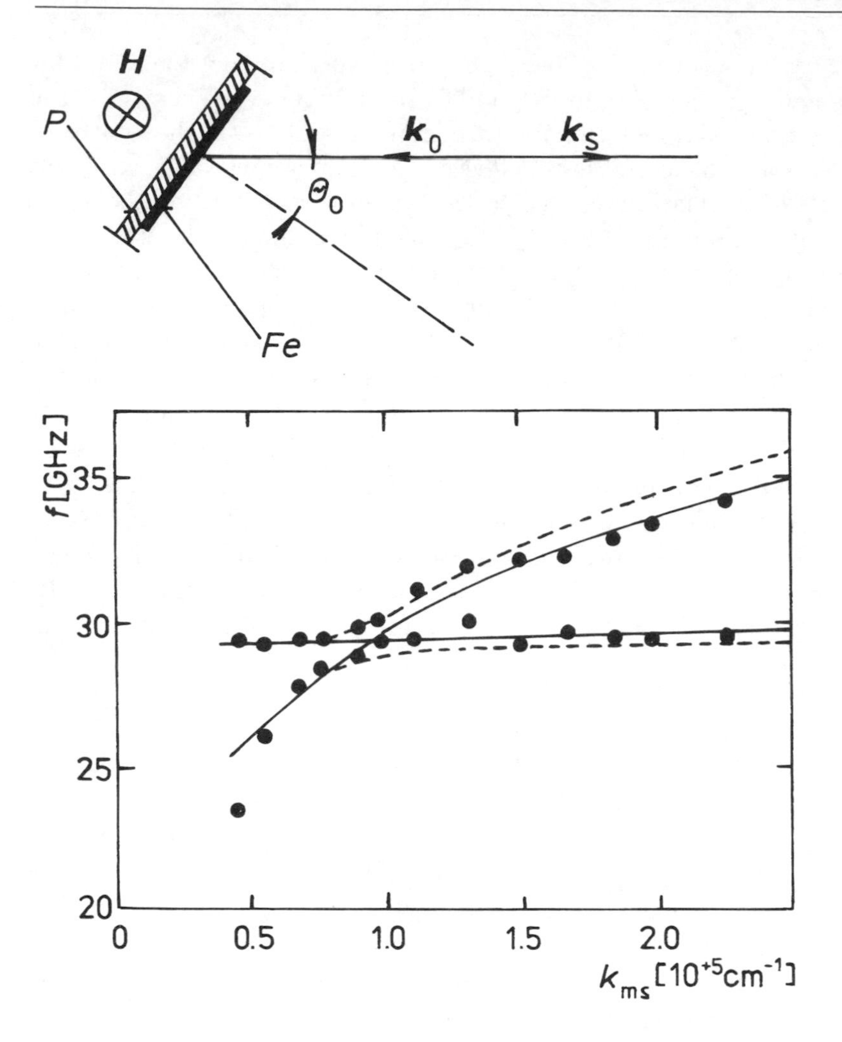

Fig. 6.19. Results of the light scattering on a thin Fe layer. a) geometry of the experiment, b) measured (points) and theoretical (lines) dependences of spin wave frequency on the amplitude of the wave vector of a surface wave.

were chosen with the intention to investigate an intersection of two different modes in the chosen range of wave numbers, which can also be clearly seen in the data obtained. The full curve in the figure represents a simple theoretical solution according to the theory of Damon–Eshbach [11] and the dashed curves represent results including exchange terms by the theory of Wolfram and DeWames [84]. A surprisingly good agreement with experiment (points in Fig. 6.19b) for displayed parameters is obtained. By the same method a sub-

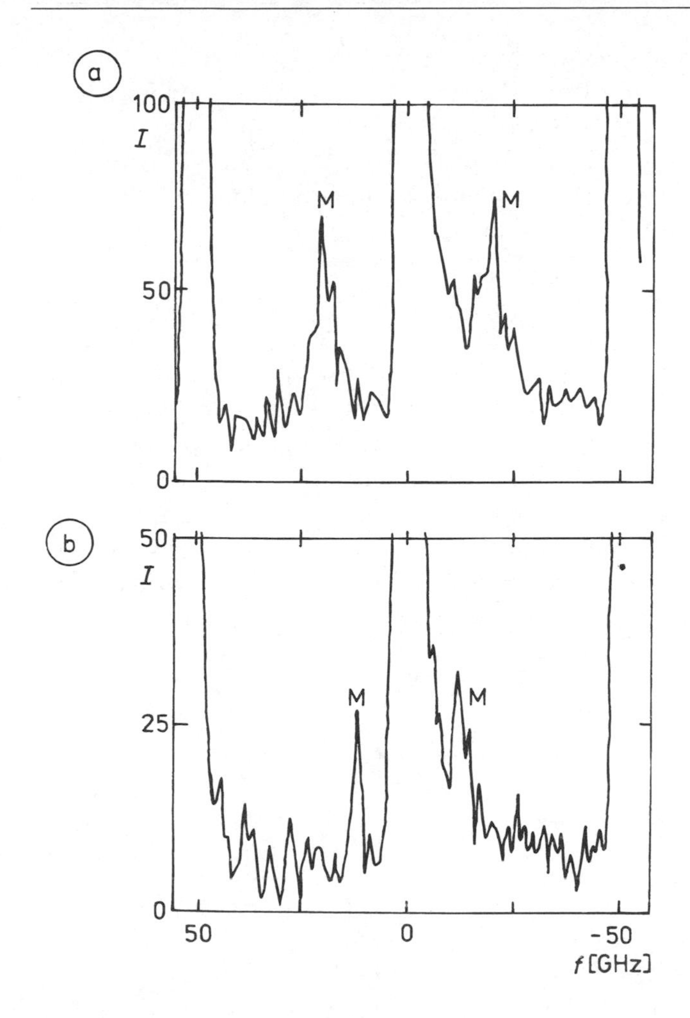

Fig. 6.20. BLS spectra for two samples of Li ferrite a) polycrystalline $x = 0.1$, b) monocrystalline $x = 0.55$. Magnetizing field $H = 320\,\mathrm{kA\,m^{-1}}$. M denotes the intensity of light scattered on spin waves. I — intensity of scattered light, f — frequency.

stituted Li ferrite was investigated in [89]. Substituted Zn passes into a sublattice A and replaces ion Fe^{3+} in a spinel structure [85], where the distribution into sublattices is given by the formula [86] $(Zn_{2x}Fe_{2-2x})[Li_{1-x}Fe_{3+x}]O_8$, where x denotes the concentration of Zn ions. The influence of Zn on static parameters of Li ferrite is well known, but up to now the measurement of the parameters of spin excitations has been indirect. Two spectra for different concentrations of Zn measured at constant magnetizing field $320\,\mathrm{kA\,m^{-1}}$ in the configuration of backward scattering are presented in Fig. 6.20a, b. The shift of the marked maxima in the spectrum of corresponding magnetic excitations obviously indicates the influence of Zn on their frequency. Spectra similar to those in Fig. 6.20a, b for various values of magnetizing field provide experimental data, which can be analyzed by the approximate formula [86]

$$\omega_k \simeq H + Dk_m^2 + M_0/6 \tag{6.18}$$

The amplitude of the wave vector k_m is given by (6.4) and the index of refraction n was chosen to be equal to the index of refraction of YIG, i.e. 2.46. The values of M_0 were obtained independently by means of a vibration sample magnetometer. From (6.18) the value of the exchange parameter D can be determined. The dependence of values of D obtained in a such a way on the substitution of Zn is shown in Fig. 6.21, in which the dashed line represents the theoretical dependence on the concentration of Zn obtained from the molecular field theory in the range when linear dependence of input variables on concentration can be expected [86].

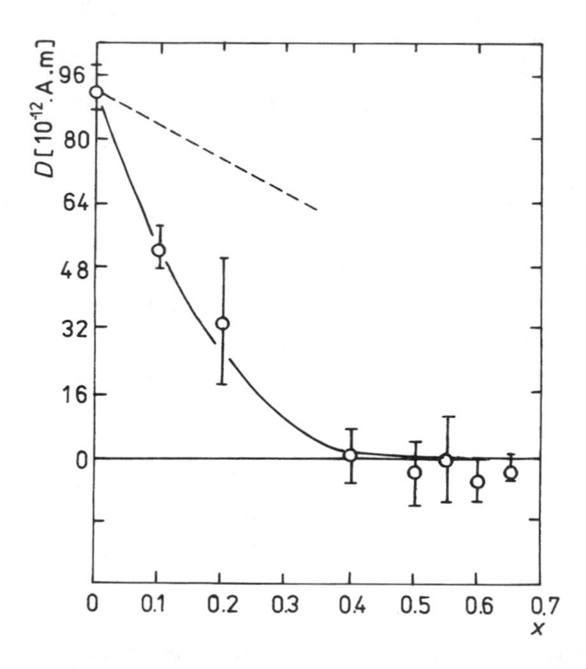

Fig. 6.21. Measured dependence of the exchange parameter D for Li ferrite on concentration. The dashed line corresponds to theoretical analysis.

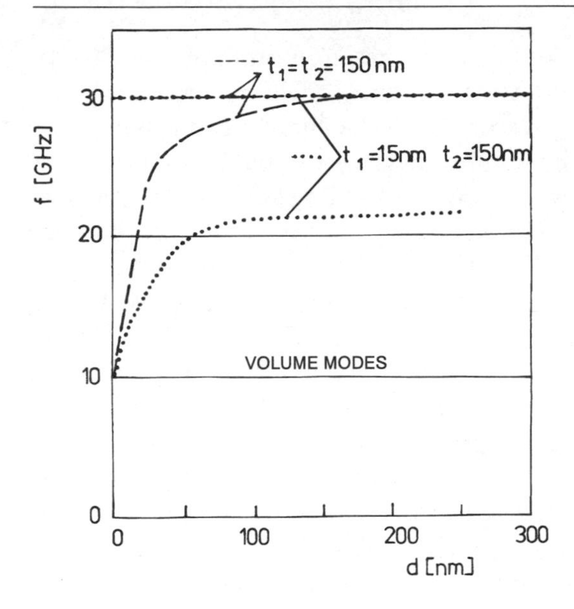

Fig. 6.22. Frequencies of modes in a two-layer structure as a function of separation layer thickness d_0 for two combinations t_1 and t_2.

Recently interest in multi-layer systems has risen considerably, is caused mainly by improvement of the technology of preparation of thin films. Typical thicknesses of individual layers range from nm to hundreds of nm, while ferromagnetic, antiferromagnetic or nonferromagnetic layers can change. Up to now almost all these artificial magnetic systems have been investigated by the method of light scattering. As a consequence of separation of individual magnetic layers the coupling between them can have different origins, one of which is the dipole interactions, effective over a large distance, which results in excitation of several types of magnetostatic waves. A theoretical approach to the solution of such a task was presented in Chapter 2. The only one collective effect which could not be explained by the theory of modified single-layer properties is the already mentioned dipole coupling between individual magnetic layers. Behaviour of a system composed of two magnetic layers separated by a nonmagnetic one was investigated by Grunberg et al. [87]. Theoretical results together with experimental ones are in Fig. 6.22 for two examples, one for $t_1 = t_2 = 15\,\text{nm}$ (full lines), and the other one for $t_1 = 15\,\text{nm}$ and $t_2 = 150\,\text{nm}$ (dashed lines). The parameters used in the calculation are listed in the figure. Experimental results in this figure are represented by points, and as can be seen in principle they prove the conclusions of the magnetostatic theory. Differences, especially for small values of the separation layer thickness, are probably the result of the fact that exchange interactions were neglected.

All measurements presented can be used also for the determination of other selected parameters occurring in the dispersion relation. A somewhat different approach should be chosen for the determination of the anisotropy constant. In this case, without regard to the experimental method used, the dependence of the dispersion relations for chosen values of wave number and of external magnetic field on the angle of rotation of the external magnetizing field in selected directions is investigated. The influence of an anisotropy on magnetostatic types of excitations was investigated in detail by Schneider [88]. In principle, this influence will manifest itself in the dependence of the components of the permeability tensor (1.42) on input parameters (in our case phenomenological) characterizing anisotropy as well as on corresponding angles between individual vectors, crystallographic axes and the chosen basic direction in an experiment. The components of the permeability tensor can in general be expressed in the form

$$\begin{aligned} \mu &= \mu(H_0, M_0, \omega, k, K_1, K_2, K_u, \dots) \\ \mu_a &= \mu_a(H_0, M_0, \omega, k, K_1, K_2, K_u, \dots) \end{aligned} \tag{6.19}$$

where K_1, K_2, K_u, ... are the constants of cubic, uniaxial or other anisotropy. The components of (6.19) occur in the dispersion relations for individual types of magnetostatic waves from which, by means of an appropriately selected experiment, the parameters of an anisotropy can then be determined. Surface MSW propagation in an anisotropic Fe layer for various directions with regard to an external magnetizing field was investigated by Kabos et al. [89, 90]. The initial position was a perpendicular orientation of the magnetic field to the direction of propagation. The measured deviation was in fact the deviation from this perpendicular direction. The experimental results obtained by means of Brillouin light scattering for two values of the wave vector $\vec{k}$ are shown in Fig. 6.23. The change of the upper spin wave band frequency as a function of the angle of propagation is illustrated in Fig. 6.23, indicated by line 3. From a comparison with the theoretical results (full lines) it follows that for small wave numbers there exists quite good apreement between theory and experiment. For larger values of the wave number ($\vec{k}$) the results are

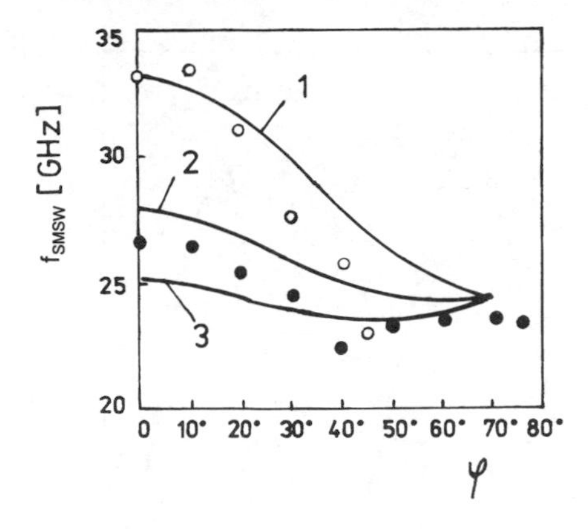

Fig. 6.23. Experimental investigation of surface MSW propagation for various directions of propagation φ with regard to external magnetic field. 1 — $k_m = 1.83 \times 10^5\,\text{cm}^{-1}$, 2 — $k_m = 0.43 \times 10^5\,\text{cm}^{-1}$, 3 — the spin wave band boundary frequency.

in agreement only for small angles of propagation. For $\varphi > 45°$ in the case investigated the surface wave almost merged with the spectrum of volume waves. However, it occurred much earlier than was expected for the theoretical critical angle which in the case investigated was 69°. Rupp et al. [91] used Brillouin light scattering and a propagation surface magnetostatic wave for the determination of anisotropic properties of materials. The dependence of the frequency of surface MSW on the angle with regard to the [100] cubic axis was measured. The mutual orientation of the magnetic field and the direction of propagation was for each 90°. The relationship obtained enabled the determination of the anisotropy constants of the material investigated from the dispersion characteristics. Reference [73] deals with the measurement of the anisotropy constants of thin YIG films used in MSW elements. With regard to low magnetic losses, the necessary information can be obtained from microwave measurements. The authors used the technique of the excitation of surface magnetostatic waves by means of microstrip transducers. Data were obtained from the measurement of the frequency dependence of the coefficient of reflection R of the transducer (Fig. 6.12) for various orientations of crystallographic axes of the thin layer with regard to the fixed positions of the transducer and external magnetizing field. The dependence of the frequency f_k (Fig. 6.12) on the angle of rotation, which was obtained in this way, is shown in Fig. 6.24. Essential directions $[\overline{1}\overline{1}2]$ and $[\overline{1}12]$ in the layer plane are depicted in the figure

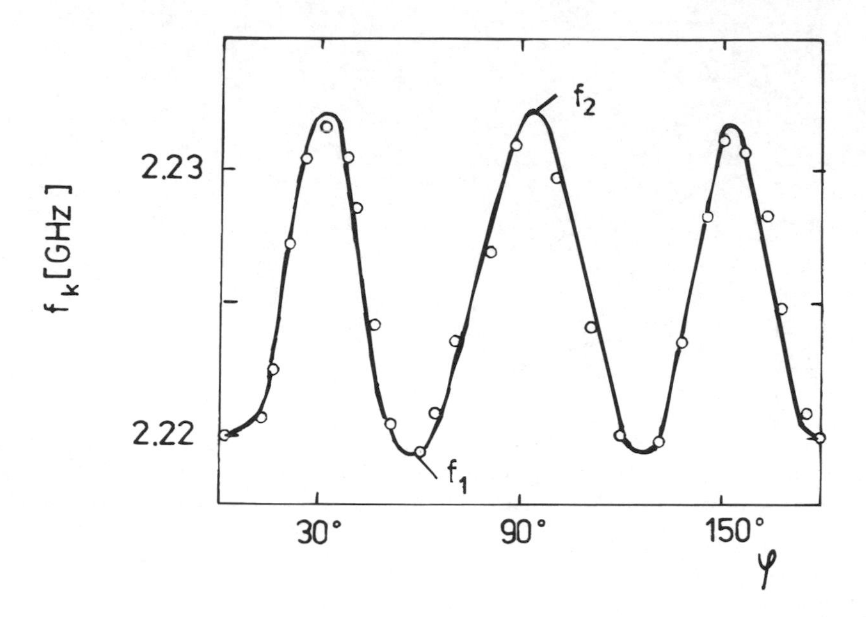

Fig. 6.24. The dependence f_k on the angle of layer rotation. Layer thickness $4\,\mu$m, $M_0 = 137.6\,\text{kA}\,\text{m}^{-1}$.

by the corresponding frequencies f_1 and f_2. If dispersion relations for surface MSW propagating perpendicularly to external magnetic field and formulae (6.19) for the components of the permeability tensor are used, and furthermore if $kt < 0,1$ (where t is the thickness of the sample) then [98]

$$\omega_1^2 = \omega_{01}^2 + \gamma^2\mu_0^2[kt.M_0(M_0 - H_A/2)/2] \tag{6.20a}$$

$$\omega_2^2 = \omega_{01}^2 + \gamma^2\mu_0^2\big[H_A^2[1/2 + 5(M_0 + 2H_e - H_A/2)/9 \div (M_0 + H_e - H_A/2)] + ktM_0(M_0 - H_A/2))/2\big] \tag{6.20b}$$

where $\omega_i = 2\pi f_i$, $\omega_{01} = \gamma\mu_0[H_0(H_0 + M_0 - H_A/2)) - H_A^2/2]^{1/2}$ is a ferromagnetic resonant frequency for $k = 0$ and $H_0 \| [\bar{1}10]$ and $H_A = 2K_1/\mu_0 M_0$. Thus the constants H_A and M_0 can be determined from frequencies ω_1 and ω_2 corresponding to the given wave number k at external field H_0 and known film thickness t. In the case that $H_0 \ll |M_0 + H_{0t} - H_A/2|$ the equations (6.20a, b) can be simplified, after elimination of k, t and H_0 we obtain [98]

$$H_A = \pm\sqrt{[(18/19)[(\omega_2/\gamma\mu_0)^2 - (\omega_1/\gamma\mu_0)^2]]} \tag{6.21}$$

or if $M_0 \| H_0$ is supposed

$$H_A = \pm\sqrt{(2[(\omega_2/\gamma\mu_0)^2 - (\omega_1/\gamma\mu_0)^2])} \tag{6.22}$$

Thus M_0 can also be determined if we know H_A. Another standard approach to the determination of anisotropy parameters of magnetic materials is the use of FMR techniques. Usually the dependence of the resonance field of FMR on the angle of rotation of the external magnetic field with respect to the selected direction is measured, and the constants of anisotropy are obtained from fitting the theoretical dependences and experimental data. Experiments of this kind were carried out by many authors using various kinds of materials. In this monograph we will not deal in more detail with the use of FMR for the determination of the parameters of anisotropy. We would like only to draw the attention of the interested reader to the publication [92] in which the FMR technique was combined with Brillouin scattering of light for the investigation of anisotropic properties of ultra-thin Ni–Fe bi-layers epitaxially grown on Ag.

6.4.2 The determination of high-frequency losses of magnetic materials

From the previous section it follows that the measurement of the dispersion relations is an effective method for the determination of magnetic material properties. The dispersion relations are not, however, very sensitive to another important parameter for high-frequency applications, i.e. to losses in the given material. Furthermore, if we take into account that $\Delta\omega$ corresponding to the usuall losses of microwave magnetic materials is of the order of 0.3 GHz and that the spectral range of FPI used for the investigation of magnetic excitations is of the order of 50 GHz, then it is obvious that a standard optical system so effective for the measurement of scattering characteristics by means of Brillouin light scattering has to be considerably modified for the measurement of losses. For this reason direct measurement of losses in ferromagnetic materials by means of optical methods has been used only in a few publications [92, 93] up to now. Therefore mainly high-frequency methods are used for the measurement of losses.

As the parameter characterizing high-frequency losses in magnetic materials an attenuation term is used from the phenomenological Landau–Lifshitz equation expressed in the form of the resonance line width ΔH. In the case that the measured resonance line width will correspond to some propagating or standing magnetostatic wave with wave number k, we will denote it by $\Delta H(k)$. It should be emphasized, however, that by $\Delta H(k)$ we will understand the variable corresponding to linear processes in magnetic materials investigated, in contrast to ΔH_k which is the parameter bound to the losses of some spin wave (usually with $k \to 0$) which is excited, however, in a nonlinear so-called parametric process. The ferromagnetic resonance line width ΔH provides information about total losses at FMR. These losses can be referred to two basic relaxation channels [94]:

(a) spin–lattice relaxation channel which represents "intrinsic" losses in perfectly polished monocrystalline or defect-free epitaxially grown samples

(b) spin–spin relaxation channel which couples the energy from uniform precession ($k = 0$) or from a high-frequency electromagnetic field to degenerated excited magnetic excitations. This coupling process is usually called two-magnon scattering and it characterizes the influence of non-homogeneities in a sample on resonance losses. With regard to the fact that propagating magnetostatic waves can be used also for the measurement of losses in polycrystalline materials we will not restrict ourselves only to monocrystalline materials or LPE films. Under the assumption that degenerate excitations do not interact with each other, the resonance line width of polycrystalline or monocrystalline ferrites with defects is expressed in the form

$$\Delta H = \Delta H_s + \Delta H_a + \Delta H_p \tag{6.23}$$

where ΔH_s corresponds to the resonance line width of a monocrystalline sample including residual losses as a result of scattering on the sample surface, ΔH_a expresses the influence of the random orientation of anisotropic axes of individual grains or of regions with different anisotropy and ΔH_p the effect of local demagnetizing fields around pores and of non-magnetic inclusions. Schloemann [94] showed that for the components ΔH_a and ΔH_p

$$\begin{aligned} \Delta H_a &\simeq p \cdot F_a(k, \theta_k) \\ \Delta H_p &\simeq p \cdot \rho \cdot F_p(k, \theta_k) \end{aligned} \tag{6.24}$$

where p denotes porosity, ρ the frequency dependence of the spin wave density of states, θ_k the angle between $\vec{k}$ and $\vec{H}_0$, and $F_{a,p}(k, \theta_k)$

the coupling functions of uniform precession ($k \to 0$) with degenerate modes. Rőschmann [95] used the measurement of the uniform mode frequency and temperature dependences of the resonance line width of individual magnetostatic modes for the separation of the influence of porosity and anisotropy in polycrystalline materials with a narrow resonance line width. This method was generalized in [96] for polycrystalline ferrites with large resonance line widths ($\Delta H \simeq 24 \div 36\,\mathrm{kA\,m^{-1}}$). Since the influence of porosity is proportional to the magnetization M_0 and anisotropy field H_A, and at higher temperatures decreases much more with temperature than M_0, these two contributions to ΔH can be separated. Moreover, measurements enable the determination also of the effective anisotropy field H_A and simultaneously show the influence of various doping materials in a ferrite. For example, Rőschmann in this way detected that In or Zn substitutions in octahedral positions of a lattice in a polycrystalline YIG reduce the effective anisotropic field to very small values and that additions of Ca result in low porosity with $p \simeq 0.05\%$. As a result of these experiments there were produced polycrystalline garnet materials with resonant line width $\Delta H \simeq 100\,\mathrm{A\,m^{-1}}$ in the X-ray frequency band. Patton showed that the effective anisotropy field in polycrystalline Li ferrites decreases at room temperature with the addition of Co and that supplementary additions of Ti cancel this influence of Co on anisotropy (Fig. 6.25). Another approach was suggested by Vrehen [97]. He introduced the concept of so-called effective resonance line width ΔH_{eff}. This parameter is determined from the frequency or field-dependent line width. The field-dependent losses, if they are interpreted as the attenuation constant, provide information about the type of scattering centres. In the case that one of the given types of centres is predominant, ΔH_{eff} can be expressed in the form of the products $\rho F_a, \rho F_p$ respectively. If it is a question of mixed types of scattering centres, then it is not possible to distinguish individual types from each other (if the temperature dependence is not made). It is, however, known that ΔH_{eff} is very sensitive to the two-magnon process and grain size [98] as can be seen in Fig. 6.26. For this reason this method was examined in [99] for the investigation of the crystallization process in amorphous ferromagnetic materials and in [100] for losses in thin permalloy films. Introduction of ΔH_{eff} also causes a problem in application to non-spherical samples. ΔH_{eff} is defined on the basis of

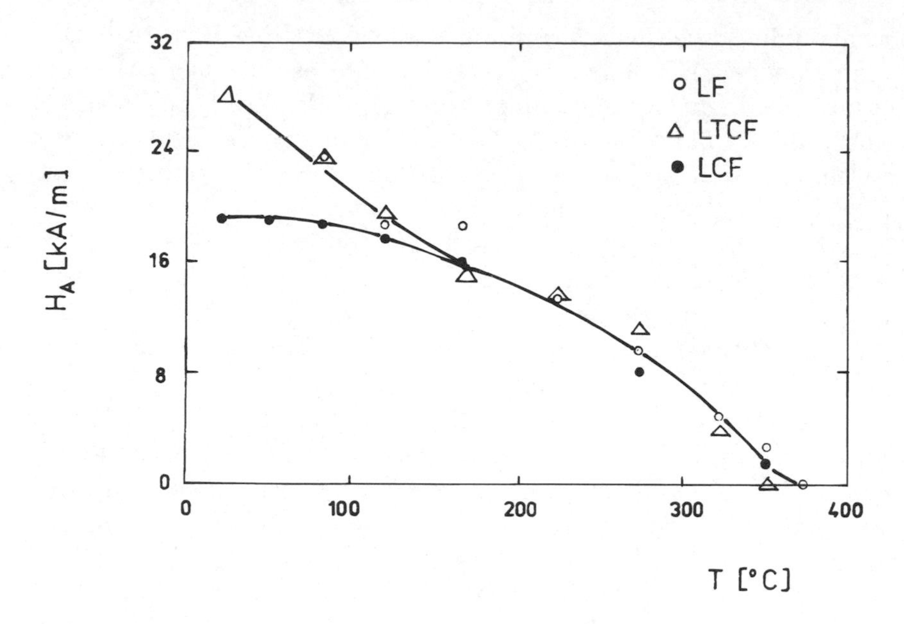

Fig. 6.25. The dependence of the anisotropy field on temperature for polycrystalline Li ferrites.

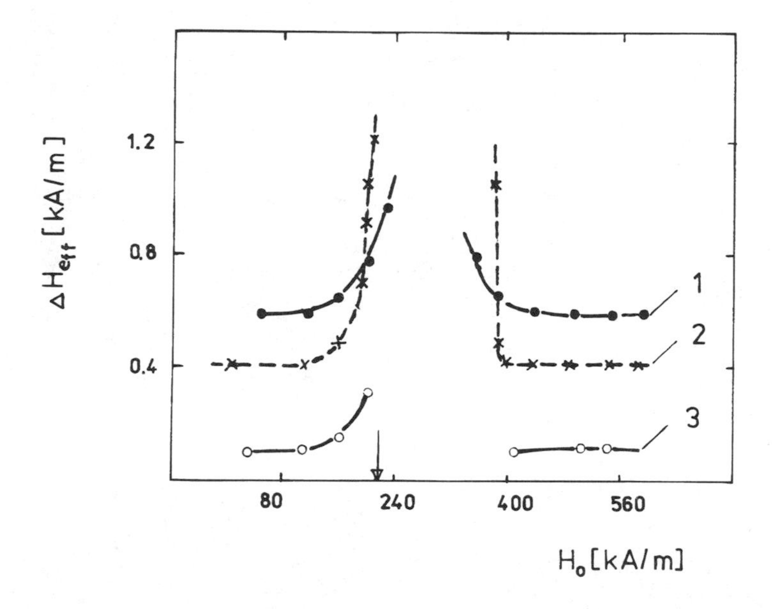

Fig. 6.26. ΔH_{eff} as a function of external magnetic field. Grain size — parameter. 1 — 1 μm, 2 — 2.1 μm, 3 — 30.1 μm.

formal analogy with the resonance line width in the form

$$\Delta H_{\text{eff}} = \frac{2\eta_0}{\gamma} \tag{6.25}$$

where η_0 is the so-called attenuation factor and γ the gyromagnetic ratio. The meaning of ΔH_{eff} for spherical samples was clarified e.g. in [101]. However, for non-spherical samples the formula for ΔH_{eff} is dependent also on the choice of the equation of motion for the vector of magnetization (its attenuation term) as well as on the way the attenuation factor is introduced. Detailed analysis of individual formulae using computer simulation of experiments showed that in spite of the differences in equations for spherical and non-spherical samples it is possible to compare mutually the results obtained for samples of different shapes. A necessary condition of this comparison is the investigation of the dependence ΔH_{eff} not as a function of external field H_0, but as a function of resonance field H_r. The analysis also showed that for non-spherical samples the so-called Gilbert's modification of the Landau-Lifshitz equation can be recommended as an initial equation. Rőschmann also showed experimentally that the effective resonance line width in the spin wave manifold is connected with the resonant line width of individual magnetostatic modes, which means that both approaches are, in some areas of applied fields or frequencies, equivalent. Unfortunately, as follows from [103], measurement of ΔH_{eff} cannot be used for thin epitaxially grown YIG films, because the influence of a paramagnetic GGG substrate cannot be neglected and it influences the results of an experiment. For this reason it was necessary to find other approaches for the investigation of the mechanism of losses in thin YIG films. With regard to their practical applications in MSW elements, direct use of MSW should be taken into consideration.

The problem of the use of magnetostatic modes for the measurement of losses in ferromagnetic materials was first solved by Adam [104] and Gusev et al. [75] for surface magnetostatic waves in thin films. Stancil [105] generalized these results for an arbitrary type of magnetostatic wave. The relationship between ΔH (or $\Delta H(k)$ in this case) and the relaxation time of propagating spin waves with the wave number k is [105]

$$\frac{1}{T_k} = \frac{\gamma \Delta H(k)}{2} \cdot \left. \frac{\partial F/\partial \omega_0}{\partial F/\partial \omega} \right|_{\omega_0 = \gamma \mu_0 H_i} \tag{6.26}$$

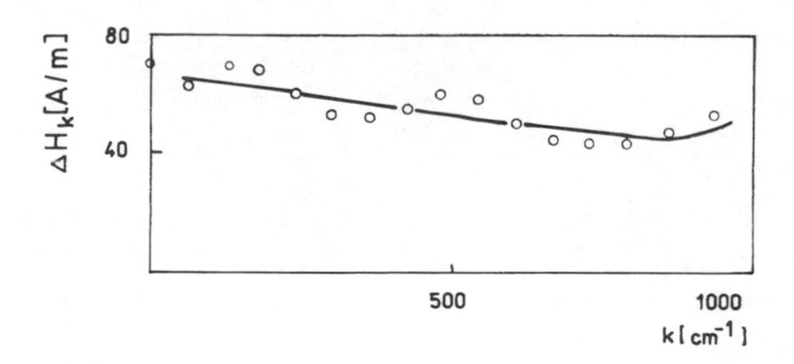

Fig. 6.27. The dependence of loss parameter on a wave number. ΔH_0 — the ferromagnetic resonance line width of measured sample.

where $F(\omega, k, \omega_0) = 0$ is the dispersion relation for the given type of excited magnetostatic wave and selected geometrical arrangement. This approach was used in [75] for a 7.3 μm thick YIG film. The measured dependence is depicted in Fig. 6.27. The scattering of experimental points in the figure is caused by oscillation of the dependence $\delta L(f)$ from Fig. 6.14, which is again caused by "seeping through" of high-frequency power at the 45 dB level. The ferromagnetic resonance line width ΔH_0 measured by means of the resonance method on a sample from the same material is also indicated in Fig. 6.27. From Fig. 6.27 follows that in the range of wave numbers 60 up to 100 cm^{-1} $\Delta H(k)$ differs very little from ΔH_0. On the other hand, however, measured values of $\Delta H(k)$ are much greater than the value $\Delta H_k = 0.16\,\mathrm{A\,cm^{-1}}$ (for $k \simeq 10^4\,\mathrm{cm^{-1}}$) measured by the parametric excitation of spin waves. The authors ascribe this difference to the scattering process of volume and surface MSW on inhomogeneities and to the influence of fast relaxing ions. Generalization of measurement to measurements at various temperatures would enable one, as in the case of standing waves, to differentiate the influences of individual mechanisms. An analogous approach was later also used by other authors [106].

Another possibility for low-loss materials is propagation of electromagnetic waves in a cut-off waveguide. The authors in [107] showed that this method can be successfully used for non-destructive measurement of parameters of thin ferrite films. The theoretical dependence of the relative band width of a cut-off waveguide on the loss parameter α is shown in Fig. 6.28. As can be seen, the width of the pass band depends considerably on the loss parameter, especially for small losses. The influence of losses in the waveguide walls

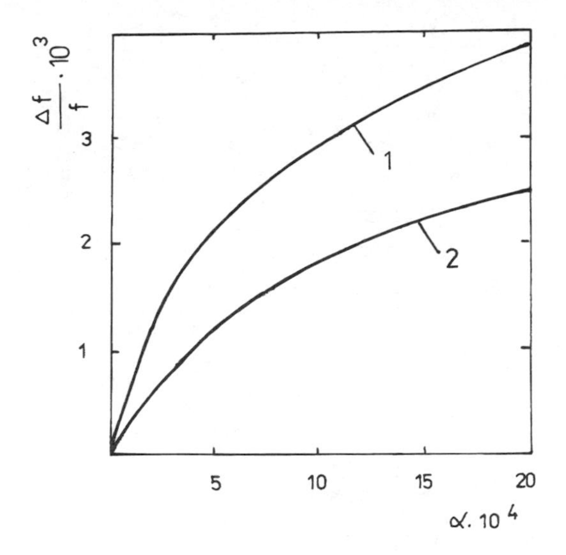

Fig. 6.28. The dependence of the relative width of the pass band on the loss parameter. $f = 30.362\,\text{GHz}$, $\omega_M/\omega_H = 0.15$. 1 — $t/h = 0.01$, 2 — $d/h = 0.5$.

can be lowered by the location of the sample layer at a sufficient distance from the side wall of the cut-off waveguide. This method is also highly appropriate for the measurement of the saturation magnetization, because the average frequency of the pass band along the cut-off waveguide is a function of it. Furthermore, if various types of magnetostatic modes are excited in the thin layer investigated, e.g. by the change of orientation of the external magnetic field with regard to the layer plane, it is possible to determine not only M_0 but also its profile through the cross-section of the layer [18].

The approach described in [106] is based on receiving MSW by means of two or more microstrip antennas, a relatively large distance from each other. With regard to the relatively large attenuation of MSW on the path of propagation, which for uniform precession is about $76.4\Delta H$ (dB μs^{-1}) (by time we mean one microsecond of delay of the magnetostatic wave propagating with a group velocity v_g) it is possible to use this method of measurement practically only for extremely pure magnetic materials with very narrow resonance line width ΔH, which, up to now, are only epitaxially grown YIG layers.

On the other hand, the technique of scanning of MSW by an induction probe also enables the investigation of characteristics in materials with higher losses [109]. The price to be paid for this

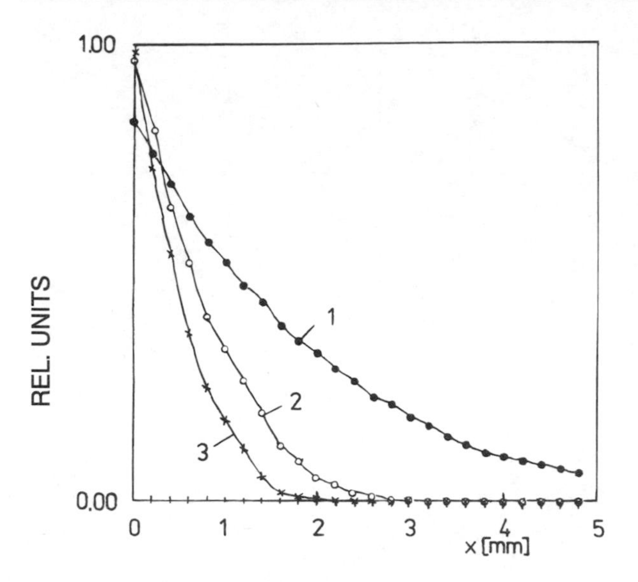

Fig. 6.29. The influence of layer thickness on attenuation of propagating surface MSW. $H_0 = 82\,\mathrm{kA\,m^{-1}}$, $M_0 = 64\,\mathrm{kA\,m^{-1}}$, $f = 3.735\,\mathrm{GHz}$, 1 — $t = 1.17\,\mathrm{mm}$, 2 — $t = 0.8\,\mathrm{mm}$, 3 — $t = 0.14\,\mathrm{mm}$.

is, however, the restriction to slabs from the material investigated which, on the other hand, restricts the range of measurable wave numbers. The influence of the layer thickness on the attenuation of surface MSW is depicted in Fig. 6.29. With regard to high attenuation only a progressive wave propagates in a ferrite slab, which can be observed also on the dependences recorded. From the measured dependences $k'' = \mathrm{Im}(k)$ is determined, from which, by means of the formula (6.30), for the recorded magnetostatic wave line width with wave number $k' = \mathrm{Re}(k)$ we have

$$\Delta H(k') = \frac{2v_g \mathrm{Im}(k)}{\gamma} \cdot \left.\frac{\partial F/\partial\omega}{\partial F/\partial\omega_0}\right|_{\omega_0=\gamma\mu_0 H_i} \tag{6.27}$$

where v_g is the group velocity of an excited magnetostatic wave. It is obvious that this approach can be used for the measurement of frequency dependences of the spin wave resonance line width with given wave number k' in a relatively wide frequency range, which creates the conditions for a partial differentiation of individual relaxation processes in polycrystalline materials investigated. Although the mentioned techniques, especially for polycrystalline materials, have not been commonly used yet, their wider use can be expected, especially in connection with the preparation of multi-layer structures. The progress of spin wave spectroscopy by means of magnetostatic waves can be expected as a result of progress in microwave applications of new types of ferrite materials. Another reason is the fact that they enable a non-destructive and approximately local measurement of the parameters of material samples in a wide frequency band (of the order from 1 GHz up to the millimetre wave frequency band). In the end we may say that spin wave spectroscopy has also become an effective tool for the investigation of the properties of magnetic materials.

6.5 The spin wave line width ΔH_k and nonlinear processes

To conclude this chapter we will briefly mention the basic concepts describing relaxation processes (losses) at higher levels of high-frequency power. If the high-frequency power exceeds the so-called threshold value, the nonlinear processes described in Chapter 5 begin to reveal themselves. While magnetic excitations of the spin system are more or less independent at lower power levels, at high power levels they are strongly coupled. As a consequence of this strong coupling there are also produced magnetic excitations which are not directly coupled with the exciting high-frequency electromagnetic field. The parameter which in a phenomenological way describes the relaxation of such magnetic excitation with the wave vector $\vec{k}$ is called, by analogy with the low-power relaxation parameter, the spin wave line width ΔH_k. The value of ΔH_k is determined from the measurement of the so-called critical field h_c, in the configuration of either parallel or perpendicular pumping, which means that the high-frequency magnetic field h is either parallel or perpendicular to the external static magnetic field H. The experimental results (see e.g. [110]) showed that ΔH_k strongly depends on the amplitude and orientation of the wave vector $\vec{k}$, especially for static magnetic fields below the field corresponding to the minimum level of applied power at which nonlinear processes occur. If the level of applied power is further increased, periodic and later chaotic oscillations of the strongly parametric excited magnetic system occur. Chaotic oscillations were theoretically predicted in [111, 112] and later experimentally verified in [113], which opened the way for an intensive investigation of magnetic excitations in nonlinear and strongly parametric excited spin systems. These problems are, however, beyond the scope of this book.

Chapter 7

FERROMAGNETIC MATERIAL AND MSW ELEMENT TECHNOLOGY

7.1 Garnets

We will mention briefly the properties of pure yttrium-iron garnets and garnets doped by some elements to modify their magnetization and anisotropy, because they are almost solely used for most technical applications of analogue signal processing.

In 1956 magnetic garnet, and especially yttrium-iron garnet, was discovered by a research group in Grenoble [114] and independently by a research group at Bell Telephone Laboratories [115]. These materials are included in the group $M^{III}{}_3Fe_5O_{12}$ where $M^{III}{}_3$ represents an element from the series of rare earths. Nowadays yttrium-iron garnet has become one of the basic magnetic materials for the study of the physical properties of ferromagnetic materials and due to its special advantageous magnetic, electrical and optical properties it is also the most widespread material for technical applications. We can say that its position in high-frequency device technology using magnetic materials is the same as the position of silicon in semiconductor technology. Although magnetic garnets are semiconductors their semiconductive properties are not interesting for technical applications because the mobility of charge carriers in oxides is lower by several orders of magnitude than in semiconductors from the groups III – V. In addition, high-frequency applications require as large a resistance as possible to reduce high-frequency dielectric losses. A summary of their properties can be found in [116].

To understand better the magnetic properties of the magnetic materials it is appropriate to use a structural analysis, since it is obvious that magnetic properties result directly from the structure, presence of individual ions and their location in the crystal. Thus we will mention the basis of structural analysis. For the structural description of crystalline solids a simple model of "tough cells", of which ionic crystals are created, is used. In this model basic elements differ from each other only by their size (ionic radius) and charge and they occupy certain geometrical positions with a predetermined configuration of surroundings (co-ordination number) which results in an unambiguous arrangement from the point of view of long-distance observation, and they create crystal lattice in this way. The smallest volume of a space lattice which contains all symmetries is called an "elementary cell". In the case that this elementary cell is a cube the lattice derived from it is called the cubic lattice. The value of the length of the elementary cell edge (lattice constant) is for most of the elements and compounds in the interval (0.3 – 1.0) nm. The lattice constant of garnets (1.2 – 1.3) nm is considered to be large. The temperature dependence of the lattice constant is one of the most important characteristics of compound materials, e.g. epitaxially grown garnet layers on non-magnetic garnet substrates, in which the differences between temperature dependences of lattice constants result in internal mechanical tensions which influence considerably the magnetic properties, e.g. losses. Yttrium-iron garnet $[Y_3][Fe_2](Fe_3)O_{12}$ or $Y_3Fe_5O_{12}$ is a prototype of ferrimagnetic garnet. The various kinds of brackets in its formula indicate three different types of co-ordination — dodecahedral or [c] side, octahedral or [a] side and tetrahedral or [d] side. It is necessary to remark that the ferrimagnetic garnets are not included in the cubic space group. In the case of the spontaneous magnetic polarization, the crystal can be centrally symmetrical (space group R3) while the magnetic moments associated with atoms or ions can be represented by axial vectors. Transformation around the centre of symmetry does not change these moments.

The "easy" direction of magnetization in YIG is the [111] direction and from this fact it follows that YIG crystalline structure is rhombohedral. It is not possible to detect this deviation from the cubic structure using classical Rőntgen diffraction and that is why the YIG crystalline structure was described by the space group Ia3d. The garnet cation location in four octants of an elementary cell can be seen in Fig. 7.1 and the location in the vicinity of an oxygen ion is in Fig. 7.2 (in a simplified way one octant of the elementary cell is depicted as a cube). In one octant, i.e. in a volume centred cube or subelement with an edge $a/2$ consisting of [a] ions, the [c] and [d] sides are located alternatively on the surfaces (see Fig. 7.1).

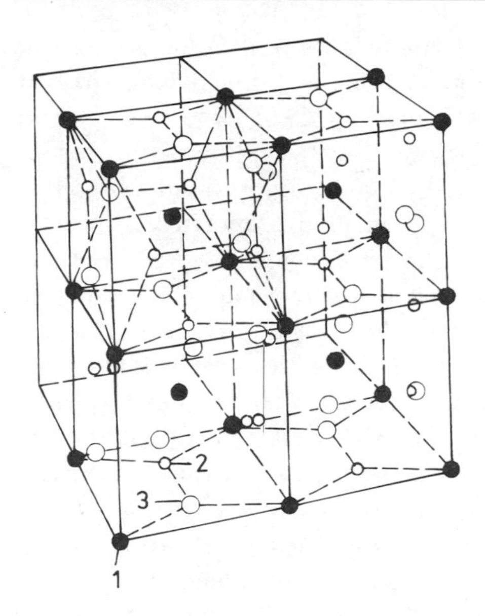

Fig. 7.1. Arrangement of cations in a garnet structure. Positions 3 — c, 1 — a, 2 — d.

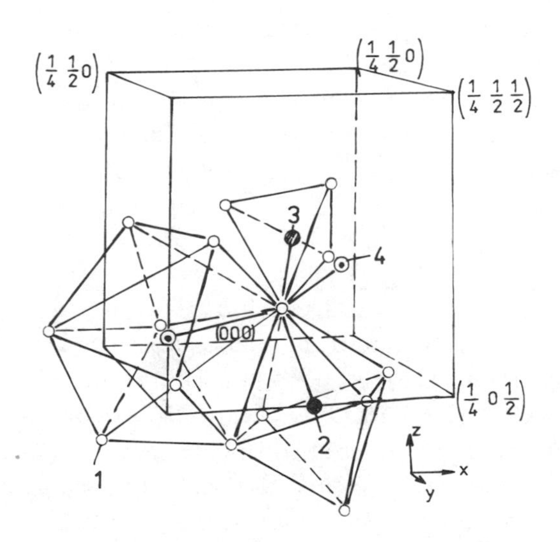

Fig. 7.2. Co-ordination around an oxygen ion in a garnet. 1 — O^{2-}, 2 — Al^{3+}, 3 — Si^{4+}, 5 — Ca^{2+}.

The position of cations is determined by the lattice constant a. [c] cations have 8 nearest oxygen neighbours, and [a] and [c] cations 6 and 4 oxygen neighbours, respectively. Inversely, each oxygen ion has 4 nearest neighbours: two on the [c] side and one each on the [a] and [d] sides. At the same time each oxygen ion belongs to two dodecahedrons, one octahedron and one tetrahedron. Geller and his co-workers [117,118] contributed to the understanding of the garnet structure in a most important way. For special material applications in electrotechnics and electronics, the optimal properties of materials are usually obtained not by using pure elements or compounds but using either extremely pure elements with special trace elements or substituted compounds. In principle both approaches are used in the case of magnetic garnets. Since all cations with ionic radii between 0.026 nm and 0.129 nm can "fit" into the garnet lattice, a number of practical combinations can exist. A crucial role in the regulation or optimization of physical parameters of these materials is played by the charge compensation. For example, if a trivalent ion (Fe^{3+}) is to be substituted, then the substituting ion must also be trivalent (e.g. Ga^{3+}, In^{3+}) or in the case that its valency is higher it must be combined with an ion of lower valency (e.g. Sn^{4+} has to be combined with Ca^{2+}). Another important problem in substitution is how substituents occupy different cation positions, mainly positions in the tetrahedron/octahedron and dodecahedron/octahedron sides. In developing basic theories it is necessary to know these distributions and their temperature dependences. In general the radius of the ion substituted determines more or less what positions will be occupied. The biggest ions (r = 0.096–0.129 nm) occupy dodecahedral positions, medium size ions (r = 0.053–0.083 nm) octahedral positions and the smallest ions (r = 0.026–0.047 nm) occupy preferably tetrahedral positions. It is obvious that this is a very simplified approach, but it provides a basis sufficient for an astonishing number of exact predictions. From one-component substitutions, whose practical applications are of great importance, the Al and Ga substitutions are mostly used ($Y_3Fe_{5-t}M_tO_{12}$) because by changing the structure they allow a change of the saturation magnetization from 140 kA m^{-1} to zero at room temperature. Both ions preferably (but not exclusively) occupy tetrahedral positions; however, with increasing concentration they also occupy octahedral positions. Since the allocation of occupied positions is extraordinarily important the tetrahedral/octahedral distribution function f_t and its thermal dependence is examined. Up to now this function has been investigated experimentally by means of RTG or neutron diffraction and Mossbauer spectroscopy. Borghese [119], having started from general thermodynamical considerations, obtained theoretically a very simple formula for the thermal equilibrium of the cation transition between the two sublattices mentioned above, but unfortunately the theory is not in agreement with experiment. The importance of this function requires further investigation concerning the individual ions creating the lattice. Distribution functions for Gd and Pb were examined in e.g. reference [131].

7.2 Garnet technology

In general, three main fields are included in the technology of crystalline inorganic electronic materials: technology of ceramic materials, massive monocrystals and epitaxial layers on monocrystalline substrates. Magnetic garnets can be prepared by all these technologies. In contrast to most of the semiconducting materials, special precautions must be taken when magnetic materials crystallize from a melt consisting mostly of oxide mixtures. The fact that, in spite of this, garnet monocrystals almost as perfect as the best silicon crystals have been prepared, is the consequence of the extraordinary interest and effort devoted to the development of these materials all over the world.

7.2.1 Preparation of garnet materials using the ceramic method

Although the elements to which this book is devoted are based exclusively on monocrystalline and epitaxially grown layers we will also mention the technology of preparation of polycrystalline garnets to make this chapter complete and because it is possible to apply these materials together with the monocrystalline ones. The basic material for the ceramic method is powder, which must be pressed to a solid form. The crucial parameters for materials prepared by the classical ceramic method are those which characterize chemical composition, crystallographic structure and so-called microstructure (including such important parameters as the size of grains originated, their distribution function with regard to their size, grain boundaries, their texture and inclusions (pores) among them). To check the influence of these parameters it is necessary to have technological devices and efficient experimental methods which enable us to follow the parameters mentioned above at the microstructural level to be able to recognize the mechanisms which give rise to certain crystallographic phases and so to give the possibility of increasing the material density when the sintering process is in progress. Under sintering we produce calcinating of the pressed sample during several hours at a temperature below the melting temperature of the given matter. Before the sintering process a one-component material or mixture of several components is usually obtained by means of chemical reactions (so-called calcination). These stages can be carried out simultaneously. If hydrostatic pressure and sintering are applied at the same time, then materials with a density value almost equal to the theoretical one can be obtained. Ceramic materials prepared in this way then differ from monocrystalline ones only by the fact that their physical properties are isotropic and their mechanical stability is less in general. The standard procedure of ceramic garnet technology begins with the mixing of component oxides in ball mills. Having dried, this mixture undergoes heat treatment for the first calcination (e.g. 4 hours at temperature 1000 – 1200°C in air). To homogenize the product of this reaction it is necessary to mill it again in a ball mill. In the case that iron balls are used in the mill it is necessary to take into account their rubbing and to decrease the input amount of iron oxide Fe_2O_3 in order to keep the stoichiometry of the resultant product. In a further stage the powder obtained is pressed hydrostatically and definitively sintered e.g. for 6 hours at 1250 – 1450°C in an oxygen atmosphere. Of course the procedure briefly described above represents only one of a number of possible modifications. Homogeneity of the basic powder at the "molecular level" can be reached, e.g. in such a way that components of the garnet material required are present in a solution with the right stoichiometric ratio and subsequently, without changing composition, they are transformed to solid state. In comparison with the standard ceramic procedure this "wet" chemical process improves the composition and homogeneity because stoichiometry is not affected by admixtures coming from the mill. In this procedure water solutions of inorganic and organic salts (nitrides, sulphates, citrates) are mostly used. The solution is dehydrated and dehydration is carried out by e.g. freezing out, spraying and immediate drying, coprecipitation or coagulation from precursors. Using one of these procedures a solid mixture is created which retains solution homogeneity. In the next stage this mixture is pyrolysed to the mixture of oxides which under these conditions represents a very fine disperse reactive system. The last stage of ceramic preparation is sintering, mentioned above. During this process individual grains grow. Condensation of material when sintering is in progress is influenced by a number of technological parameters, e.g. the pressure used, the size of grains, shape of grains, sintering temperature and duration of sintering, composition and pressure of atmosphere, impurities or doping elements used to influence the sintering process, etc. In a simplified way the sintering process can be divided into three phases. In the first so-called initial phase the creation of the interparticle bridges begins at the atomic level, bringing them nearer to each other; the second so-called transition phase occurs when the density reaches (60–70) % of its theoretical value and when grains begin to grow. The system still has a continuous pore phase which begins to decay when the density is (85–95) % of its theoretical value, and this results in the third phase, when pores are separated and they are in the matrix of grains. Migration of grain boundaries and the grain growth associated with it are the major processes in which the material sintered tries to achieve a minimum of energy. In the case that non-continuous grain growth occurs a great number of pores remain closed in the grains and that is why they are isolated from the grain boundaries and the sintering process is terminated. But then scattering centres arise whose influence is disadvantageous, mainly on losses.

7.2.2 Preparation of massive monocrystalline materials

The wide application of garnet materials in microwave and recording electrotechnics results from the fact that monocrystals of high quality can be prepared. The basis for the preparation technology of monocrystals growing from the melt is the work of Remeika [120], Nielsen [121–124] and Van Uitert et al. [125, 126] published in the late sixties. Remeika demonstrated that monocrystals of many types of orthoferrites of the perovskite type can be obtained by growing from a melt of lead oxide. Since many of these compounds contain the same components as magnetic garnets, garnets were considered to be able to crystallize from a lead oxide melt, keeping the right conditions. First these conditions were found out empirically using such oxides as PbO, Fe_2O_3 and Y_2O_3. Considerable development was achieved by adding other components to the melt, e.g. PbF_2 and B_2O_3, which resulted in a decrease of crystallization temperature and viscosity of the basic melt. Using an appropriate temperature profile, i.e. the temperature of the higher part of the crucible is 8–10°C higher than the temperature of the lower part of the crucible, we can secure the creation of nuclei on the melt surface, which results in an intensive increase of size and also in an improvement of quality of the crystals obtained. The disadvantage of this procedure consists on the one hand in the fact that the pressure of PbF_2 vapour is high at the melting temperature 1250°C and for this reason it is necessary to seal the crucible in order to prevent losses caused by evaporation and the subsequent change of composition while grain growth is in progress, and on the other hand that while crystals are separated from the melt expensive basic solution is lost. In addition to this there are also problems with the change of composition. These shortcomings were overcome elegantly by the method independently introduced by Bennet [127] and Tolksdorf [128, 129]. In a closed platinum crucible which can rotate and lower the pressure, crystals are separated from the melt at the end of the slow period of cooling when the crucible rotates in the furnace in such a way that it turns bottom upwards and crystals stuck to it are torn from the melt. In so doing almost no melt loss occurs. A special design of the deck enables one to take out the garnet crystals, supply corresponding basic material and to continue the crystal growth.

The overpressure of the oxygen atmosphere prevents the evaporation of PbF_2/PbO during the crystallization process and it also suppresses the creation of Fe^{2+} ions in crystals. In another procedure the so-called technique of accelerated rotation of the crucible is used. The rotation changes periodically, e.g. over 20 s from zero to 35 revolutions per minute and back to zero, with subsequent 10 s rest. In co-operation with localized cooling (e.g. a flow of cold air directed to the bottom of crucible), keeping sufficiently quick melt flow to the boundary with the crystal and preventing undercooling, this enables the solution of the problem of controlling the creation of nucleation centres (nucleation).

A typical course of production of YIG crystals from a cooled melt is as follows. The mixture is heated for several hours until it is melted at a temperature of about 1280°C. Then the temperature decreases at a rate of $40°C h^{-1}$ to 1190°C when the system of the accelerated rotation technique is switched on, together with a cooling air flow directed at the bottom of the crucible with capacity $170 l h^{-1}$. While nucleation and crystallization are in progress the

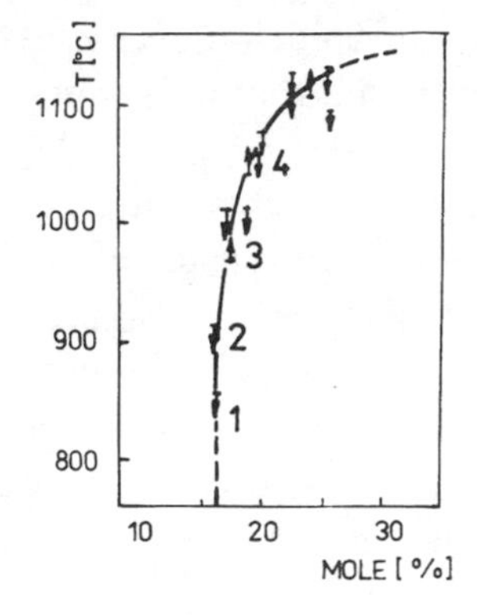

Fig. 7.3. Saturation concentration of YIG in a melt of type B (PbO — 36.3, PbF_2 — 27, B_2O_3 — 5.4, CaO — 0.1, Fe_2O_3 — 20.78, Y_2O_3 — 10.42 mol %). Arrows down–cooling with regard to required temperature, arrows up–heating-up by 10°C with a rise of $0.5°Ch^{-1}$.

temperature decreases at a rate of only $0.5°Ch^{-1}$. As can be seen from Fig. 7.3 [130] for the composition in the caption saturation for YIG is reached at the temperature $1130 \pm 10°C$. Below this point crystals begin to be created by spontaneous crystallization with subsequent growth during continuing cooling. The temperature decrease after nucleation is $0.2°Ch^{-1}$ during the first 175 hours and $0.5°Ch^{-1}$ during the last 80 hours. To accelerate the crystallization process a nonlinear programme of cooling can be used. While the process of crystal growth is in progress the temperature in the crucible must be maintained with accuracy 0.02°C at least. The residual content of lead in crystals is crucial for optical applications (see e.g. [131]). For this reason lead-free melts were looked for. It turned out that one of the possible melts is Ba–B_2O_3–YIG. Using the appropriate weight ratio yttrium garnet is a stable phase between $1190 \pm 3°C$ and $1000 \pm 3°C$. A disadvantage of the procedures of crystal growth from the melt mentioned above is the fact that, as a result of multinucleation, the crystals are relatively small. This problem can be overcome by using nuclei, stirring the melt in an appropriate way and checking thoroughly the temperature distribution in the crucible. Under these conditions crystals weighing 0.6 kg have been obtained. Unfortunately, up to now quantitative criteria for stable growth are missing. Till now, almost the whole technology of crystal growth has been based on the experience of a qualified worker. That is why it is still necessary to develop the theory of crystal growth from high-temperature salt solutions, although the process of crystal growth is controllable in practical applications and can be divided into four phases: (a) transport of solution to the crystal boundary, (b) diffusion through the boundary layer, (c)surface diffusion through the crystal boundary, (d) coagulation of solution particles into crystals. A concise review of crystal growth theories was made by Parker in 1970 [132].

To end this section we will mention briefly the characterization of massive crystals in which, especially, the chemical composition and structure are included. The composition of crystals is investigated using standard chemical analytical methods for solutions. The most effective and exact method is local non-destructive analysis by means of Rőntgen fluorescent analysis, using which surfaces to a depth of 3 mm are analyzed while calibration standards (polycrystalline) of approximately the same composition are used.

Initially, the structure properties can be checked also by simpler means (e.g. microscope) and later by more complicated magneto-optic, magnetic or Rőntgen methods. There is also a big variety of structural defects in the garnets obtained from the melt, beginning with inclusions in the basic solution and ending with dislocations and growth defects (growth zones, zoning, etc.). It has been found experimentally that the creation of inclusions depends on the growth velocity. At values below 15 mm s^{-1} defect-free growth occurs in unstirred solutions. The density of dislocations and their distribution depend on the level of deviation from the conditions of homogeneous crystallization. Published results differ from each other and the results obtained are more or less empirical. This results from the fact that up to now reliable, unambiguous and theoretically based procedures to influence defects in magnetic crystals do not exist. On the other hand we must say that the empirical technology of preparation has been developed to such an extent that crystals of extraordinarily high quality prepared in a standard way contain such small amounts of defects that they are suitable also for optical applications.

7.2.3 Preparation of epitaxial monocrystalline layers

Magnetic recording applications and special bubble memories have forced the development of the monocrystalline thin-layer technology. It was conditioned mainly by the fact that there is a direct relation between the size of bubble domains (their diameter) and the thickness of the magnetic crystal used. To obtain small bubbles with sizes of the order of a few μm it is necessary to produce magnetic layers with this thickness and of high quality. In principle this problem results in several mutually connected tasks. To create high quality layers it was necessary to find a suitable substrate on which the thin layer with the parameters required could be laid. Almost at the same time it turned out that the thin-layer technology used for the creation of magnetic layers also provides a number of possibilities for microwave applications. We will describe briefly the preparation of thin monocrystalline layers. In general they are prepared by the process of epitaxial growth, i.e. by oriented crystal growth on the surface of the crystal already existing — substrate. This represents one of the most important conditions for obtaining high quality layers and that is why it must fulfil certain criteria. They are [133]:

(a) the substrate must be monocrystalline with the crystal structure coincident with the structure of the epitaxial layer required

(b) the difference between the lattice constants of the substrate and the layer must be minimal

(c) the density of defects in the substrate must be lower than that in the layer

(d) the substrate must have the prescribed orientation with regard to the crystallographic axes

(e) if magnetic garnet layers are to be grown, then the substrate must be dia- or paramagnetic.

A world-wide long-term investigation in this field has shown that for the epitaxy of magnetic garnets the most universal substrate material is gadolinium-gallium garnet (GGG). Its chemical formula is $Gd_3Ga_5O_{22}$, it crystallizes in a cubic system with space group $O^{10}{}_h$ — Ia3d — and its lattice constant is $a = 1.2383 \pm 0.0002\,\mathrm{mm}$. With respect to further treatment the difference between the lattice constants of the substrate and the layer Δa has to fulfil the following condition:

$$|\Delta a| \leq 0.0013\,\mathrm{nm}$$

Massive monocrystals GGG with the quality required (i.e. a density of defects smaller than $5\,\mathrm{cm}^{-2}$) can be prepared only by the so-called Czochalski method of growth from the melt at temperature about 1750°C in an N_2 atmosphere with 2% addition of O_2. The physical and chemical parameters of GGG are in Table 7.1.

For industrial applications GGG slabs cut in the (111) plane are used. For research purposes orientations (110), (100) and (211) are also used. The first epitaxial ferrite layers were obtained by the so-called chemical vapour deposition (CVD) process. In the CVD

Table 7.1

Chemical formula	$Gd_3Ga_5O_{22}$	Colour	colourless
Molar weight	1012.34	Toughness	7.5(Mohs) 1098(Knoop)
Type of crystallo-graphic structure	garnet	Melting temperature	1750°C
Elementary crystallographic cell cell	cubic ($a = 1.2383$ nm)	Index of refraction	1.97 ($\lambda = 632.8$ nm)
Space		Thermal expansion	
Space group	$O_h{}^{10}$ — Ia3d	coefficient	$9.5 \times 10^{-6}\,K^{-1}$ (300 K)
Density	$7090\,kgm^{-3}$	ε_r	10

technique [134] garnet components are transported in the form of vapour (chlorides) and mixed in the reaction chamber in which the substrate is placed at sufficiently high temperature. Although using the CVD technique high quality YIG layers of various compositions have been obtained this process has now begun to be replaced by the liquid phase epitaxy (LPE) technique. First, the LPE equipment is relatively simple and it can also be built under laboratory conditions. Secondly, thin garnet layers of complex composition can be prepared without independent vapour sources for each component which should dope the garnet. In the LPE technique (Fig. 7.4) oxides of garnet components Y_2O_3 and Fe_2O_3 are diluted in the solvent of mixture PbO and B_2O_3, and in this way the same melt is created as for the preparation of massive YIG crystals. The well-prepared melt is maintained for a long time (several hours) under the saturation temperature (up to 50°C, $T_s \simeq 1000$°C depending on mixture) without spontaneous crystallization. During this period the substrate on which the garnet layer is to be grown is placed in the furnace and heated to the furnace temperature and then plunged into the melt. In so doing the substrate behaves as a nucleus on which the garnet layer is created. The process of thin layer growth is terminated when

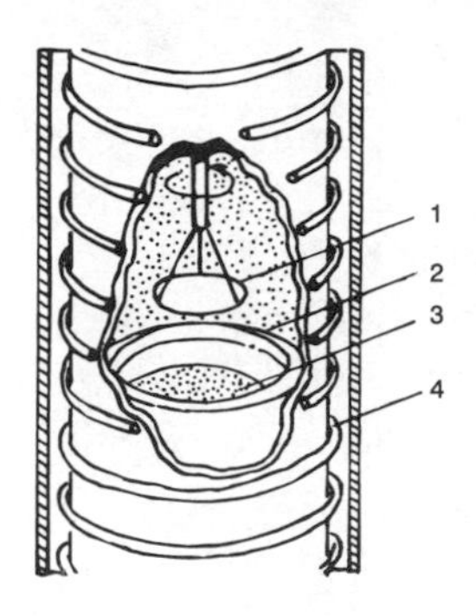

Fig. 7.4. The isothermal method for obtaining epitaxial YIG layers by dipping into a melt. 1 — substrate, 2 — platinum crucible, 3 — melt PbO–B_2O_3, Y_2O_3, Fe_2O_3, 4 — resistance furnace.

the substrate is pulled out from the melt. During the growth period the melt temperature is kept at a constant value. In addition, if the amount of melt is big enough, then the grown layer represents only a small part of it and the composition of the melt is considered to be almost the same as before. Rotation of the substrate during the growth of the layer provides an equable liquid flow with good stirring of the melt and its equable distribution on a prevalent part of the substrate. It results both in homogeneity of thickness and composition of the layer. On top of this special demands are put on the layers prepared for microwave applications. They are: low losses at FMR (represented by the resonance line width ΔH) and the thickness of the layer which should be as far as possible in the interval (10–100) μm or more. The resonance line width in low-loss materials is extraordinarily sensitive not only to physical lattice defects but also to impurities and point defects [135]. For example a small lack of oxygen results in turning up Fe^{2+} ions, with a subsequent increase of conductivity losses. Typical LPE growth periods for YIG layers are about $1\,\mu\mathrm{m\,min}^{-1}$. Problems occur when the thickness of the layer is above $50\,\mu$m. For these thicknesses special conditions are required to keep the quality of the layer high [136, 137]. Another difficulty in growing thicker layers is the problem of accommodation of garnet and substrate lattice constants. YIG and GGG substrate lattice constants are 1.2376 nm and 1.2383 nm, respectively. Although this difference is very small (7×10^{-4} nm), in comparison with many epitaxially prepared semiconducting materials, it can have considerable consequences. When the growth is in progress this difference causes a tension in the plane of the layer and at a certain critical thickness it is energetically more favourable to release this tension by fracture [138]. The critical thickness experimentally observed for YIG on GGG substrate is about $15\,\mu$m. The difference between lattice constants can be reduced by changing the lattice constant of either layer or substrate. A simpler way is to change the composition of the thin layer substituting La for a small amount of Y or building in Pb as an impurity from the melt [138, 139, 140]. Since La is non-magnetic and its valence is the same as for Y, its building into the lattice should cause a change of lattice parameter only. However, it is necessary to be careful because La can be contaminated with other ions of rare-earth elements which contribute directly to the relaxation mechanism. At first sight the building in of Pb ions seems to be disadvantageous. Its valence is mostly 2+, although it can also be 4+. Both cases change the charge equilibrium, which contributes to the conductivity mechanism. It has been found experimentally that, depending on the content of lead, the minimum ΔH exists and this minimal loss is just near the concentration needed to compensate for the difference between lattice constants [152]. To elucidate this surprising fact the authors assumed that first of all Pb^{2+} compensates for another impurity, mainly Pt^{4+}. It is obvious that the true explanation will be more complex. To confirm this we can use the results of other authors working in this field and we can see that they differ considerably from each other.

Magnetic properties of YIG layers can be modified changing their composition. E.g. we have mentioned already that Ga substitution changes values of saturation magnetization in a wide range. The corresponding dependence is depicted in Fig. 7.5. However, to keep the match between lattice constants it is necessary, in the case of thin lay-

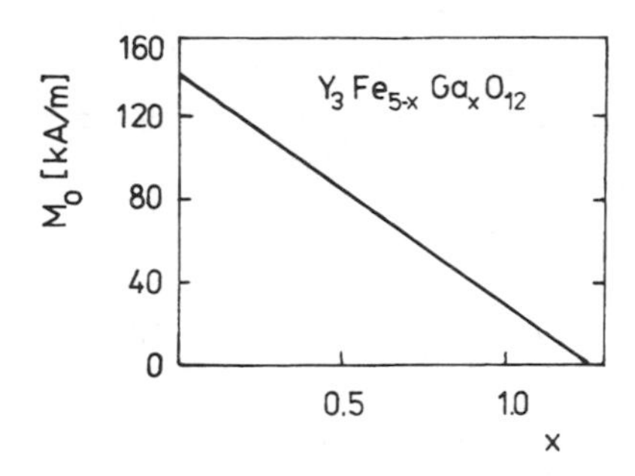

Fig. 7.5. The change of saturation magnetization of YIG by substitution of Ga.

ers, to combine Ga substitution with La. On one hand, at combined substitutions the possibility of the creation of non-homogeneities with subsequent broadening of the resonance line increases, and this limits application of such layers in microwave elements. On the other hand, La, Ga, YIG layers are used for the preparation of multilayer materials, or in other words materials with changing properties in the cross-section of the layer, which are begining to be used in MSW devices to improve or modify dispersion dependences ([140]). Substitutions can also be used for other purposes. Bi^{3+} ion substitution results in a considerable increase of the Curie temperature, by 38 K per Bi^{3+} ion. Similar results are obtained when Ca^{2+} and V^{5+} ions are substituted in garnet, and this system can even have a higher Curie temperature than pure YIG. Several experiments have been carried out to explain this effect but their presentation is beyond the scope of this book and the reader is therefore referred to [116] and the references presented there. Another group of the ions substituted are paramagnetic ions. In connection with them, the ions Mn, Fe and Ni have been investigated. In these cases electrostatic equilibrium is reached by simultaneous substitution of either Ge^{4+} or Si^{4+} ions. The Mn^{2+} ion occupies octahedral positions and contributes to the magnetization with each ion. However, on the other hand it weakens a–d interactions which results in the deflection of the moments of Fe^{3+} ions in d-positions (canting). It is obvious that Ni^{2+} ions preferably occupy octahedral positions and Co^{2+} tetrahedral positions. Each of them influences some magnetic parameters in a specific way, but detailed description beyond the scope of this book.

One of the shortcomings of LPE-YIG layers and devices which use them is the change of resonant frequency with temperature. This results from the temperature dependence of the saturation magnetization, which then results in the change of resonant frequency. In devices using massive YIG this effect can be compensated for by using spherical monocrystalline samples with an appropriate orientation with regard to the external field. In LPE layers temperature drift is compensated for either by external means (e.g. using feedback in a magnetic circuit, choosing permanent magnets with inverse temperature dependence, deflecting magnetic field) or appropriate substitution in YIG, layers which decreases temperature contribution from magnetization and in addition gives rise to the contribution which acts inversely through the mediation of induced anisotropy [145]. In spite of the problems we have mentioned, the low-loss LPE-YIG layers are prepared in a routine way. Their quality is sufficient for MSW applications and it is possible to prepare them with thicknesses up to 150 μm.

To end this Section we will mention briefly general trends in the preparation of thin layers not from YIG materials. Several experiments have been carried out to adapt the LPE technique to the growth of both spinel and hexagonal ferrite thin layers. Analogously with YIG layers on GGG substrates, non-magnetic crystals of the same structure as final ferrite have been tried as substrates. Up to now results are not satisfactory. Thin epitaxial films have been obtained but their quality is not sufficient for use in practical applications. It seems that these problems are caused mainly by the substrates used. Up to now the most common substrates for spinel ferrites have been monocrystals MgO, $ZnGa_2O_4$ or $Mn(In,Ga)O_4$. These substrates have not reached the quality of GGG. Another problem related to the spinel growth is that they prefer growth in the [111] "facet". This causes, at slow cooling, crystals to grow in the shape of octahedrons and if the substrate orientation differs from [111] orientation, then small octahedral crystals arise and on the substrates with [111] orientation hollows and terraces arise in contradistinction to the smooth surface of YIG garnets. For this reason, up to now Li ferrite layers of the highest quality have a resonance line width of the order of $100\,\mathrm{A\,m^{-1}}$ [146]. If we take into account that no special attention has been given to Li layers, then we can expect that if they are needed in practice, intense efforts will bring improvement of their quality. A similar situation is in the case of hexagonal ferrite thin layers which have potential applications in the millimetre wave frequency range. But the layers obtained have a relatively high value of resonance line width — about $3.5\,\mathrm{kA\,m^{-1}}$ at 60 GHz [147].

In the near future we can expect considerable efforts in preparation of not only monocrystalline but also polycrystalline thin ferrite layers because ferrite elements which are compatible with monolithic integrated circuit technology are being looked for. Current technological procedures in preparation of monocrystalline ferrite layers require high temperatures and chemically reactive surroundings which calls into question their possible combination with semiconductors such as GaAs. The possibility in this field seems to be to use epitaxial ferrite layers together with epitaxial GaAs layers on a suitable substrate. The epitaxy of GaAs and InP on GGG has been carried out already [148,149]. Another possibility is to use the so-called ferrite patching [150] which enables the deposition of ferrite layers from water solutions at not too high temperatures above room temperature. Promising results have also been obtained recently by a new emerging technique of laser pulse deposition. Of course other techniques are possible, but descriptions of them are beyond the scope of this brief summary.

7.3 Preparation of MSW excitation, receiving and control elements

7.3.1 Substrates used for the implementation of microwave integrated circuits

In principle the elements for MSW excitation, receiving and control are realized in two ways, either directly on the surface of a thin ferromagnetic (YIG) layer or producing the structure required on a different substrate and then putting the ferromagnetic layer on a structure. In both cases procedures of standard technology which are used for the implementation of microstrip hybrid integrated circuits, are employed. Taking into consideration that the relation between transducer and MSW is strong (in some cases it can cause undesirable effects) in practice it is often reduced by increasing the distance between the transducer and the magnetic medium. This can be done in a simple way when MSW control elements are separated from the ferromagnetic layer itself. For this reason we will pay attention to the methods of preparation of these structures for the latter case mentioned above. But the principles for the creation of control structures are similar in both cases.

The structures required are produced on an appropriate substrate. In this case the substrate is a dielectric wafer on which the system of conductors is placed and this creates the required electric circuit for MSW excitation, receiving and control. In estimating substrate suitability for microwave circuits it is necessary to realize that the substrate has to serve three different purposes. They are:

(a) mechanical support for conductors and other elements (including the ferrite layer) and simple mechanical coupling with the other parts of the device

(b) medium for transmission of electromagnetic waves

(c) technological medium for the implementation of conducting structures, resistors, capacitors and in the case of combination with semiconductors it is also the medium for the production of semiconducting structures.

Of course, it is necessary to take into consideration other mechanical, temperature and chemical criteria in order to choose an appropriate substrate. The demands put on the mechanical stability of the substrate are based not only on the necessity to build them safely into the holder, but also the influences of the surroundings in which these circuits will be operating must be partly taken into consideration. These further demands can be operation at high or low temperatures, resistance against temperature change, moisture, surrounding aggression, mechanical vibrations, mechanical quakes and exposition to radiation. In the case of its function as an electromagnetic field guiding structure the value of the relative permittivity ε_r of the substrate, dielectric loss factor represented by $\tan\delta_\varepsilon$ and the thickness h of the substrate play an essential role and they influence the basic parameters of the resultant line — wave impedance, wave propagation velocity and losses in the line.

From the technological point of view it is necessary to distinguish whether metallized substrates will be treated exclusively by photochemical etching or whether thin- or thick-layer technology will be applied to non-metal-coated substrates. In the former case it is sufficient for substrates to be resistant to a temperature of about 250°C, which is valid for many substrates made from plastics, e.g. teflon strengthened by glass fibres. In the latter case only inorganic materials are suitable (e.g. ceramics, glass), since for thin-layer technology temperatures up to 400°C are necessary for evaporation and sputtering. To secure good adhesion of layers, ceramic substrates are heated to a temperature of about 1000°C in order to remove impurities from the process of metallization. In the thick-layer technique temperatures of about 850°C are necessary to calcinate pastes for individual structures.

Besides the demands mentioned above we cannot forget economic factors which suggest simultaneous production in one process. It is noteworthy that an ideal material for substrates which would meet all the demands for all possible applications does not exist and that is why it is necessary to find an optimal compromise for each specific technical application.

From the material standpoint substrates can be divided into two groups: (a) plastics and organic materials and (b) inorganic materials. Organic materials are used in the pure form only occasionally, e.g. teflon, polyolefin. They have low permittivity $\varepsilon_r = 2 - 3$ and low loss factor $\tan\delta_\varepsilon$. But they are unstable mechanically. To suppress the mechanical instability, glass fibres, for example, are used as supports. In so doing ε_r increases only a little but losses increase considerably. In addition to that, the material becomes anisotropic. Another possibility is to build in as a support ceramic materials and in this way materials are obtained with permittivity $\varepsilon_r = 5 - 20$ which can be easily mechanically processed.

Inorganic materials are also divided into subgroups: ceramics, glasses, monocrystalline materials such as silicon and sapphire, ferrites and semiconductors. The most commonly used one is alumina (Al_2O_3 ceramics) which is the most important substrate for thin-layer technology (from the standpoint of microwave hybrid integrated technology) and that is why it is also the most used substrate material for MSW excitation, receiving and control structures. Sapphire is a monocrystalline form of Al_2O_3. It is characterized by a surface of high quality, low loss factor $\tan\delta_\varepsilon$ and a very low dispersion of relative permittivity ε_r. Unfortunately, it is dielectrically anisotropic. The reason for the wide application of alumina as a construction element in microwave integrated circuits is its relatively simple production and a number of applications in a wide frequency range. For this reason, nowadays, alumina has become one of the most important materials for substrates. It is produced with thicknesses 0.254, 0.508, 0.635, 1.27 and 2.54 mm and size 25.4×25.4mm (1×1inch) or 101.6×101.6mm (4×4inches) by a foil-pulling process. In this process first 99.9% pure Al_2O_3 is mixed with a small amount of magnesium and silicon. This material is ground, then solvent and binding agent are added and a thick but still liquid mass is produced which is poured into the porous mould. After drying a flexible foil arises (so-called green ceramics) from which plates of the size required are obtained. Then these plates are calcinated to final hard ceramics. Magnesium oxide within ceramics produces a glassy mass among grains of Al_2O_3 on the surface which secures the adhesion needed for the processes of thin-layer technology. The resultant material is 99.9% alumina ceramic with relative permittivity $9.8 \pm 2.2\%$ (influenced by porosity of ceramic), $\tan\delta_\varepsilon \simeq 0.0001$ (at 25°C and 10 GHz) and specific resistance $> 10^{12}\Omega$m.

7.3.2 Preparation of control structures

To implement the control structures required, which are represented by the elements of hybrid integrated circuits, various techniques can be used. They differ from each other by the material used for substrates and by the technological procedures employed for the implementation of conductors, resistors, capacitors and also by the density of integration. In the case of alumina substrates either thin-film or thick-film technology is used. We will describe them only briefly. We would like to remark that thin- or thick-film technology is defined with regard to the technological procedures used for implementation of conductors, capacitors and resistors and without regard to the thickness of the film created.

7.3.2.1 Preparation of conducting structures using thin-film technology

Nowadays, the most important technology for implementation of hybrid high-frequency integrated circuits is thin-film technology (see e.g. [151, 152]). It makes use of the processes of sputtering and evaporation which are carried out in vacuum, and using them films with thicknesses up to 1 μm can be realized. We are interested most in the implementation of conductors. They must meet the following criteria: low specific resistance, conductor thickness $t > 3\delta$ where δ is the skin depth, good accuracy of edges, resistance against oxidation and aggressive gases, as e.g. SiO_2, solderability and the ability to make various types of contacts, good adhesion to substrates in the case of mechanical strain and resistance against the ageing process. In the case of multifilm conductors there is an additional demand for a weak mutual effect and negligible diffusion of conductors. Various types of conductor materials used in thin-film technology are shown in Table 7.2. As follows from this table, all good conductors have bad adhesion to alumina substrate. That is why in order to create well-conducting and adhesive films with thickness 5–10 μm it is necessary to deposit a very thin holding layer of Cr, Ta or Ti on the surface and then to deposit the conducting film (by evaporation or sputtering, with thickness about 0.1 μm) of Ag, Cu, Au or Al, which is further strengthened galvanically to the required thickness (5–10 μm). But it is necessary to realize that diffusion between individual films can occur and this results in a decrease of electric conductivity. In order to prevent this, combinations of such metals are chosen which mutually diffuse only a little, or another intermediate film is added, e.g. Pt, to suppress undesirable diffusion. The thicknesses of the holding layer and intermediate layer are to be so small (about 0.05 μm) that the increase of attenuation resulting from the increase of resistance is negligible. Among conductors priority is given to gold because its resistance against influences of the surroundings is the highest and it is suitable for all types of contacts.

Table 7.2

Conductors	Relative specific resistance compared with Cu	Penetration depth at 2 GHz (μm)	Coefficient of temperature expansion (μm m^{-1} °C^{-1})	Adhesion to Al_2O_3
Ag	0.95	1.4	21	bad
Cu	1.0	1.5	18	bad
Au	1.36	1.7	15	bad
Al	1.6	1.9	26	bad
	Materials for adhesive intermediate films			
Cr	7.6	4.0	8.5	good
Ta	9.1	4.5	6.6	good
Ti	27.7	7.8	9.0	good
	Materials for separating films			
Pt	6.2	3.6	9.0	
Pd	6.2	3.6	11.0	

The technological procedure itself begins with the production of an adhesive layer on which a thin conducting film is deposited. These films are deposited on the substrate. Two procedures are used — evaporation technique and cathode sputtering. Both of them work in vacuum (evaporation in high vacuum — pressure ca 10^{-4} Pa, sputtering in low vacuum — pressure ca 1 Pa). The evaporation technique is depicted schematically in Fig. 7.6a. The material evaporated is put into the electrically heated holder which must be from a material (usually tungsten or molybdenum) whose melting temperature is above the evaporation temperature of the material evaporated. In high vacuum (10^{-4} Pa) the material is heated to a temperature of about 1000°C. It is evaporated and particles produce the vapour deposit on the substrate after straight-line motion.

In the case of cathode sputtering, whose schematic depiction is given in Fig. 7.6b, a protective cylinder is filled with inert gas (e.g. argon). In the working space there is a system of electrodes which serves as anode and cathode and it is fed with a voltage of several thousands volts which causes a glow discharge. In this way a plasma of Ar^+ ions is created between anode and cathode. The substrates on which films are to be deposited are placed on the anode. Ar^+ ions are accelerated to the cathode and they fall and strike it with the velocity gained in the cathode fall region. The cathode (target) is made from material which is to be sputtered. When the cathode is bombarded with Ar^+ ions the material particles from the cathode are released; they move in the space between the anode and cathode and deposit on the substrate. The amount of material sputtered per unit time under constant conditions is inversely proportional to the gas pressure and the distance between the anode and cathode and as in the evaporation technique it is not very big. That is why both basic films — the adhesive one and the basis for the conducting film are made up to thicknesses 0.01–1 μm and the thickness needed for microwave applications is obtained by subsequent electrodeposition of the conducting film. In the last stage, to create the structure required, photolithographic procedures using so-called photoresists are used. The structure required is produced using a photomask. Photomasks are films or glass plates on which the structure which is to be created on the substrate to the scale 1:1 is depicted with high accuracy (deviation ca 1–5 μm). The masks are made either by hand or automatically. When they are ready, it is possible to realize the structure required on the substrate. In the process either a negative procedure (when the surface except the conductors required is covered) is used or a positive procedure (when conductors are covered) together with the technique of so-called selective etching which consists in etching the film from radiated (unradiated) material by an appropriate agent while the film from unradiated (radiated) material remains intact. This description is, of course, simplified very much. Problems arising in connection with keeping as high an accuracy of the edges as possible, are solved using special procedures and the reader can obtain further information in the specialized literature [153]

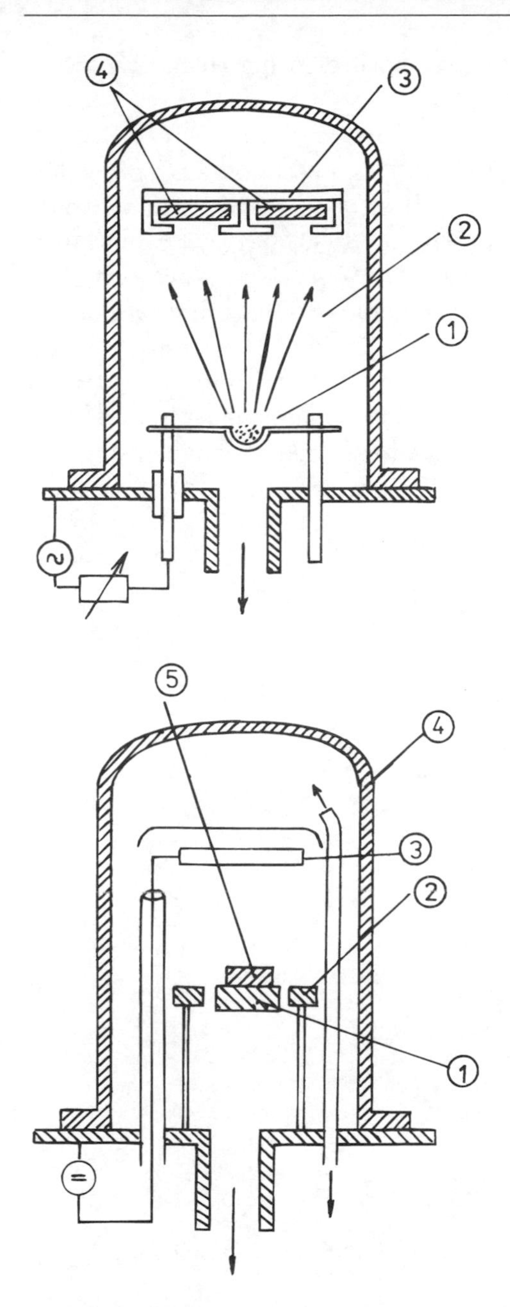

Fig. 7.6. Thin-film technology a) evaporation method, 1 — evaporated material, 2 — vacuum $\simeq 10^{-6}$ Torr, 3 — substrate holder, 4 — substrate, b) cathode sputtering, 1 — substrate holder (pre-heated), 2 — anode, 3 — cathode, 4 — recipient, 5 — substrate.

7.3.2.2 Preparation of conducting structures using thick-film technology

The thick-film technique [154, 155] is a technological procedure for implementation of conducting structures, resistors and capacitors on inorganic substrates, mostly on Al_2O_3. Here, it represents an alternative technique to the thin-film one. There are two standard procedures in the thick-film technique. The first one is the technique of sieve pressing (or direct deposition of pattern) and the second one the etching of thick films. The first one is used more frequently and is not very expensive. Its principle is as follows. The basic material in the form of a paste is pressed through a sieve (or is deposited directly in the form of the pattern required) on the sub-

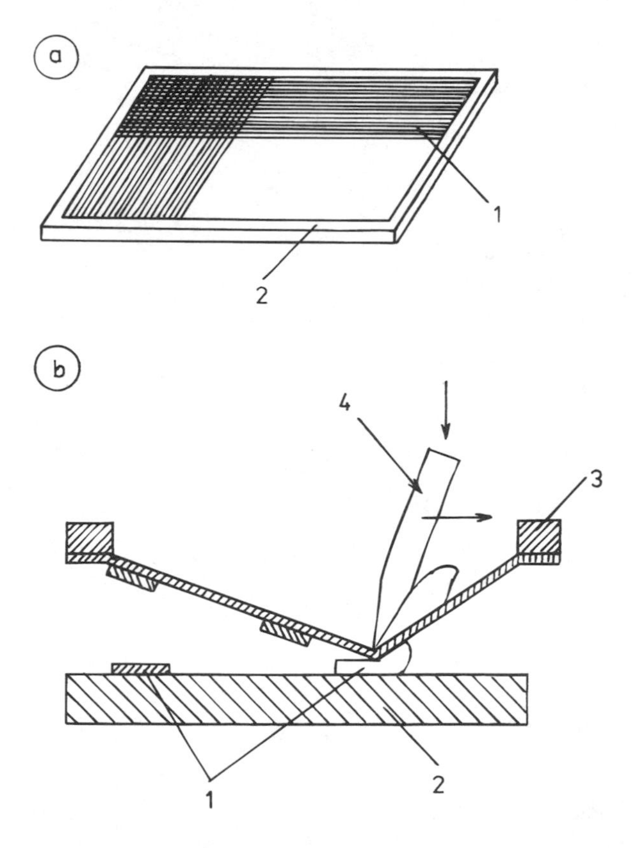

Fig. 7.7. Thick-film technology a) 1 — sieve, 2 — frame, b) printing by paste on a substrate, 2 — created conductive structure, 2 — substrate, 3 — sieve with frame, 4 — spatula.

strate. The grid of very fine plastic or metallic fibres is tightened on the frame (Fig. 7.7). The grid contains 50–150 fibers per cm. On this sieve, by means of photosensitive emulsion, the picture of the required conducting pattern which is to be deposited on the substrate is created (by exposure of the emulsion through the mask with the structure required). After exposure the exposed parts are washed out and the sieve has free regions in which conductors are to be placed. Then the sieve is put on a ceramic film and using a spatula conducting paste is pressed through the prepared pattern on the sieve on the substrate. In this way conductors are deposited on the substrate. The paste deposited is dried at a defined temperature profile at temperature 100–200°C and then calcinated at temperature 700–1000°C. Conducting pastes are mixtures of metal, glass powder, organic solvent and binding agent. As a matter of course gold, gold–palladium, silver–palladium and copper pastes are manufactured for conductors. As has been mentioned already this technique is comparable with the thin-film technique but there are some differences: in particular, lower investment in technological equipment is needed, running costs as a result of the smaller number of operations in the course of manufacture are lower, productivity is higher and final conductors are thicker (about 25 μm and more). The disadvantages are that the least practically reachable gap between conductors is 100–200 μm, edges have an accuracy of only 20–30 μm (as a result of the grid structure) and there are restricted possibilities of contacting by soldering and sticking (except gold pastes). The thick-film technique can be used only in a relatively low-frequency range (up to now ca 20 GHz) in contrast to the thin-film technique which is applicable for frequencies up to $\simeq$ 50 GHz. For the broad-band excitation of all types of magnetostatic waves (surface and volume MSW) a narrow microstrip transducer (10 μm) has to be used. In narrow-band devices meander, interdigital and mesh type transducers are used. All types can be realized with impedances currently used in microwave device technology, i.e. in the range 25–75 Ω, in relatively broad frequency bands. To reduce contingent losses which result from mismatch, matching circuits are realized. By considering the advantages and disadvantages of all the techniques mentioned above, one of them can be chosen for a specific technical application.

7.4 Permanent magnets and electromagnets

Ferrite microwave devices for most applications require the presence of an external magnetizing field. That is why the realization of an external magnetic circuit is an inseparable part of the design of ferrite devices operating in certain frequency bands. The realization of the external magnetic circuit is usually massive, which considerably increases the device size and complicates (at least up to now) efficient integration of non-reciprocal ferrite elements, in spite of their unique properties, into monolithic integrated circuits that will predominate in the frequency band from 1 to 300 GHz in the coming 10–15 years. That is why ways are being looked for as to how external magnetic circuits in monolithic integration can be implemented, saving all the advantages of the ferrite materials used. It is noteworthy that nowadays many prominent laboratories all over the world participate in research and development concerning this field and they are intent on applying non-reciprocal ferrite elements in microwave circuits. Mainly we will address the possibilities of realization of magnetic circuits in hybrid integration when the corresponding non-reciprocal ferrite element is an independent part of a more complex device. Predominantly we are interested in a special group of these devices which take advantage of MSW. Problems concerning the production of external magnetic circuits are, of course, more complex, but in principle they are the same for all types of non-reciprocal microwave devices which take advantage of the properties of gyrotropic media. In spite of that we will outline concisely the approaches which appear in connection with monolithic integration. As follows from the previous chapters, an external magnetizing field is one of the crucial external parameters which is present in dispersion relations for all types of excited MSW. A unique property of microwave ferrite devices is just the possibility of their "retuning" in a relatively wide frequency range by changing the external magnetic field. Although the velocity of this "retuning" is relatively low (a magnetic circuit represents a strong inductive load with corresponding time constants), for many technical applications it is possible, using an appropriate combination of stationary magnetization by permanent magnet and an auxiliary electromagnetic circuit, to obtain sufficient operating velocities near the operating point (of the chosen frequency).

The last but not the least important demand put on the external magnetizing circuit is its ability to compensate for undesirable changes of properties of the ferromagnetic material used which result from the change of temperature conditions. The ability of "retuning" mentioned above can be used in this case to finish the adjustment of the state required. It can be done by choosing appropriate parts for the external magnetic circuit, e.g. permanent magnets, in such a way that their temperature characteristics compensate automatically for e.g. changes of magnetization caused by the change of temperature of the ferromagnetic material, by a corresponding change of magnetizing field. Feedback circuits with corresponding sensors regulating the current in the electromagnet used are also employed. These solutions require the properties and corresponding characteristics of the hard- and soft-magnetic materials used to be known as well as a basic knowledge about magnetic circuits. We will pay attention to this in the following sections.

7.4.1 Hard-magnetic materials

High-frequency applications of ferrites require relatively high intensities of external magnetic field. For elements operating near ferromagnetic resonance (e.g. in devices taking advantage of MSW), in the first approximation Kittel's form is valid for resonant frequency [2], in which the geometry of the sample used is taken into account by means of demagnetizing factors. For common applications, e.g. in the X band, the magnetic field intensity is of the order of $100\,\mathrm{kA\,m^{-1}}$ and depends on the parameters and shape of the ferromagnetic material used. When we take into consideration that in realizing the magnetic circuit we try to minimize its size we come to the conclusion that the parameters required can be obtained only using hard-magnetic materials (permanent magnets) of last generations with high so-called energetic products $(BH)_{\mathrm{max}}$. Intermetallic compounds based on cobalt and metals of rare earths, and alloys of systems $SmCo_5$ or NdBFe characterized by high uniaxial magnetocrystalline anisotropy, high saturation magnetization and reasonable Curie temperature are included in high performance ferromagnetic materials of last generation with extraordinarily high values of remanent induction B_r, coercivity H_c (or H_{cB} — coercivity determined from hysteresis loop at $B = 0$) and maximum value $(BH)_{\mathrm{max}}$.

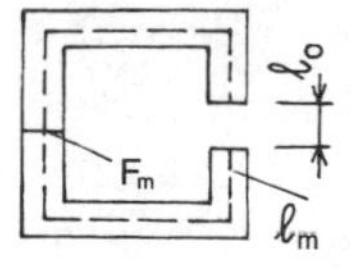

Fig. 7.8. Simple magnetic circuit with permanent magnet.

To elucidate why the energy product is so important let us consider, as an example, a simple magnetic circuit of toroidal shape with cross-section F_m, length l_m, remanent induction B_r and air gap l_0 (Fig. 7.8). If we neglect the edge effects, then the induction B_0 in the air gap will be the same as in the sample. From Ampere's law

$$-H_m l_m + B_0 l_0/\mu_0 = 0 \tag{7.1}$$

and from this

$$l_m = B_0 l_0/(\mu_0 H_m) \tag{7.2}$$

Since the volume of the magnet is $V_m = l_m F_m$ and $B_0 = B_m$, if the stray flux is neglected (small air gap), then for the volume of the magnet the following holds:

$$V_m = l_m.F_m = (B_0 l_0 F_m)/(\mu_0 H_m) = (B_0^2 l_0 F_m)/[\mu_0(H_m B_m)] \tag{7.3}$$

From (7.3) it follows that for given parameters B_0, l_0, F_m the volume of the permanent magnet V_m will be a minimum if the product $B_m H_m$ is maximum. This product is a measure of the appropriateness of the use of a certain permanent magnet in magnetic circuits. The historical development of the highest $(BH)_{\max}$ which have been successively achieved in the materials used in practice, is shown in Fig. 7.9, in which hard-magnetic materials are marked by numbers 1–3, alloys AlNiCo by 4–8, hard-magnetic ferrites by 9–13, alloys Fe–Cr–Co by 14 and 15 and Mn–Al–C materials by 16. From the early seventies the highest values of energy product $(B{\cdot}H)_{\max}$ have been provided, as has been mentioned, by materials based on rare-earth metals. Their development continues and materials of $R_2(Co,Fe)_{17}$ type, in which R is an element from the rare earth series. In particular, the newest group of materials of the R–Fe–B type, which are overcoming current record values, are the most auspicious. Research and development on materials whose parameters are below the connecting line of the highest values continue, as they are based on cheaper and more accessible elements.

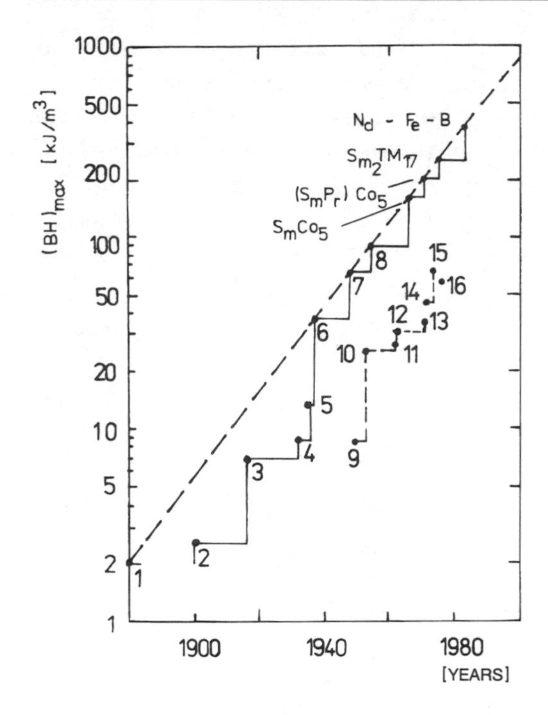

Fig. 7.9. Historical development of maximal values of the energy product.

7.4.2 Reasons for magnetic hardness

Although the demands put on materials for permanent magnets can be changed in detail, the basic demand — magnetic hardness of the material characterized above all by coercivity H_{cB} — remains. Let us imagine a magnet composed of a set of one-domain particles. If these particles are oriented in space randomly, then the resultant remanent induction B_r is equal to $J_S/2$ (J_S is the saturation magnetization). But if these particles are ordered in such a way that their easy magnetization directions are parallel, then the remanent induction B_r increases to $B_r = J_S$. So, the orientation of particles or grains influences significantly H_c and B_r of the whole system. Doing more detailed analysis we can find out that magnetic hardness is always connected with the presence of some kind of anisotropy which prevents the arbitrary ordering of magnetic moments and accordingly prevents the demagnetization of ferromagnetic materials. We can claim then that without anisotropy, permanent magnets would not exist. Various kinds of anisotropies can work, but the most frequent ones are shape and crystallographic anisotropies.

As follows from the analysis of B–H characteristics, materials with a high product $(BH)_{\max}$ have to have high values of H_c and B_r at the same time. Coercivity depends on several factors, e.g. non-magnetic inclusions present in magnetic material (Kondorsky's model [156]), but predominantly on various anisotropies. A magnetization process is carried out by either domain wall displacement or the rotation of the vector of magnetic polarization. From knowledge of the magnetization process in magnetic materials it follows that high values of H_c cannot be achieved if magnetic reversal by domain wall displacement prevails. That is why for permanent magnets it is necessary to exclude this process as much as possible and to force the magnetization by the rotation of the M vector, for which a crucial factor is magnetic anisotropy. Now, at least briefly we will present possible contributions to the magnetic anisotropy. The contribution of magnetoelastic anisotropy will be

$$H_A^{\sigma} = c\lambda_s\sigma_i/\mu_0 M_0 \tag{7.4}$$

where λ_s is the coefficient of shape magnetostriction, σ_i internal tension and c a function dependent on the magnetization direction. Substituting the usual values of these parameters in (7.4) we will find out that this contribution to the anisotropy field is too small to be a crucial factor for permanent magnets. Materials with shape anisotropy are characterized by higher coercivity. Let us assume that the material is a slim bar in the shape of an elongated rotational ellipsoid with the longer axis in the x-axis direction; then the shape anisotropy field will be

$$H_A^F = M_s(N_z - N_x) \tag{7.5}$$

where N_x, N_z are demagnetizing factors in the x-, z-axis directions, respectively. E.g. for small size needles which can be considered to be one-domain formations, $N_x = 0$, $N_y = N_z = 1/2$ and $H_A^F = M_0/2$. In the case that magnetization is carried out, as we assume, by the coherent rotation of the magnetization vector, then the proper coercivity H_{cJ} will be equal to H_A^F. From this it follows that even in this case extreme values cannot be reached.

It seems then that only materials with high crystallographic anisotropy can fulfil the demands put on an ideal magnet. However, to secure the correlation between the proper coercivity and the field of crystallographic anisotropy, i.e. $H_{cJ} = H_A^K$, the permanent magnet must consist of fine particles of "subcritical" size (one-domain formations) oriented parallel to the easy magnetization direction. Only under these conditions can the magnetization process be carried out by the coherent rotation of the magnetic moment vectors. The description of individual physical processes, from which the relation between coercivity and the magnetization mechanism can be seen, exceeds the scope of this book and the reader can obtain further information about them in relevant publications (e.g. [157]). However, it is obvious that the demands mentioned above are very heavy and they can cause technological problems in permanent magnet manufacture. Since the crucial factor is the size of particles, in the production of these materials methods of powder metallurgy are used or new techniques, such as rapid cooling of the melt, originally developed for manufacturing of amorphous soft-magnetic materials. Nowadays the demands mentioned above (high anisotropy field) are fulfilled by intermetallic compounds of elements from the rare-earth series. We cannot present here the whole range of materials prepared so we will pay attention only to the most important representatives and their properties.

7.4.3 Permanent magnets based on cobalt and rare-earth metals

Intermetalloids of group RCo_5

This group of magnets has been developed as the first one and up to now it is the most used group of magnets based on rare earths. Equilibrium diagrams of rare-earth metals are similar to each other and there are phases whose formulae are stoichiometrically identical. This results from the fact that rare-earth metals have similar chemical properties and the ionic radii of rare-earth metals have approximately the same values $\cong 1.8 \times 10^{-8}$m. The highest magnetocrystalline anisotropy among all possible phases has the phase RCo_5 which crystallizes in the $CaZn_5$ type hexagonal structure and the easy magnetization axis is the c crystallographic axis. A typical representative, for which the best results have been achieved, is the alloy $SmCo_5$ with average magnetic properties: $(BH)_{max} \simeq 150$–$180\,\mathrm{kJ\,m^{-3}}$ and $H_c = 1.2\,\mathrm{MA\,m^{-1}}$, $T_c = 720°\mathrm{C}$. These values are dependent on the technological procedure (structure) and up to now the highest value for this compound $(BH)_{max} = 224\,\mathrm{kJ\,m^{-3}}$ is pre-

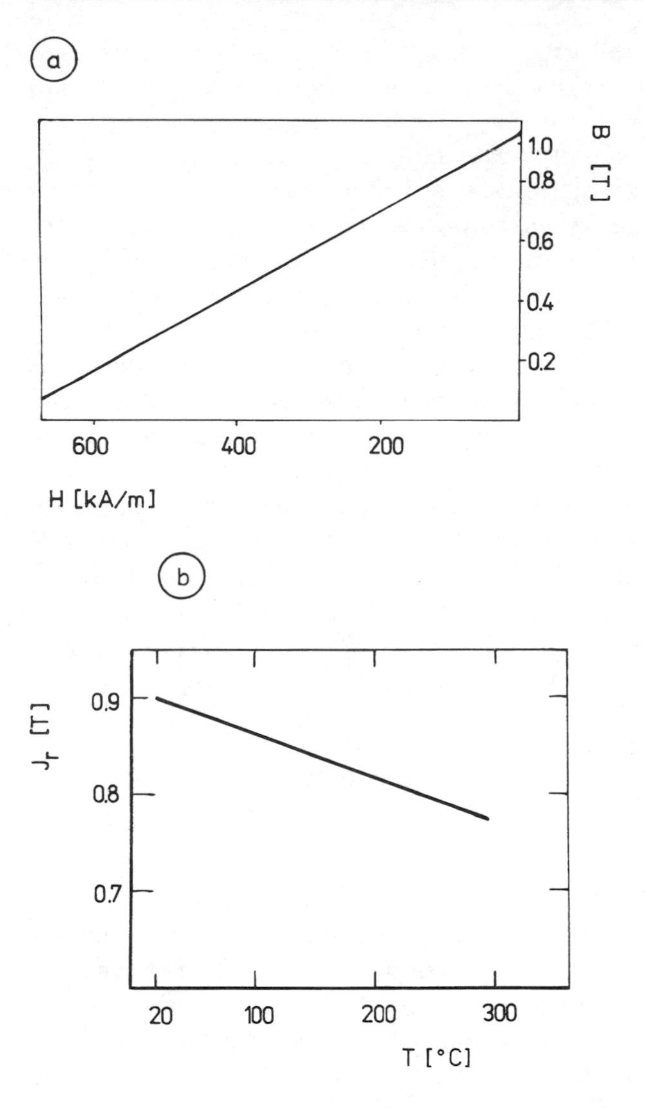

Fig. 7.10. Permanent magnet $SmCo_5$: a) demagnetization characteristic, b) temperature dependence of magnetic polarization.

sented in [158]. The corresponding demagnetization characteristic is in Fig. 7.10a and the temperature dependence of saturation magnetization is in Fig. 7.10b.

Magnets of R_2T_{17} *type*

These magnets have been developed from R_2Co_{17} phases as a further generation of permanent magnets based on rare-earth metals. Similarly, as in magnets of RCo type, in these the element samarium has an exclusive position and it is present in all compounds practically used. At room temperature, the phase Sm_2Co_{17} shows, relatively in comparison with other phases of R_2Co_{17}, the highest uniaxial magnetocrystalline anisotropy with K_u value $K_u = 3.52\,\mathrm{MJ\,m^{-3}}$, saturation magnetization $M_0 = 1.024\,\mathrm{MA\,m^{-1}}$ and Curie temperature $T_c = 922°\mathrm{C}$ [159]. It turned out, however, that the phase R_2Co_{17} itself is not suitable for preparation of sintered permanent magnets. Suitable alloys for preparation of sintered magnets have been found by substituting a part of the cobalt by iron and in order to increase the coercivity a small amount of other elements (e.g. Cu, Zr, Ti, Nb and other) has been added. Using procedures of powder metallurgy permanent magnets of these compositions and with higher energy product than $SmCo_5$ are manufactured. It is assumed that later, by finding out the optimum composition and heat treatment, values up to $320\,\mathrm{kJ\,m^{-3}}$ will be achieved [160].

Magnets of R–Fe–B *type*

Development of these magnets has been stimulated by a considerable increase in the price of Co. In the field of R–T alloys, interest has been focused on the alloys of R–Fe type (T represents an element from the 3d transition metals). Alloys $Nd_2Fe_{14}B$ and $Pr_2Fe_{14}B$ have the most favourable properties for use in the manufacture of permanent magnets. B is added as a glass-making element. In 1984 Sagawa et al. [161] presented a new way of manufacturing R–Fe–B magnets using powder metallurgy; magnets had energy product $(BH)_{max} = 290\,\mathrm{kJ\,m^{-3}}$ ($B_r = 1.23\,\mathrm{T}$, $H_c = 960\,\mathrm{kA\,m^{-1}}$, $T_c = 312°\mathrm{C}$) while predicted average values $(BH)_{max}$ are 330–340 $\mathrm{kJ\,m^{-3}}$. The major disadvantage of R–Fe–B permanent magnets — low T_c value and from this a resulting disadvantageous temperature dependence — can be eliminated by substituting iron for part of the cobalt [161]. The demagnetization characteristics are depicted in Fig. 7.11a. The R–Fe–B permanent magnets are expected to become the most important kind besides cheap ferrite magnets, since their parameters have high values and raw materials for their manufacture are relatively easily accessible. However, in the near future a number of problems have first to be solved, e.g. improvement of temperature stability, new magnetic circuit designs, economic problems, etc.

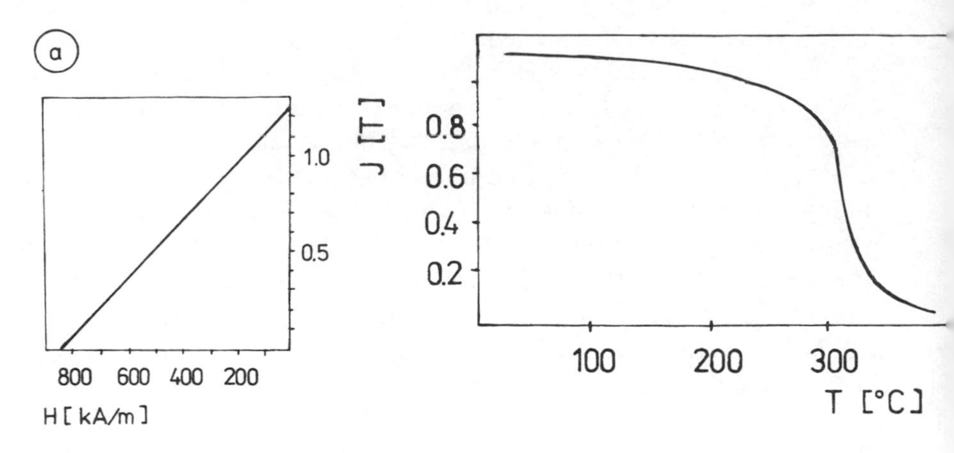

Fig. 7.11. Permanent magnet of R–Fe–B type. a) demagnetization characteristic, Neomax 30 H, b) temperature dependence of magnetic polarization, material $Nd_{15}Fe_{78}B_7$.

7.5 Thin film permanent magnets

In connection with possible integration of ferrite elements in monolithic integrated circuits, new ways of manufacturing high quality permanent magnet layers are being looked for. F. J. Cadieu [162] used special sputtering methods to synthesize magnetic layers with high coercivity, high saturation magnetization and special anisotropies. If various anisotropies are present, then the axis of easy magnetization in different layers can be oriented in or outside the plane of a layer or it can have a significant direction in the plane of a layer. High coercivity values were achieved using intermetallic alloys mentioned in the previous chapters. It was found out that magnetic properties are extraordinarily sensitive to the texture of the layer. So, e.g. for the composition $Nd_2Fe_{14}B$, layers with easy axes both in the plane of the layer and perpendicular to the plane of the layer were realized depending on the magnitudes and signs of the crystallographic anisotropy constants. The energy product reached the value $168\,kJ\,m^{-3}$ without any further treatment. A possible configuration of the magnetic circuit with a thin film permanent magnet is depicted in Fig. 7.12. The volume of the air gap is determined by the length, width and distance between the two thin layers with large coercivity in the plane of the layer. In this configuration two

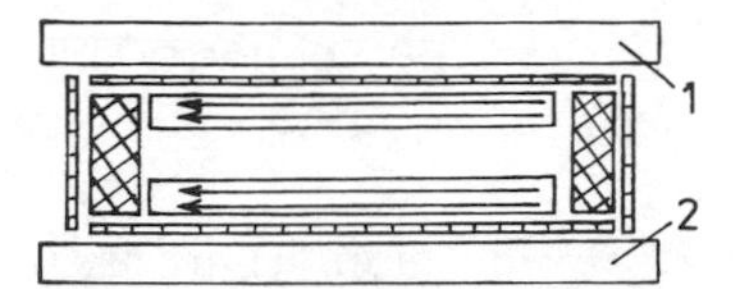

Fig. 7.12. Geometrical arrangement of magnetic circuit using thin film permanent magnet. 1, 2 substrates. Hatched areas represent a magnetic circuit for magnetic field formation.

layers of permanent magnet are magnetized in the same direction, which makes it easy to magnetize layers in the course of manufacture. Closing the magnetic layers enables the concentration of the magnetic field in the gap of the magnet. The magnetic field created in the gap can be extraordinarily uniform, with induction values up to 1–2.5 T.

7.6 Design procedure for circuits with permanent magnets

When designing a circuit with a permanent magnet the following characteristics of this magnet are of great importance: the shape of the demagnetization characteristic, i.e. the part of hysteresis loop in II quadrant, remanent induction B_r, coercivity H_{cB} (or H_{cJ}) and maximum energy product $(BH)_{\max}$. The value of remanent induction determines primarily the cross-section of the permanent magnet needed to reach the required induction in the air gap. The ability to demagnetize the permanent magnet depends on the coercivity, and the energy product is crucial for the volume of permanent magnet for a given application. The progressive permanent magnets mentioned in the previous sections are characterized by the linearity of their demagnetization characteristics in the II quadrant. The magnetic circuit has to be designed in such a way that the operating point of the permanent magnet has an optimum position on the demagnetization curve. The properties of materials mentioned already differ so distinctly from the properties of classical ferrite and AlNiCo materials for permanent magnets that it was necessary to find new approaches for the design of magnetic circuits. The use of a correctly designed magnetic circuit can be a significantly greater task than only the use of a maximum energy product and the subsequent size reduction. Construction of devices using these materials is simplified considerably and much better characteristics can be achieved than by using classical materials. Reliability, performance and efficiency also increase.

In magnetic circuits for high-frequency applications there is, in most cases, a combination of a permanent magnet and an electromagnet. It is obvious that individual specific circuits demand a special design of the whole magnetic circuit corresponding to the given demands (above all spatial demands) but in spite of that basic principles can be elucidated using a simple circuit depicted in Fig. 7.13a. When analyzing we will assume that the magnetic circuit is outside

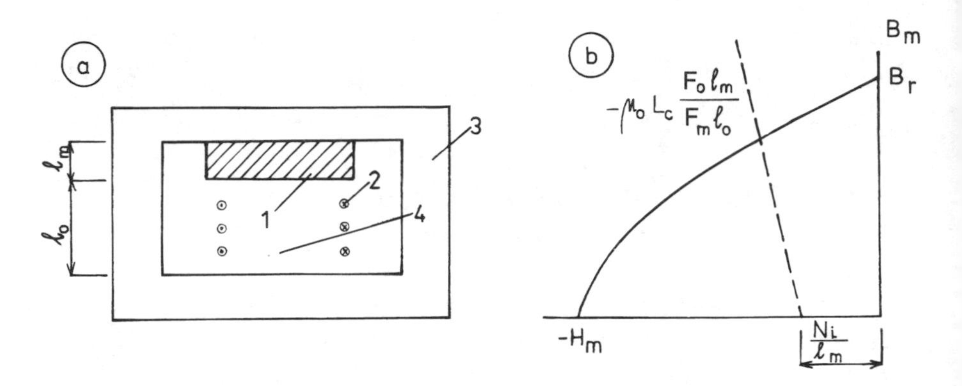

Fig. 7.13. a) Magnetic circuit scheme, 1 — permanent magnet, 2 — coil, 3 — magnetic yoke, 4 — air gap, b) its graphical solution.

the permanent circuit, coil and air gap and magnetically is ideally conducting, i.e. the magnetic core has infinite permeability. Then applying Ampere's law and considering continuity of magnetic flux we get

$$Ni = H_m \cdot l_m + H_0 \cdot l_0 \tag{7.6}$$

$$B_m \cdot F_m = B_0 \cdot F_0 \tag{7.7}$$

where H_m, B_m and H_0, B_0 are the magnetic field intensity and induction in the permanent magnet and the air gap, respectively, and l_m, F_m and l_V, F_V represent the length and cross-section of the permanent magnet and the air gap, respectively. If we introduce the stray coefficient k_0, then $F_0 = k_0 S_m N$. N is the number of turns of the coil and i is the current flowing through the coil. In the air gap

$B_0 = \mu_0 H_0$ and using (7.6) and (7.7) we get

$$B_m = \mu_0 k_0 \left[\frac{Ni}{l_0} - \frac{H_m \cdot l_m}{l_0} \right] \tag{7.8}$$

If the demagnetization characteristic is known, then equation (7.8) can be solved graphically (Fig. 7.13b) or numerically. The current change causes the change of position of the operating point on the demagnetization characteristic and consequently the change of induction in the air gap. For dynamical application the position of the operating point is determined by the demands put on the change of induction in the air gap with respect to the maximum current, the rate of change of magnetization and also by the demand to optimize the size of the whole magnetic circuit. Starting from equations (7.7) and (7.8) and other boundary conditions the problem can be solved by using a computer, while we can take into account the finite permeability of the soft-magnetic parts of the circuit. In principle it is necessary to realize that the energy product is not the only criterion for the use of permanent magnets but we must take into consideration also other demands put on the resultant magnetic circuit.

In conclusion, let us remark that there are efficient and comprehensive professional computer program tools for the solution of the shape of the field in the air gap for various configurations of soft-magnetic parts of the circuit and they also provide possibilities to optimize circuits with respect to given demands. Some details can be found in e.g. [163].

Chapter 8

HIGH-FREQUENCY MSW ELEMENTS

Before we proceed to the description of specific applications we would like to make two remarks.

First of all it is necessary to emphasize that although there is a theoretical possibility of implementing the interaction of MSW with active media (flows of charge carriers) in order to amplify MSW, up to now this has not achieved practical results. All investigations done until now have been in the field of implementation of passive high-frequency elements with the trend to modify the parameters of MSW elements, as e.g. frequency band, phase of device output, delay, phase and group dispersion (forward, inverse, none), noise suppression, etc.

The second remark has to do with work materials (ferrites). As has been mentioned several times, almost all experimental results have been obtained using thin YIG layers. Other ferrite materials (spinels, hexagonal ferrites etc.), although some of their parameters are better than YIG ones (broader frequency band or higher internal fields), cannot compete with YIG because of losses. That is why all information in this chapter concerns the use of thin YIG films.

First, we will pay attention to linear MSW elements, i.e. elements whose parameters do not depend on the signal amplitude. Filters, delay lines and magnetostatic resonators are included in them. All these elements represent MSW transmission lines with various characteristics. Nonlinear applications will be dealt with at the end of this chapter.

8.1 Properties of the pass band and dispersion of MSW transmission lines

First, let us examine the band properties of MSW transmission lines, i.e. we will evaluate the possibilities of various types of MSW from the standpoint of their use in high-frequency bands. The question is that MSW in magnetically ordered matter have no limitations in principle up to frequencies given by $\lambda \cong a$ (a is the lattice constant). However, the situation in practice is different.

As has been presented in Chapter 2, MSW transmission lines are band filters in principle, i.e. they transmit a high-frequency signal in a defined frequency band. This frequency band is determined both by the external magnetic field $\omega_H = \mu_0 \gamma H$ and by the magnetic material properties characterized by means of $\omega_M = \mu_0 \gamma M_0$. Volume MSW exist in the band $\mu_0 \gamma H \leq \omega \leq \mu_0 \gamma \sqrt{[H \cdot (H + M_0)]}$, surface MSW in the band $\mu_0 \gamma \sqrt{[H \cdot (H + M_0)]} \leq \omega \leq \mu_0 \gamma \cdot (H + M_0/2)$ for a free ferrite slab, and $\omega \leq \mu_0 \gamma \cdot (H + M_0)$ for a metal-coated layer (let us note that the last case is not interesting in practise because in metal Joule losses increase intensively). On increasing the frequency and subsequently changing the external magnetic field, surface and volume MSW have different band characteristics. From theoretical relationships of the operating frequency with the external magnetic field for surface and volume MSW in a free thin YIG layer ($M_0 = 139\,\mathrm{kA\,m^{-1}}$) it follows that with increasing frequency the surface MSW band in a free slab narrows and by contrast for increasing frequency the volume MSW band broadens. From dependences analogous to the dependences in Fig. 5.7 it follows that MSW excitation and propagation begin approximately at 1 GHz. At the frequency $f_{op} = 1.5\,\mathrm{GHz}$ the volume MSW band represents (at $H = 28\,\mathrm{kA\,m^{-1}}$) about 1.5 GHz and the surface MSW band (for fields $< 24\,\mathrm{kA\,m^{-1}}$) also about 1.5 GHz. At $f_{op} = 3\,\mathrm{GHz}$ the volume MSW have a band width about 2 GHz ($H \cong 64\,\mathrm{kA\,m^{-1}}$) and the surface MSW about 1 GHz (at $H \cong 32\,\mathrm{kA\,m^{-1}}$). At $f_{op} = 6\,\mathrm{GHz}$ the volume MSW have a band width $\simeq 2.4\,\mathrm{GHz}$ and the surface MSW about 0.5 GHz. At $f_{op} = 10\,\mathrm{GHz}$ for the volume MSW we get a band width of 2.5 GHz and for the surface MSW only 0.25 GHz. On increasing the frequency further, the band width of volume MSW remains almost the same and the band width of surface MSW is, in the limit, close to zero. So, this is the behaviour of the maximum possible band characteristics of the free transmission line for surface and volume MSW. However, such estimates have a relatively low practical value. In practice we are usually not interested in the whole transmission band of the transmission line but in the band in which the dispersion characteristic has a certain defined shape. Explicitly, the lines we are concerned with have to serve as tunable delay lines. Workers dealing with the problems of the design of such lines are interested in delay lines (DL) with constant delay in the transmission band or with delay that changes linearly (increases or decreases). Estimation of MSW properties made from this standpoint results in slightly different results. Let us make again some introductory remarks. The line signal delay is determined by the magnitude of the group velocity and along the length L is given by the form $\tau(\omega) = L/v_g(\omega)$, where $v_g(\omega) = \mathrm{d}\omega/\mathrm{d}k$ is entirely determined by the dispersion dependence $\omega(k)$. If the relation $\omega(k)$ is linear in the operating band, i.e. $\omega = \text{const.} \times k$, then $\tau(\omega) = \text{const}$ and does not depend on the frequency. If the relation $\omega(k)$ is quadratic, i.e. $\omega^2 = \text{const.} \times k$, then the delay $\tau(\omega) = \text{const.} \times \omega$ is linear and can increase or decrease with frequency depending on the shape of the MSW dispersion.

The dependence of the output phase (the difference between the phases of input and output signals in the line $\varphi(\omega)$) on the frequency is of great practical importance. Since for the output phase φ_{out} holds:

$$\varphi_{\text{out}} = \varphi_{\text{in}} + k(\omega) \cdot L,$$

then

$$\varphi(\omega) = \varphi_{\text{out}} - \varphi_{\text{in}} = k(\omega) \cdot L \tag{8.1}$$

For small frequency deviations an expansion in Taylor's series can be used:

$$\varphi(\omega) = \varphi(\omega_0) + \frac{\partial \varphi}{\partial \omega}\Delta\omega; \quad k(\omega) = k(\omega_0) + \frac{\partial k}{\partial \omega}\Delta\omega$$

Inserting this in (8.1) and taking into account that $k(\omega_0).L = \varphi(\omega_0)$, we obtain

$$\frac{\partial \varphi}{\partial \omega} = \frac{\partial k}{\partial \omega} \cdot L = \frac{L}{v_g} = \tau(\omega), \quad \varphi(\omega) = \int_{\omega_1}^{\omega_2} \tau(\omega)\mathrm{d}\omega \tag{8.2}$$

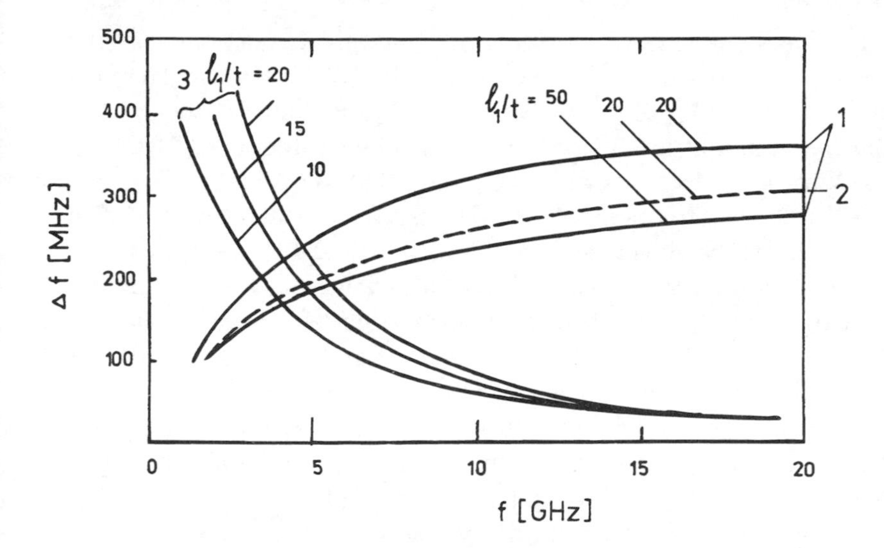

Fig. 8.1. Frequency dependences of the band of non-dispersive delay for $n = 1$, 2 — backward volume MSW, $n = 1$, 3 — surface MSW, $l_1/t = \infty$.

From (8.2) it follows that if $\tau(\omega)$ does not depend on the frequency, then the phase changes linearly with frequency and if $\tau(\omega)$ depends on the frequency linearly then the output phase (strictly speaking the output phase difference) depends on the frequency quadratically.

After these preliminary remarks we get to the problems of computation of MSW transmission lines with specified given properties. As band and MSW dispersion control elements, metallic boundaries separated from YIG by dielectric layers have been investigated. Limit values for both the magnitude and band of non-dispersive delay, and also the bands and gradients of retuning of transmission lines with linear frequency delay modulation and limit shift values of controlled delay lines on YIG films in the very high frequency range have been computed. The program for computation has been designed in such a way that on the $\tau(\omega)$ curves it automatically looks for the parts with zero or linear dependence of delay τ on the frequency while at the band edges the requirement $\tau(\omega) \leq \pm 10\%$ $(\Delta\tau/\tau)$, which is valid for various values of parameters l, t_d, t in various points of the frequency band, is not fulfilled.

In Fig. 8.1 there are depicted frequency dependences of non-dispersive delay bands Δf (i.e. bands with linear dependence $\omega(k)$) for surface and volume MSW for various distances between metallic and YIG films. As can be seen from Fig. 8.1 the non-dispersive delay band for surface MSW is at first quite wide, but when the frequency increases it narrows relatively quickly. On the contrary, for volume MSW the non-dispersive delay band broadens at first and then remains almost constant. Absolute values of Δf for surface MSW at centre frequencies 2–3 GHz are 300–400 MHz and for volume MSW at mean frequencies 10–15 GHz, $\Delta f/f$ has the same values. Let us notice that the relative band width $\Delta f/f$ decreases finally, but for volume MSW this decrease is slower than for surface MSW. We can conclude that for implementation of broad-band tunable DL on YIG films in short-wave frequency bands, volume MSW show better prospects than surface MSW.

Now, let us compare absolute values of delay τ (ns cm^{-1}) in surface and volume MSW delay lines. In both cases absolute values of delay τ can reach 150–200 ns cm^{-1}. But for surface MSW, on increasing the frequency the delay frequency band narrows and for volume MSW it remains almost constant up to very high frequencies. Theoretically, this also benefits volume MSW for non-dispersive delay lines at high frequencies. On the other hand, the use of volume MSW has its own specific difficulties, e.g. high magnetic fields, limitation of amplitude frequency characteristic, etc. That is why in each specific case it is necessary to reconsider these problems again.

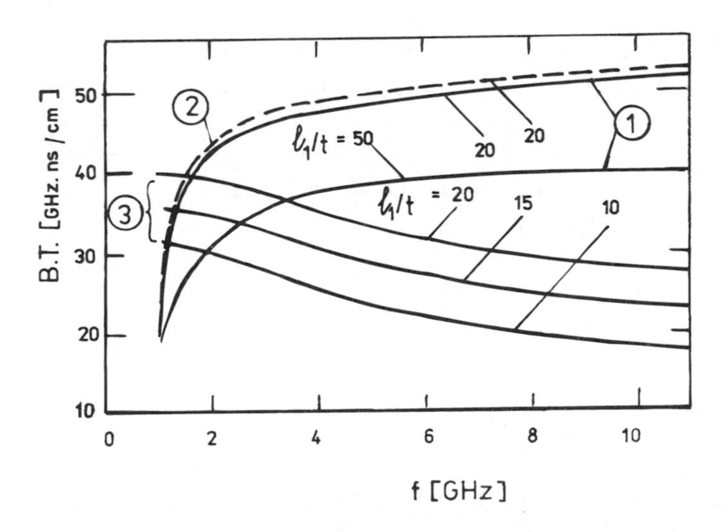

Fig. 8.2. Frequency dependence of generalized BT parameter for non-dispersive volume waves, 1 — forward volume MSW, $n = 1$, 2 — backward volume MSW, $n = 1$, 3 — surface MSW, $l_1/t = \infty$, $t = 10\,\mu$m, material YIG.

In the literature DL are usually characterized by the generalized parameter BT equal to the delay band (GHz) multiplied by the delay time (ns cm^{-1}) (BT = *band* × *time*). Fig. 8.2 shows the dependence of this parameter on the frequency for thin YIG film with $t = 10\,\mu$m and various values l_1/t at $l_2/t = \infty$ for both surface and volume MSW. As can be seen, values of the BT parameter for surface MSW are in the range 20–40; they are slightly dependent on geometry and decrease with frequency. For volume MSW, BT is of the same order; it is a little more dependent on geometry and increases with frequency. Later we will see that these theoretical results are in agreement with experimental results for non-dispersion DL.

An important parameter of frequency tunable DL is sensitivity of the output phase, i.e. its dependence on frequency and sensitivity to the change of external parameters, e.g. geometry of structure, magnetic field, etc. Since we examine regular MSW lines in which the dependence $k(\omega)$ has no singularities, the DL output phase cannot have singularities in the pass band either. Fig. 8.3 shows the dependences

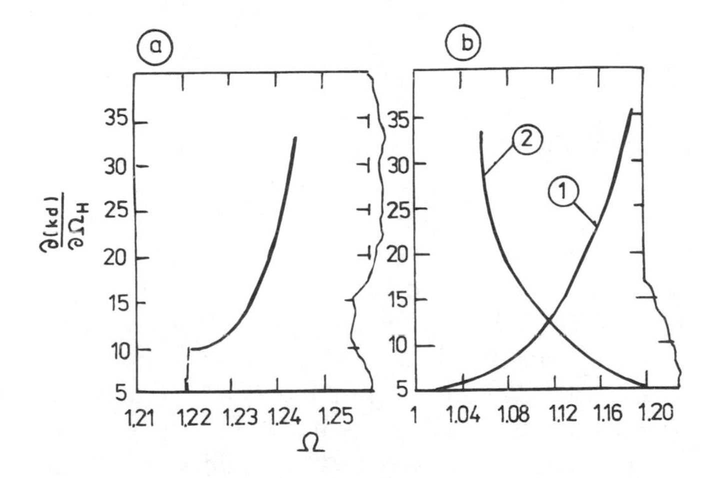

Fig. 8.3. The dependence of phase sensitivity on Ω a) for surface, b) for volume MSW: 1 — forward, 2 — backward. $l/t = l_1/t = \infty$, $\Omega_M = 0.5$.

of phase sensitivity $\partial(kd)/\partial\Omega_H(\Omega, \Omega_M)$ on Ω, where $\Omega = \omega/\omega_H$, $\Omega_H = \omega_H/\omega_M$, $\Omega_M = \omega_M/\omega_H$, $\omega_H = \mu_0\gamma H$, $\omega_M = \mu_0\gamma M_0$ for surface, forward and backward volume MSW in a free ferrite slab without metal. We can see that in this case the phase dependences have a monotonic character. As has been expected, the behaviour of the output phase for forward and backward volume MSW is different.

For surface and volume MSW it has a normal dispersion — phase sensitivity increases with frequency. For backward volume MSW it has anomalous dispersion — phase sensitivity decreases with frequency. But we can observe these simple dependences only in regular homogeneous ideally matched transmission lines whose dependences $k(\omega)$ have no extraordinary features. But already a simple consideration of the reflection at the DL ends leads to a complex dependence of the output phase on the frequency and to possible frequency discontinuities of the output phase.

To conclude this part, let us remark that the presented estimations of the band, dispersion and phase properties of transmission lines with surface and volume MSW at various frequencies, although they have an illustrative character, can be useful for orientation when choosing construction of the delay line with the properties needed. Further, we will show that there are theoretical estimations close enough to the experimental results obtained on DL types mentioned above.

8.2 Delay lines

8.2.1 Delay lines with linear delay

Now, let us treat the results of practical investigations of MSW delay lines. A lot of papers and books deal with this problem (see e.g. [164–167] and corresponding bibliography). We will present only some important results. In [164, 165] there are described delay lines for the 3 cm and 10 cm bands with linear frequency dependence in the pass band. The constructions of these DL are typical and they include all the basic features of MSW delay lines. That is why they deserve adequate attention. A characteristic speciality of the elements described is the use of outer metallic boundaries to control dispersion. An automatic search for linear sections in dependences $\omega_H d/v_g(\omega)$ with deviation from linearity $\Delta\tau = \pm 5\,\mathrm{ns\,cm^{-1}}$, for YIG film with the thickness $t = 10\,\mu\mathrm{m}$ for various parameters l_1/t and $l_2/t = \infty$ in a broad frequency band (up to 20 GHz and higher) for surface and volume MSW, leads to in the results in Fig. 8.4, in which dependences of the linear delay bands $\Delta f(\mathrm{GHz})$ on frequency for various values of l_1/t are depicted. As can be seen, the nearer the metallic boundary to the YIG surface, the broader the linear delay

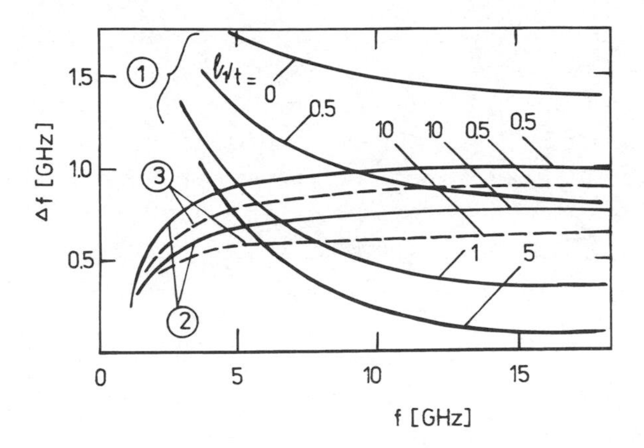

Fig. 8.4. The dependence of bands of linear delay on frequency. 1 — surface $\tau_{\mathrm{str}} = 30 - 50\,\mathrm{ns\,cm^{-1}}$, 2 — forward volume $\tau_{\mathrm{str}} = 300 - 400\,\mathrm{ns\,cm^{-1}}$, 3 — backward volume MSW. $t = 10\,\mu\mathrm{m}$, $l_1/t = \infty$.

band will be. For surface and volume MSW at the centre frequency $f \simeq 10\,\mathrm{GHz}$ the band width can reach values about $\simeq 1\,\mathrm{GHz}$ and more. On increasing the frequency the linear delay band narrows for surface MSW and broadens for volume MSW. However, from the standpoint of absolute value of delay τ, volume MSW are considerably more favourable because they give an average delay value τ_s of about $\tau_s \simeq 300$–$400\,\mathrm{ns\,cm^{-1}}$ while for surface MSW it is only $\tau_s \simeq 30$–$40\,\mathrm{ns\,cm^{-1}}$, i.e. it is lower by one order of magnitude. Elements are realized for forward volume MSW which, in a similar way as before, seem to be more favourable at higher frequencies. Typical construction of these elements is shown in Fig. 8.5. As a matter of fact, in this case the control boundary represents the basis of the microstrip line and that is why the input and output of the high-frequency signal have been realized by means of coplanar waveguides etched in the same conducting layer. A dielectric substrate (glass) with thickness $20\,\mu\mathrm{m}$ has been deposited on the conducting layer. Then microstrip transducers connected with central conductors of coplanar lines have been realized on the substrate. From above, YIG film grown on GGG substrate (for illustration it is separated in Fig. 8.5) has been placed on the glass substrate. The size of the YIG film was: length 25 mm, width 5 mm, thickness $t = 20\,\mu\mathrm{m}$, distance to the conducting plane (thickness of dielectric layer) $l_1 = 20\,\mu\mathrm{m}$. As analysis for surface and volume MSW shows this ratio ($l/t \simeq 1$)

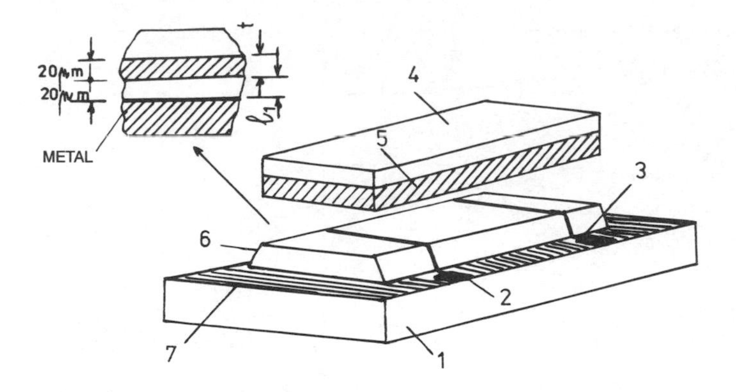

Fig. 8.5. Typical construction of a delay line. 1 — alumina, 2 — input of a coplanar line, 3 — output of a coplanar line, 4 — GGG substrate, 5 — YIG, 6 — glass, 7 — metallic layer.

creates the best conditions for the rise of linear sections in delay line characteristic. Input and output microstrip conductors have been narrow enough (up to 50 μm) to secure the inevitable band width for MSW excitation. As follows from theoretical analysis done for forward volume MSW the band of linear delay for frequencies in the 3 cm and 10 cm bands should be of the order 700–800 MHz, the band of linearity ($\Delta\tau$) of the order 250–300 ns cm^{-1}, and the parameter $BT \simeq 150$–$200\,\mathrm{cm}^{-1}$. Owing to the reasonably constructed element, the experimental parameters have been close to those computed theoretically. So, e.g. at wavelength 3 cm they correspond to the theoretical values mentioned above. The intensity of external perpendicular magnetic field was about 360 kA m^{-1}, the linear delay band $\simeq 1$ GHz, the parameter $BT \simeq 200\,\mathrm{cm}^{-1}$, and the band of linearity 50–250 ns cm^{-1}. However, in consequence of the fact that the metallic boundary was placed near enough to the YIG film the attenuation was large — of the order of 40dB (which was also sufficiently exactly predicted by the theory). Recently, in order to achieve the required shape of dispersion characteristic, there has been a tendency to use control metallic boundaries with a gradual change in gap, in which the depth and length of steps change according to a certain regularity. In reference [166] the authors assume that in this way the DL frequency range can be broadened and the parameter BT increased. There are also described DL in the 3 cm frequency band with a rising linear characteristic (up-chirp) in the 500 MHz range for surface MSW, with delay difference 80 ns cm^{-1} and insertion attenuation $\simeq$ 20 dB. In [167] there are described DL using backward volume MSW with a decreasing linear characteristic (down-chirp). The linear delay, with a good linearity $\simeq \pm 5$ ns, has been obtained for $\lambda = 3$ cm in the range 800 MHz. Attenuation values have been in the range 20–30 dB. Since control metallic boundaries, as a result of their finite conductance, have big ohmic losses, other ways to influence the dispersion characteristic have to be taken into account.

Another approach, which enables smaller losses, starts from the change of material properties. It uses e.g. the influence of the change of thickness of a layer on the dispersion characteristics or multilayer structures. The influence of the thickness of a layer on the dispersion characteristics of surface, forward and backward volume MSW has been investigated in Chapter 2. In principle two approaches have been developed for multilayer structures. The first one enables control of the dispersion characteristics by layers with different saturation magnetization, i.e. the medium is magnetically non-uniform in the direction perpendicular to the direction of MSW propagation. The second one consists in the production of a structure of two magnetic layers with the same magnetization separated by a non-magnetic dielectric layer. The theoretical solution of this structure has also been obtained in Chapter 2. In addition, the approach consisted of the introduction of surface permeability, which enables the formation of a simple algorithm and the use of a computer to obtain corresponding dispersion relations. This considerably simplifies the analysis of possible cases and enables effective computer simulation. As follows from this analysis the linear delay demanded can be obtained by the change of magnetization of individual layers or by the change of distance between the same two magnetic layers. However, up to now a certain disadvantage of multilayer structures is the simultaneous excitation of undesirable modes which causes undulation in the characteristic measured. Practical use of these structures requires the solution of the problem of suppression of these undesirable modes.

8.2.2 Non-dispersive delay lines

Non-dispersive broad-band delay lines are potential candidates for the replacement of phase shifters in the microwave frequency band. The demand put on non-dispersive delay lines is the existence of a delay tunable by means of an external magnetic field but independent of frequency in a certain frequency range. One of the possible approaches to the realization of delay lines with the properties mentioned above is to use the dispersion dependences of basic MSW modes by on appropriate choice of operating point. In reference [174] the authors tackled this problem in a different way. They presented a cascade connection of two delay lines with different dispersion characteristics. One of them operates in the range in which the phase delay increases with the increasing frequency. The second line operates with backward volume MSW with the inverse slope of the dependence presented. Corresponding frequency relations were linearized for both types by changing the distance of the ground plane under the magnetic layer. The resulting frequency dependence of the delay is given as the sum of the dispersion dependences of individual delay lines. This procedure is depicted schematically in Fig. 8.6. The element realized consisted of the two delay lines with YIG layers with thickness 25 and 50 μm for forward MSW and backward volume MSW, and with the gradual change of distance of the ground plane in the range ca 250 μm, its band width was 270–370 MHz with phase error 8.7–12.2° at central frequency 3 GHz. The electronic change of the delay was 47 ns. The line mentioned above was not optimized and that is why its attenuation in the operating band was relatively big (ca 35 dB). The other techniques recently used represent the use of multilayer structures [175, 176], non-uniform magnetization [177] or reflective microstrip fields [178].

8.3 High-frequency MSW filters

Filtration of electric signals is usually examined as a frequency selective process analyzed in the frequency domain. An element in which this filtration is carried out is called the frequency filter. In principle any delay lines can be used as filters, but it is necessary to modify their corresponding amplitude and phase characteristics. The basic demands put on filters are:

(a) as low a value of damping as possible in the pass band

(b) good suppression of the signal outside the pass band, needed for the minimization of deformation of the electric characteristics of the filter ($>$ 40 dB)

(c) in some special cases there are certain demands on the phase transmission characteristic

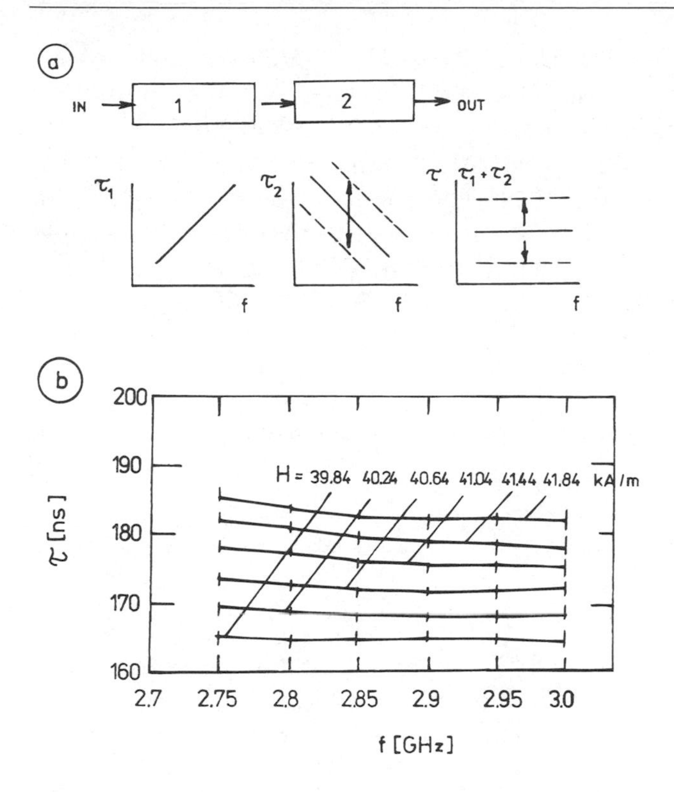

Fig. 8.6. The implementation of the non-dispersive line by connection of two dispersive lines a) schematic representation of the device, 1 — forward MSW delay line, 2 — backward volume MSW delay line. b) corresponding dispersion relation.

According to the relative width of the pass band it is possible to divide MSW filters into very narrow-band, narrow-band, medium-band and broad-band. It is obvious that the various kinds of filter will differ considerably in their construction and structure.

In classical filters the pass band or rejection band are obtained as a result of resonance in series and parallel branches of the circuit. But if MSW are used a different physical process occurs. Filtration is carried out by the transition of the input signal through a certain space structure. It is possible to model this by a certain number of delay lines connected in series and by the subsequent "addition" of delay signals. Under certain conditions these delay signals "add" themselves in a constructive way (pass band) and under other ones in a destructive way (rejection band). Filters in which the filtration process is in progress in the way mentioned above are called transversal filters. The transversal filter is modelled by means of branches separated by the delay lines with delay τ_k, while in principle the signal from each branch is multiplied by a weighting coefficient a_k. Modified signals from individual branches are added and they create the output signal [179]. The complex transmission function of such a filter can be obtained if we assume that at its input there is a function $e^{j\omega t}$. Then the output voltage will be

$$u_L(t) = \sum a_k e^{j\omega(t-\tau_k)} \tag{8.3}$$

where τ_k represents the total signal delay in place k. The expression

$$H(f) = \sum_{k=1}^{n} a_k e^{-j2\pi f \tau_k} = \sum_{k=1}^{n} a_k e^{-j2\pi f k \tau_0} \tag{8.4}$$

then represents the complex transmission function of the transversal filter. From equation (8.4) it follows that the transversal filter is characterized by two variables: the weighting coefficient a_k and delays $\tau_k = k.\tau_0$, in the case that there is a constant delay τ_0 between two branches. Their values determine the filter transmission function $H(f)$. This approach is often used in MSW filter design in which modification of individual coefficients (predominantly weighting coefficients) is achieved by a change of the electrode length — so-called apodization. With regard to the strong coupling between the transducer and MSW generated, apodization is seldom used in MSW elements in the way that it is used in surface acoustic wave (SAW) devices. Apodization or modification of weighting coefficients can be realized by changing the width of transducers, changing the distance between individual microstrip conductors or by changing the distance between the transducer and the layer. In addition, an appropriate choice of the width of the sample can shift the pass band of higher modes out of the pass band of the basic mode. The influence of circuit structures themselves on the resultant characteristics, because of strong coupling, is more significant and that is why in the case of MSW it is necessary to examine the response of the whole structure. It can be expressed by means of the so-called "transmission efficiency" defined as the ratio P_L/P_G where P_L is power delivered to the load and P_G is the maximum power supplied by the generator. In the case of MSW the power delivered to the load means the total power transmitted by MSW generated. If we neglect contributions of reflections, then the following holds:

$$P_L/P_G = |H_1(\omega) \cdot H_2(\omega)|^2 \tag{8.5}$$

This is proportional to the squared product of the transducer transmission functions. For a simple transducer the attenuation can be introduced as follows:

$$A(\mathrm{dB}) = -10\log(P_L/P_G) = 20\log|S_{21}| \tag{8.6}$$

where S_{21} is the coefficient of the scattering matrix of a given system understood as a two-port network which is commonly measured by standard network analyzers. The expression for S_{21} depends on the equivalent circuit of the transducer with ferrite slab used. E.g. for the equivalent circuit model with lumped parameters [180], assuming reciprocity of the coupling coefficient, for a pair of identical transducers the following equation can be obtained [180]:

$$A(\pm) = 20\log\left[\frac{(R_g + R_m)^2 + X_m^2}{4R_g R_m^{\pm}}\right] \tag{8.7}$$

where $R_m = R_m^+ + R_m^-$ is the total radiation resistance of the transducer, X_m is its reactance and R_g the impedance of the generator. Also attenuation as a result of MSW propagation must be included in (8.7), adding the following term

$$\simeq 76.4\tau_g \Delta H(\mathrm{dB}) \tag{8.8}$$

where τ_g is MSW delay on its propagation path and ΔH is the resonance line width of the material used. This approach, enlarged by the description of other circuit parts using scattering matrices and the principle of decomposition [181], enables effective analysis and design of MSW band-pass filters. An analogous procedure was used in [182] to analyze MSW filters with various types of transducer. Results are in agreement with experiments. However, recently volume MSW have been used more frequently for construction of filters because of a simpler magnetic circuit (narrower air gap) and the shape of dispersion characteristics of this magnetostatic mode, in spite of the fact that these MSW are reciprocal. Efficient means have been found to suppress the influence of reflections as well as the influence of higher MSW modes.

8.3.1 Filters of other types

The use of the change of the dispersive properties of MSW on their propagation paths as a result of the change of frequency or external magnetic field for implementation of filter elements attracts attention of workers from several firms. So, e.g. Adam [184] designed a 10 channel filter for the X-ray band with the following parameters:
— the pass band of each channel at level $-3\,\mathrm{dB}$ is $50\,\mathrm{MHz}$ and at level $-50\,\mathrm{dB}$ is $100\,\mathrm{MHz}$
— suppression of side bands $55\,\mathrm{dB}$
— insertion losses $\simeq 20\,\mathrm{dB}$
— pulsation in pass band $1\,\mathrm{dB}$

The filter consists of 10 narrow-band delay lines with forward volume MSW creating individual channels which have a common input and different output transducers consisting of idle microstrip ($635\,\mu\mathrm{m}$ wide) lines. Distance between layers and microstrips was $160\,\mu\mathrm{m}$ and this secured that frequency characteristics of channels were narrow-band (Fig. 8.7). The position of each delay line referring to the end of line was determined by the form $(N - 1/2)(\lambda/2)$ where N is the channel number and λ the wavelength corresponding to the central frequency of the Nth path band of the filter. In order to suppress higher modes 18 aluminium ribbons with thickness $26.4\,\mu\mathrm{m}$ were deposited. The central frequency of each channel was determined by the magnitude of the magnetic field, which changed

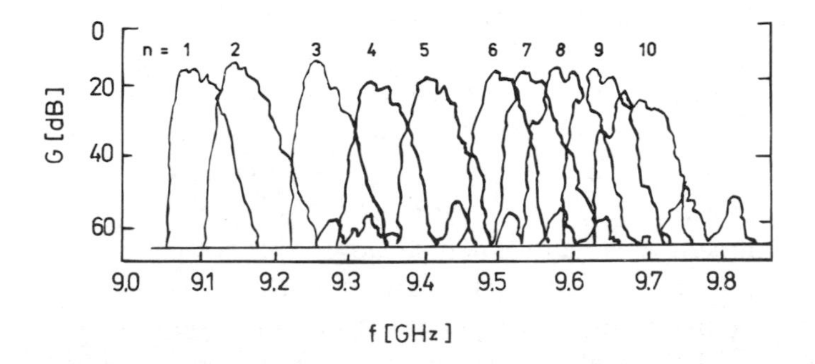

Fig. 8.7. Characteristics of 10-channel filter.

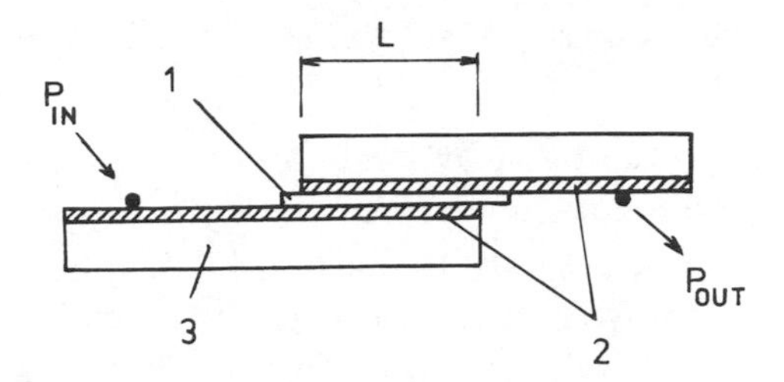

Fig. 8.8. The configuration of directional coupler used as a filter. 1 — mylar foil, 2 — YIG, 3 — GGG.

linearly starting from the first channel by means of the appropriately shaped outer magnetic circuit.

N. S. Chang [185] suggested the formation of narrow-band filter amplitude frequency characteristics by taking advantage of the thickness of ultrathin YIG layers to regulate the gradient of dispersion characteristic $\omega(k)$ at constant magnetic field. At the central frequency $f_c = 3258\,\text{MHz}$ a quality factor of about 1040 and attenuation 23 dB were achieved. An interesting idea has been developed by T. Koike [186] who used for filtration a directional coupler implemented by means of a double-layer waveguide structure (Fig. 8.8).

8.4 MSW resonators and their properties

Although the parameters of MSW band-pass filters discussed above, are useful for many applications, there are others which require considerable signal selectivity, i.e. as narrow a pass band width as possible. To satisfy this the construction of the filter has to be based on different principles. One of the possible ways is to use the properties of resonator systems. Their principle of operation consists either in the creation of a resonator by the appropriate location of reflection systems or in the use of the MSW properties. The coupling needed in reflection systems is created by placing one or several transducers in the resonator cavity.

8.4.1 Resonators consisting of reflection elements

With respect to the large influence of boundary conditions on various types of MSW propagating, various techniques are used to construct this type of resonator cavity.

8.4.1.1 Surface MSW resonators

The grooves created either by etching or by ion bombardment are used predominantly for the implementation of the resonator structure for surface MSW [187–191]. The resonator consists of a couple of wavelength selective periodic grooves separated by spaces in which transducers are placed. A schematic depiction of this resonator is given in Fig. 8.9a. The grooves created in this way behave as mirrors of a planar Fabry–Perot resonator. The step between grooves is chosen in such a way that it corresponds to a half wavelength of the corresponding resonant frequency. From this it also follows that the wavelength of the spin wave must be chosen in such a way that the YIG surface used is small and grooves of quarter wavelength width can be realized simply also by a photolithographic technique. In addition, losses associated with MSW propagation increase at very short wavelengths. From all the results mentioned above the optimum wavelength of spin waves is found to be in the range 40–500 μm.

The basic theoretical work devoted to MSW propagation in periodic structures is given in reference [187]. Although the analysis

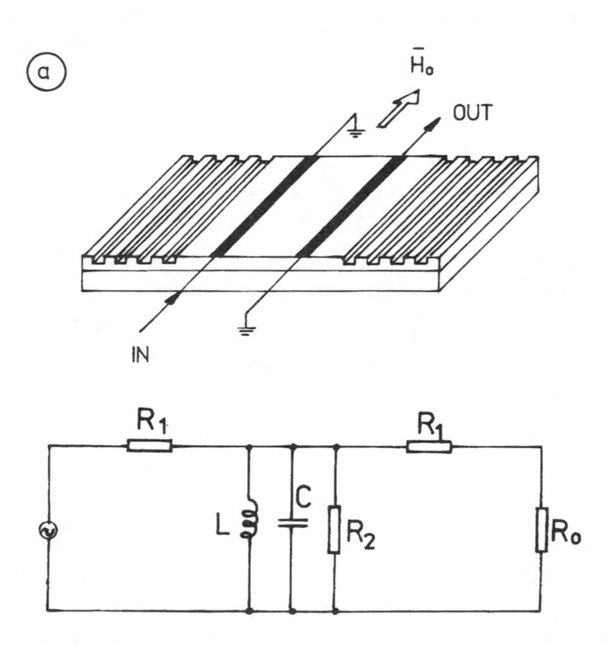

Fig. 8.9. Resonator on surface MSW: a) schematic representation, b) its equivalent circuit.

carried out there was done using strong simplifying assumptions such as e.g. a loss-free medium and neglect of the influence of boundaries, it has served as the basis for all subsequent work modelling the system of spaced out reflectors. The equivalent circuit of the resonator of this type is in Fig. 8.9b with parameters which can be obtained from experiment and theory in the forms [193]

$$\begin{aligned} R_1 &= R_0(1-r)/(1+r) \\ R_2 &= 2R_0 r/(1-r) \\ R_0 &= 50\,\Omega \\ Q &= 2R_0 C\omega/(1+r)^2; \qquad L = 1/(\omega^2 C) \end{aligned} \tag{8.9}$$

The parameter r represents the resonator attenuation given by $A = -20\log r$.

Obviously, in designing the resonator, it is necessary to know the dependence of the quality factor Q, attenuation A, suppression of side bands R, etc. on the physical parameters of basic resonator elements, i.e. the layer thickness t, groove depth d_h, number of grooves N, distances between grooves and distances between mirrors a. It has been found experimentally that the groove width can be expressed in the form

$$g = \lambda(1 - d_h/t)/4$$

where λ is the wavelength of the MSW generated. The distance between mirrors is also proportional to λ.

The number of grooves N and the ratio of the groove depth to the layer thickness t determine the width of the reflection band. The experimentally found dependences for surface MSW at a wavelength of 300 μm are depicted in Fig. 8.10, where the ratio d_h/t is taken as the parameter. From the dependences obtained it follows that the band width broadens with the ratio d_h/t and for $N \geq 20$ the width of the reflection band does not depend on the groove number.

Another group of MSW resonators consists of resonators in which reflectors are not created by the groove system but by a metallic strip system [195]. The perpendicular incidence of MSW on the system of metallic strips with finite conductance was dealt with theoretically in detail by Brinlee et al. [195] who examined the two-port network Fabry–Perot resonator with surface MSW, in which the reflectors of

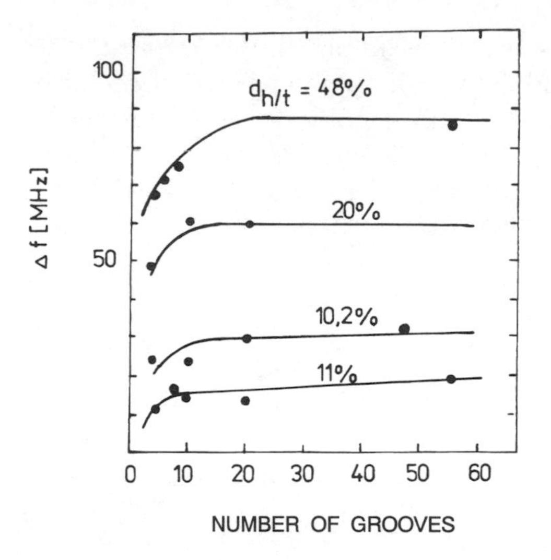

Fig. 8.10. The reflection band of resonator as a function of the number of grooves with the ratio d_h/t as a parameter, d_h — the groove depth.

the resonator cavity were created by means of metallic strips. They took into account losses in the ferrite, ferrite load caused by metallic strips and also matching of transducers. They constructed a resonator with an unloaded quality factor of 897 while the theoretically computed value was 956.

However, the disadvantage of resonators with surface MSW is that mainly at lower frequencies (under ca 4 GHz) their performance is limited to low power levels due to nonlinear effects which occur as a result of the coupling between MSW and spin waves of the system. Another disadvantage is that the external magnetic field applied in the plane of the layer requires a relatively large magnetic circuit because of the big air gap, and in addition the negative temperature coefficient of surface MSW cannot be sufficiently compensated for by the temperature change of the temperature coefficient of magnets.

8.4.1.2 Volume MSW resonators

The magnetic field of forward volume MSW is perpendicular to the YIG layer and MSW propagation in the plane of the layer is considered to be isotropic. For this reason the conventional configuration of a two-port network resonator used for surface MSW is not convenient. In this case the double-direction transducer response results in a slight suppression of the signal outside the band required. That is why a new configuration of resonator with volume MSW has been looked for. A schematic depiction is given in Fig. 8.11. The resonator consists of two cavities (in which mirrors R1, R2 and R3, R4 are placed) and a reflection mirror R5 with 45° slope which changes the beam path. The reflection coefficients of the mirrors R1 and R2 are chosen to be large and reflection coefficients of the other ones are smaller.

Suppression of the signal outside the pass band and the quality factor of the resonator (or its single-mode ability) depend on filtration properties of lattice reflectors. Similarly, as for the case of surface MSW, the magnitude of attenuation is a linear function of the groove depth d_h/t with the slope increasing with the groove number. So, e.g. for the ratio $d_h/t = 0.02$ the peak depth is 10.14 or 34 dB for groove numbers 10, 20 and 66, respectively. The reflection peak width, i.e. the pass band, is a function of the groove number as well as the ratio d_h/t. In this way implemented resonators arc tunable in a relatively broad frequency band (e.g. 2–11 GHz) with attenuation 20–32 dB and quality factor in the range 290–1570.

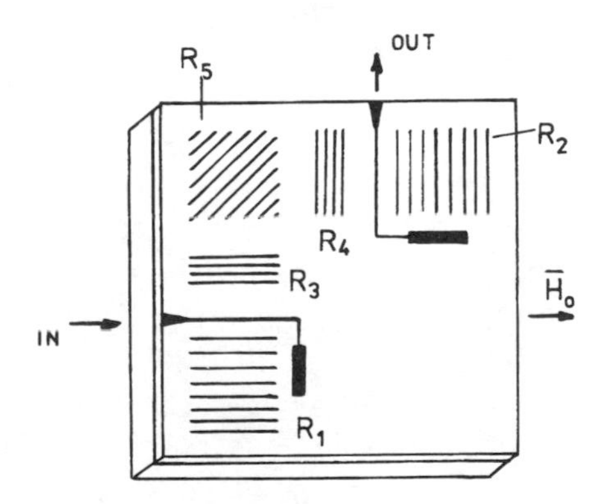

Fig. 8.11. Schematic representation of the resonator on volume MSW.

Magnetostatic resonators with volume MSW have very good prospects because they have several advantages in comparison with resonators with surface MSW, e.g.:

— above 0.5 GHz they have a higher threshold level for the onset of nonlinear effects

— the magnetic field applied is perpendicular to the plane of the layer and that is why the air gap of the magnet used is small (typically 1 mm), which simplifies implementation and reduces the size of the whole magnetic circuit

— they have a positive frequency temperature coefficient which can be compensated by the temperature shift of the properties of commonly-used hard magnetic materials.

8.4.1.3 Grid-free resonators

Up to now we have dealt with resonator structures, in which complex reflection surfaces (mirrors) have been used, which have been realized by means of a groove system which created the resonator cavity required. But implementation of these reflectors is complicated from the technological point of view and that is why other ways have been looked for to achieve the resonance effect required. In this section we will deal with resonance structures which use the features of MSW at boundaries, i.e. resonant structures which use the results described in Chapter 4.

The first investigators who were concerned with the implementation of resonators of this type were Ishak and his co-workers [196, 197]. The edges of YIG layers serve directly as reflectors for such resonators. A typical configuration of the resonator with surface MSW is illustrated in Fig. 8.12. It consists of a ferromagnetic res-

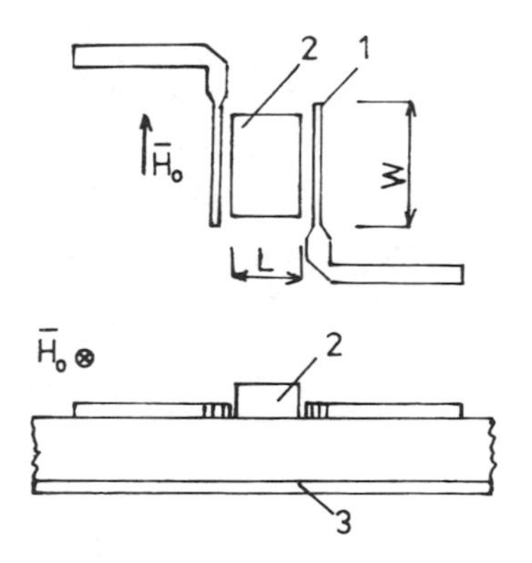

Fig. 8.12. Surface MSW resonator with layer edges as reflectors, 1 — microstrip, 2 — YIG/GGG, 3 — conductive plane.

onator cavity placed on the microstrip transducer structure. The resonator cavity consists of a thin YIG film on GGG substrate in the shape of a rectangle. The external magnetic field direction is also shown in Fig. 8.12. Its orientation enables the coupling between high-frequency power and surface MSW by means of one of the transducers. The waves excited propagate on the surface of the YIG layer and reflect at the edges to the other side, which makes the resonance possible if the following condition is fulfilled:

$$(k_+ + k_-)l = 2n\pi, \quad n = 1, 2, \ldots \tag{8.10}$$

where k_+ and k_- are the wave numbers at the top and bottom surfaces respectively of the layer and l is the distance between the edges. This condition is valid only in the case when the width of the layer w is infinite. If the width of the layer is finite, then it is necessary to include it in this condition because of the possibility of excitation of stationary modes across the layer [198]. Taking into account these modes the resonance condition (8.10) can be expressed as

$$(k_{m+} + k_{m-}) = 2\pi n \tag{8.11}$$

where

$$\begin{aligned} k_{m\pm}^2 &= k_\pm^2 + (m\pi/n)^2/\mu \text{ inside YIG layer} \\ k_{m\pm}^2 &= k_\pm^2 + (m\pi/n)^2 \text{ outside YIG layer} \end{aligned} \tag{8.12}$$

$n = 1, 2, 3, \ldots$, $m = 1, 2, 3, \ldots$ and μ is given by (1.43). However, there is still the problem of undesirable modes and shift of frequency at tuning. It is also possible to use edges as MSW reflectors for forward volume MSW. One such possibility was presented in [199]. The authors started from the configuration of a rectangular resonator, which uses edges as mirrors, and analyzed the possibilities of excitation of higher modes in the resonator. The idea of construction of such a resonator consists in the location of transmitting and receiving transducers in such places that one higher mode of MSW is selectively excited. To suppress undesirable modes, e.g. the basic and other higher ones, it is possible to use the location of grooves in the ferromagnetic layer. In this way a resonator with forward volume MSW was implemented with insertion loss 3.1 dB and side-band suppression higher than 20 dB.

The basic arrangement of another possibility is in Fig. 8.13 [200]. In this resonator, mainly the part which is linked with the input transducer is excited. Energy transmission into the input transducer region, and in so doing also excitation of the whole volume

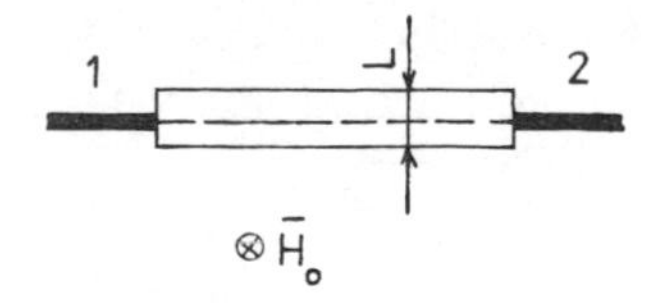

Fig. 8.13. Volume MSW resonator with layer edges as reflectors, 1 — input microstrip, 2 — output microstrip.

of the resonator, occurs as a result of the existence of waves with wave vectors which have no zero components in the resonator axis direction. Calculations on such a resonator are based on the standard theory of the Fabry–Perot resonator extended by introduction of the influence of losses. It can also be shown [200] that another of the basic resonator parameters — its quality factor — is directly proportional to the product of the operating frequency and group wave delay on the path L and inversely proportional to the quantity which represents propagation losses, losses as a result of the coupling between resonator and transducers and losses caused by reflections at the boundaries of the layer creating the resonator. An essential difference between electromagnetic F–P resonators and MSW resonators consists in the dispersion and attenuation of MSW and in special features of the MSW group velocity (Chapter 4). MSW group velocity can differ substantially from the phase velocity with regard to both its amplitude and direction, and in addition these velocities are much smaller than the electromagnetic wave propagation velocity. This elucidates the fact, that although in comparison with traditional resonators MSW resonators have relatively small coefficients of reflection and relatively big losses, they can achieve a high quality factor. The configuration depicted in Fig. 8.13 also provides a slight direct coupling between transducers. In contrast to resonators with surface MSW, resonators of this type have a sharper amplitude frequency response. As an example, in Fig. 8.14 there is the experimental amplitude frequency response of a resonator with a YIG layer size $8 \times 1.5 \times 0.3\,\mathrm{mm}$. Transducers consist of simple microstrip lines with width $100\,\mu\mathrm{m}$ and the gap between lines and a layer is $200\mu\mathrm{m}$. A magnetic field of intensity $248\,\mathrm{kA\,m^{-1}}$ was applied perpendicularly to the plane of the layer which secures the excitation of forward volume MSW. An experimental examination showed that it is possible to construct the resonator to the size $3 \times 0.5 \times 0.03\,\mathrm{mm}$ without substantial worsening of its basic characteristics, which en-

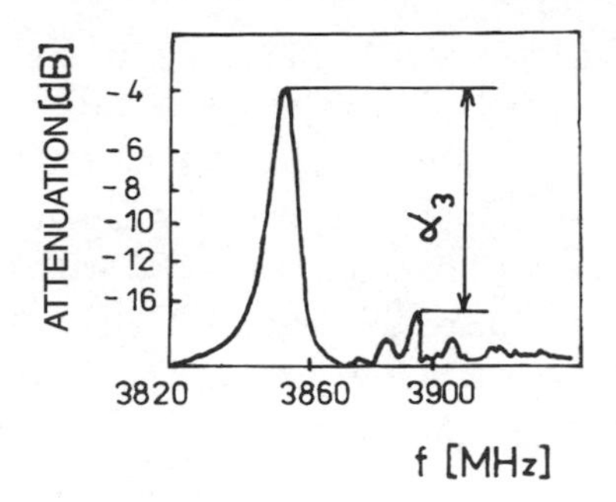

Fig. 8.14. Experimental characteristic of the resonator in Fig. 8.13. Material — $M_0 = 140\,\mathrm{kA\,m^{-1}}$, layer size $8 \times 1.5 \times 0.03$ mm. Width of microstrips 100 μm.

ables reduction of the size of the resonators implemented.

Poston and Stancil [201] suggested a new type of resonator with surface MSW which uses the possibility of localization of surface MSW energy on a circular orbit in a non-uniform magnetic field. In this way a ring resonator was created. Its structure is depicted in Fig. 8.15.

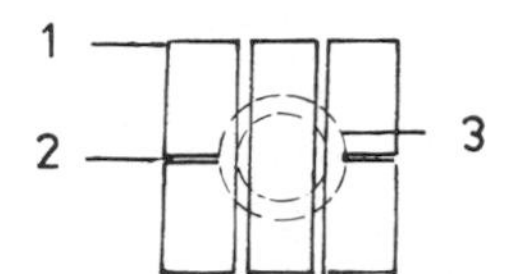

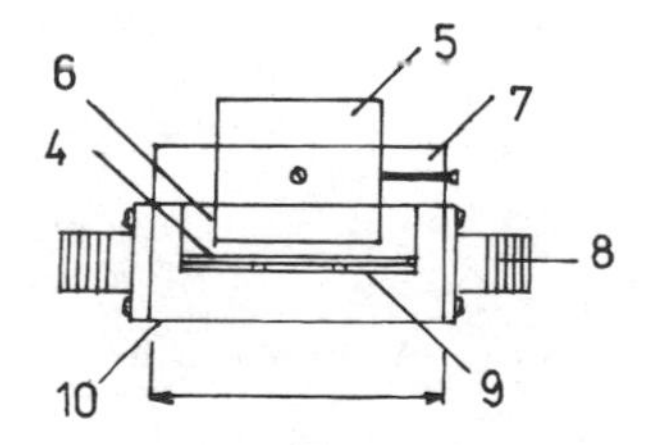

Fig. 8.15. Ring resonator. 1 — alumina substrate, 2 — microstrip line, 3 — beam path formed by magnetic field, 4 — YIG on GGG substrate, 5 — permanent magnet, 6 — foam rubber for fixing of the ferrite layer, 7 — plastic cover, 8 — OSM connector, 9 — alumina substrate with microstrips, 10 — aluminium holder.

A currently produced permanent magnet with concentric south and north poles separated by a 0.19 mm wide gap was used to create a non-uniform magnetic field. Microstrip coupling structures were realized by the standard photolithographic technique on 635 μm thick alumina substrate. Transducers of width 50 μm are short-circuited and supplied from the 50 Ω line. The dashed circular lines in Fig. 8.15 indicate the placement of the magnet air gap. The frequency distance between neighbouring resonant frequencies is

$$\Delta\omega = (\Delta\omega/\Delta k)\Delta k = v_g \cdot \Delta k = 4\pi/\tau \tag{8.13}$$

where Δk is the change of wave number between resonances and τ is the group delay for the wave propagating along the ring. From equation (8.13) it follows that the separation of individual resonances can be increased if the group delay decreases by means of e.g. the reduction of element size. The quality factor obtained was 1100 with an insertion loss of 12 dB, or 1900 with an insertion loss of 23 dB.

Another non-traditional but highly promising method of creation of reflection structures for MSW resonators is based on the total internal reflection of MSW at the boundary of the two multilayer structures of the type metal-dielectric-ferrite-dielectric-metal (Chapter 4). An advantage of this approach is the fact that for the realization of the reflection structures required classical photolithography is sufficient. The construction of such a resonator is depicted in Fig. 8.16 [202]. The oblique incidence of MSW at the resonator mirror created by the boundary ferite–metal-coated ferrite, is used in

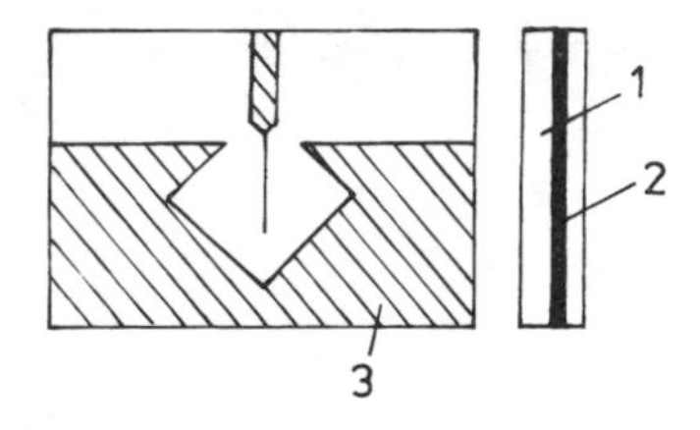

Fig. 8.16. Resonator using total internal reflection of MSW. 1 — YIG/GGG, 2 — alumina substrate, 3 — metallic structure.

this configuration. If we choose a more complicated resonator configuration and appropriate ratio l_1/t where l_1, t are the thicknesses of the dielectric and the layer, respectively, then it is possible to observe the phenomenon of total internal reflection also at perpendicular incidence of MSW at the boundary. The effect described occurs for all types of MSW and can be used usefully for the implementation of simple MSW resonators.

8.5 Optomagnetic MSW elements

Recently, elements which use the interaction between MSW and light have been in the limelight. It turns out that these elements have the greatest potential to take off as the first currently produced elements with wide practical use as high-frequency spectral analyzers, frequency shifters, microwave filters, deflectors, light modulators,

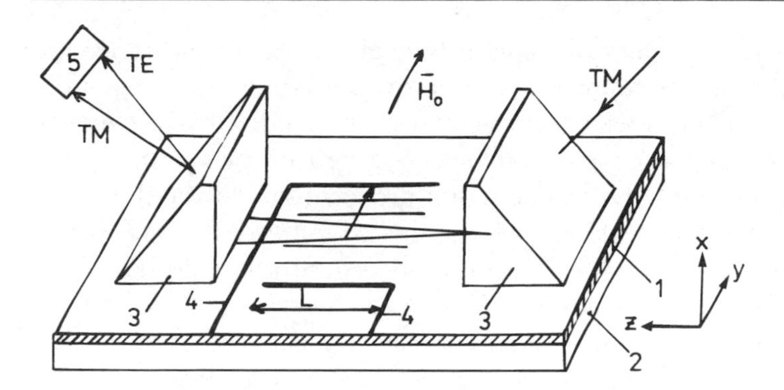

Fig. 8.17. Bragg light scattering on MSW. 1 — BiLuIG film, 2 — GGG, 3 — GaP prism, 4 — exciting and receiving MSW antennas, 5 — analyzer.

etc. For the investigation of individual possibilities the arrangement schematically depicted in Fig. 8.17 is used. Semiconducting laser diodes are usually used as light sources. According to the optical band chosen, prisms are selected for light guidance into the optical waveguide created by a thin YIG layer. To achieve good optical and at the same time high-frequency properties, these layers are usually highly doped, mainly by bismuth (see e.g. [144]). The orientation of the light and MSW beams with respect to one another can be either parallel or perpendicular. In the literature both configurations are investigated. Similarly, asalso for acousto-optical elements the optomagnetic elements described are based on anisotropic Bragg scattering of optical modes at surface or volume magnetostatic waves [209]. As a consequence of the interaction between the guided TE and TM polarized light and MSW, conversion of individual modes occurs, which can be used for element construction. In [192] there was demonstrated the possibility of light beam scattering by means of forward volume MSW, and light scattering by 4° was achieved at the change of frequency from 2.22 GHz to 3.06 GHz. Another possibility is to use the interaction presented for the spectral analysis of high-frequency signals in real time. In the work [202] mentioned above the frequency distinction 30 MHz at the carrier signal 10.925 GHz was demonstrated. From the results obtained concerning these elements and published in the literature it follows that the present situation in the development of optomagnetic elements based on the interaction between MSW and light corresponds to the results obtained on acousto-optical elements in the late seventies. However, in comparison with acousto-optical elements, future optomagnetic elements provide the following unique properties:

(a) a much broader range of carrier frequency tuning e.g. 2–20 GHz which can be carried out by changing the stationary magnetizing field. This enables direct signal processing without the necessity to process signals indirectly by conversion to lower frequencies

(b) a broad band which can be achieved by a simpler geometry of transducers

(c) the possibility of achieving much higher modulation, switch off/on and scattering velocities because MSW propagation velocity is higher than the propagation velocity of surface acoustic waves.

8.6 Nonlinear MSW elements

The basis of MSW elements operating in the nonlinear regime is a microstrip line loaded by a ferrite slab, which was examined in Section 5.5.4. There we showed that the nonlinear problem of the parametric excitation of spin waves can be transformed to the linear problem of the excitation of magnetostatic waves in a medium with loss, but in so doing it is necessary to consider the dependence of losses on the input power. This simple analysis enables us to understand phenomena which result in special behaviour of MSW elements in the range of their nonlinear excitation. Two of these elements have taken off up to now. Although they have similar construction their functions in the signal processing are antagonistic. They are frequency selective limiters which have a low attenuation for low-level signals and high attenuation for high-level signals, and elements for noise suppression or elements with inverse dynamic nonlinearity which suppress low-level signals more strongly than signals with a large amplitude.

8.6.1 Signal-to-noise enhancers

The typical construction of a transmission line with inverse dynamic nonlinearity which is used for the construction of a signal-to-noise enhancer is depicted in Fig. 5.8 [203, 204]. A metallic base and an exciting microstrip with width 25 μm were deposited by evaporation on the bottom and top sides of a dielectric substrate (thickness 635 μm and $\varepsilon_r = 70$), respectively. The wave impedance of the microstrip line was 50 Ω and the impedance of the feeding coaxial cable was also 50Ω. A YIG layer was attached to the microstrip from above. The layer was in the shape of a disc with radius 8–24 mm. The thickness of the YIG layer was changed in the range 20–50 μm. To reduce reflections at the layer boundaries, either the edges of the layer were ground off or a thin aluminium or NiCr layer ($\simeq$ 60 nm) was deposited on the dielectric substrate. The magnetic field direction was parallel to the microstrip. Hence surface MSW were excited in the YIG layer and propagated perpendicularly to the strip.

The following results were obtained for this element. In Fig. 8.18 are presented the dependences of absorption coefficient of the input signal G on the input power P_{in} amplitude for various values of the diameter L of the YIG layer (L is the length of the microstrip line). The level of input power at which the coefficient of signal suppression begins to decrease is denoted as the threshold level P_{thr}. Further increase of the input power ($P_{\text{in}} > P_{\text{thr}}$) causes that suppression coefficient to decrease and on reaching the level $P_{\text{in}} = P_{\text{sat}}$ (saturation power), the suppression has the minimum value (G_{min}). On further increasing the input power $P_{\text{in}} > P_{\text{sat}}$, the suppression remains almost constant. From the dependences in Fig. 8.18 it follows that for the given element the threshold value of input power was $\simeq$ 1 mW. Increasing the length of the microstrip line (diameter of the layer) the difference between suppression at low-level ($P_{\text{in}} < P_{\text{thr}}$) and high-level ($P_{\text{in}} = P_{\text{sat}}$) signals increases. So, for $L = 24$ mm, the quantity

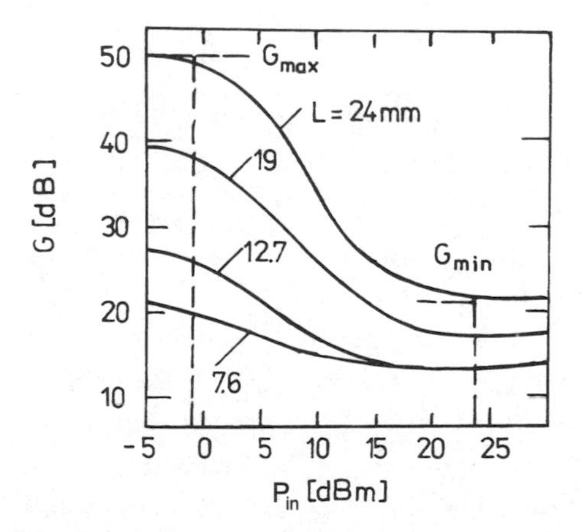

Fig. 8.18. The dependence of the coefficient of signal absorption on the level of input power. $t = 45.7\,\mu$m, $w = 25\,\mu$m, $H = 46.4\,$kA m^{-1}, $f = 4\,$GHz. Parameter — the line length.

$\Delta G = G_{\max} - G_{\min}$ was about 30 dB. For the central frequency $f = 4$ GHz the operating frequency band in which the suppression of signal occurs was about 800 MHz. For low-level signals the suppression coefficient was $\simeq$ 40 dB and for high-level signals it was $\simeq$ 10 dB.

The length of the microstrip line can be increased considerably if it is realized in the shape of a meander. In this case, the length of the meander can even be decreased in comparison with the length of a simple microstrip line, which results in a decrease of the width of the YIG layer in the magnetic field direction. This, of course, makes it easy to realize a uniform magnetic field. In the literature (see e.g. [205]) constructions of such elements are described. The increase of "effective" microstrip length enabled values $\Delta G \cong$ 30–40 dB to be obtained with a considerable decrease of the high-level signal suppression ($G_{\min} = (0-2)$ dB). To broaden the operating band of the element, the YIG layer can be metal-coated or coupled YIG layers can be placed on opposite sides of the microstrip. In the latter case the fast and slow MSW modes propagate in the coupled structure and the coupling with the fast wave considerably broadens the operating band of the element (see e.g. [206]). In addition to that the suppression level increases. An important characteristic of the signal-to-noise enhancer is the frequency selectivity in a broad frequency band, i.e. the possibility of suppressing the signal at a given frequency independently of the presence of signals of other frequencies. This problem was examined in [203], [204]. The element input was fed with two signals at the same time — a low-level one ($P_{\text{in}} < P_{\text{thr}}$) as noise imitation and a high-level one ($P_{\text{in}} > P_{\text{thr}}$) whose frequencies were different. Fig. 8.19 depicts the level of the output power of the low-level signal ($P_{\text{in}} - 20$ dBm) as a function of the frequency difference between low- and high-level signals at various values of the input power of the high-level signal. When the high-level signal was absent, the output power of the low-level signal was -55 dBm, i.e. suppression was 35 dB. On increasing the input power of the high-level signal (frequency 3.3 GHz) the suppression level of the low-level signal near this frequency began to decrease. In so doing, on increasing P_{in} of the high-level signal the mistuning of the high-level signal Δf, at which the decrease of low-level signal suppression was observed, increased. So, at P_{in} of the high-level signal $\simeq -3$ dBm it was ± 5 MHz, at $P_{\text{in}} \simeq 10$ dBm, $\Delta f \simeq \pm 10$ MHz, at $P_{\text{in}} \simeq 20$ dBm, $\Delta f \simeq \pm 40$ MHz. So, we can conclude that the presence of a high-level signal improves the characteristics of the noise suppressor near this signal. It is necessary to take this into consideration when using noise suppressors on YIG layers in the feedback circuit of a high-frequency generator.

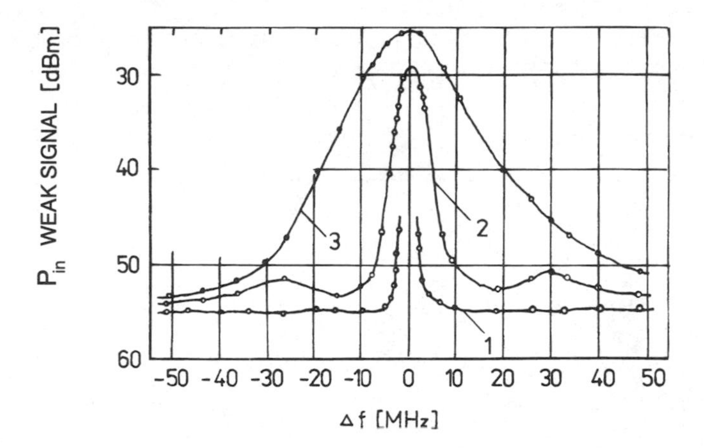

Fig. 8.19. The dependence of output power of a weak signal on the difference between low- and high-level signal frequencies. Input power of low-level signal $= -20\,\mathrm{dBm}$. Input power of high-level signal 1 — 3 dBm, 2 — 20 dBm, 3 — 20 dBm. $f = 3.3\,\mathrm{GHz}$, $H = 34\,\mathrm{kA\,m^{-1}}$.

To verify this assumption forward experiments were carried out [205] to follow the noise properties of the high-frequency generator near the carrier frequency. As a power source a magnetron was used which was adjusted by voltage with the low noise–signal ratio. The P_{out} signal directly from the magnetron was observed on the spectral analyzer screen and it is depicted in Fig. 8.20 using a logarithmic scale (dotted line). There is also the signal P_{out2} (full line) from the same high-frequency generator but after passing through the element (the levels of regular signals at the basic frequency $f = 2.7\,\mathrm{GHz}$ became equal to each other). We can see that maximum noise suppression is observed at the frequency which differs from the basic one approximately by $\pm 20\,\mathrm{MHz}$, and is approximately 40 dB in comparison with the previous level. The signal at the basic frequency is then suppressed by $\simeq 20\,\mathrm{dB}$. It is obvious from this picture that at a small difference from the basic frequency ($< \pm 5\,\mathrm{MHz}$) the noise suppression effect does not show up. This is in good agreement with the results for two regular signals which have been described above.

So, we can conclude that the frequency selectivity of a noise

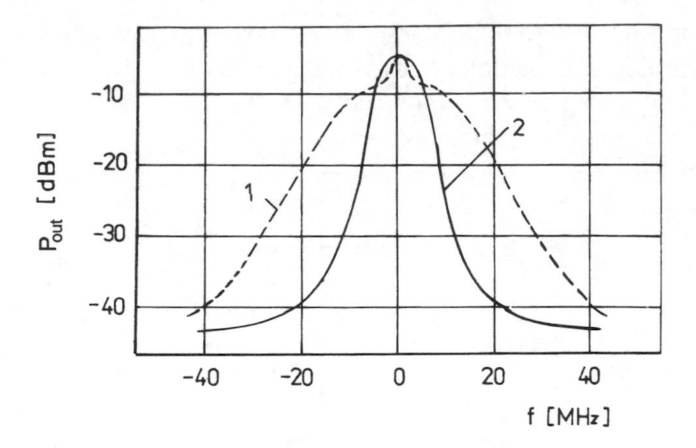

Fig. 8.20. Frequency dependence of the generator output power. 1 — before and 2 — after noise suppression.

suppressor on YIG is not lower than ±20 MHz, i.e. it is considerably broader than the Doppler frequency band.

8.6.2 Very high-frequency amplitude correctors

Let us examine other possibilities to use lines with inverse dynamic nonlinearity. One of them is the possibility to use them as amplitude correctors [207]. For the transmission of information without distortion in the microwave band, broad-band amplifiers are needed which have linear amplitude characteristics for a broad range of the input signal amplitude change. Permactrons or travelling wave tube (TWT) amplifiers comply well with these demands. However, linearity of their amplitude characteristic occurs as a rule at low amplitudes of the input signal. One of the possibilities for correcting the amplitude characteristic of a permactron amplifier is to connect it to an element with inverse dynamic nonlinearity at the amplifier input.

In [207] a corrector was examined which consisted of a microstrip line loaded with ferrite. To decrease the damping difference for low- and high-level signals a relatively short length of the microstrip line was chosen. The corrector operated in the band $\simeq 3$ GHz, the effective length of the microstrip was (3–5) mm, the low- and high- level signals were suppressed by 2–4 dB and $\simeq 1$ dB, respectively and the magnitude of the threshold power was about 100 mW.

As an amplifier the O-type TWT was used, which operated in

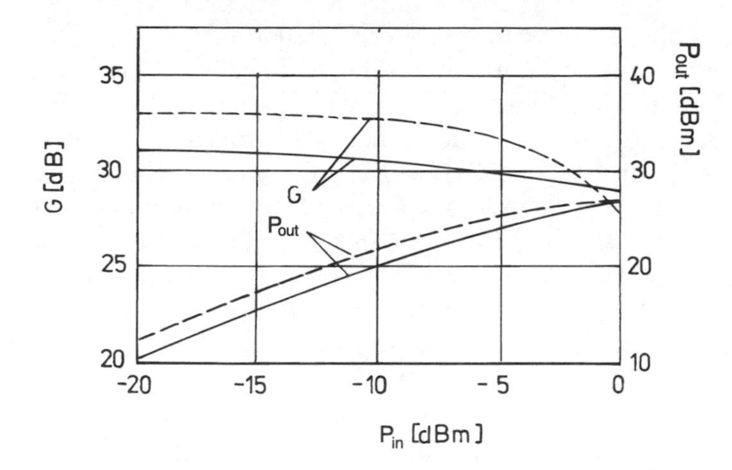

Fig. 8.21. Experimental dependences of amplification and output power of amplifier without and with corrector.

the band 2.5–3.6 GHz with gain factor 33 dB in the linear regime and output power $\simeq$ 500 mW in the saturation regime. Fig. 8.21 depicts (dotted lines) the dependences of the output power P_{out} and gain coefficient G of the amplifier on the input power P_{in} without a corrector. Characteristics of the amplifier with a corrector are marked by full lines. We can see that by changing the input power by 20 dB the decrease of the gain factor without a corrector is 5 dB and with a corrector -2 dB. Nevertheless, in the linear regime with a corrector the gain of amplification decreases by the attenuation of the signal in the correcting element. It is also necessary to note that recently, external ways of correcting high-frequency device characteristics have found a wide range of applications.

8.6.3 MSW power limiters

The existence of nonlinear losses in ferromagnetic material related with the parametric excitation of spin waves enables the realization of various types of broad-band power limiters based on the transmission line with a YIG layer. Special characteristics of these elements are a low power threshold level and the dependence of the threshold value on the magnetic field. The value of limiting power is determined by the losses in the layer and usually is of the order of a few μW. The dynamic range of these elements is 7–10 dB because on further increasing the input power in the MSW transmission line auto-oscillation phenomena can be observed [206].

An interesting power limiter is presented in [208]. It is based on the fact that in a transmission microstrip line loaded by a ferrite layer at frequencies higher than the upper frequency limit of parametric excitation ($f > 4.2\,\mathrm{GHz}$) usual dynamic nonlinearity is observed, i.e. the attenuation increases with increasing input signal. Such an element has the shape of a meander structure with a total length $L = 170\,\mathrm{mm}$, the width of the microstrip $w = 25\,\mu\mathrm{m}$ and the thickness of the YIG layer $t = 57\,\mu\mathrm{m}$. A constant magnetic field $H \cong 16\,\mathrm{kA\,m^{-1}}$ is applied perpendicularly to the meander line. The initial attenuation in the line was 7 dB.

The experimental dependences of the output power on the magnitude of the input power in such a line for various signal frequencies are depicted in Fig. 8.22.

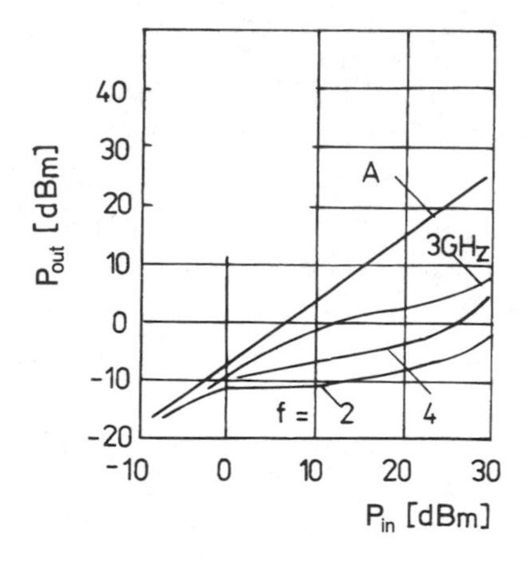

Fig. 8.22. The dependence of the output power of a limiter on the level of input power. $t = 57\,\mu\mathrm{m}$, $w = 25\,\mu\mathrm{m}$, $L = 170\,\mathrm{mm}$, $H = 16\,\mathrm{kA\,m^{-1}}$. A — line of initial losses $-2\,\mathrm{dB}$.

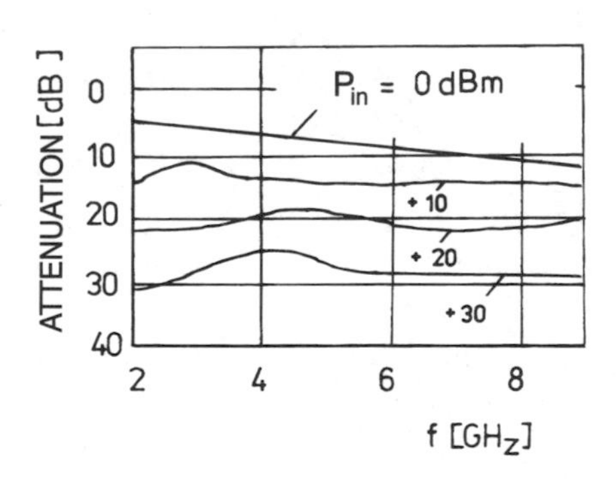

Fig. 8.23. The dependence of insertion loss on frequency for various levels of input power. Input signal 1 — 0, 2 — +10, 3 — +30, 4 — +30 dBm.

At frequency $f = 2\,\mathrm{GHz}$ and changing the input power in the range from zero to $+20\,\mathrm{dBm}$ the output power is kept at the level $-10\,\mathrm{dBm}$, so the dynamic range of the limiter is 20 dB. In the frequency band 2–4 GHz the threshold power $P_{\mathrm{thr}} \simeq 1\mathrm{mW}$ and dynamic range $\simeq 17\,\mathrm{dB}$. Fig. 8.23 depicts the dependences of the damping on the signal frequency for various input power levels. From this figure it follows that the attenuation is almost independent of the frequency in a two octave band (from 2 GHz to 8 GHz). It is necessary to add that up to now the mechanism of operation of such a limiter has not been explained in a satisfactory way.

REFERENCES

[1] Vashkovsky, A. V., Stalmachov, V. S., Sharaevsky, Yu. N.: Magnetostatic Waves in High-Frequency Electronics, SGU, USSR, 1991 (in Russian).

[2] Gurevitch, A. G.: Magnetic Resonance in Ferrites and Antiferromagnetic Materials, Moscow, 1973 (in Russian).

[3] Kvasnica, J.: Theory of Electromagnetic Field, SNTL (in Czech).

[4] Kittel, Ch.: Introduction to Solid State Physics, New York, Wiley, 1986.

[5] Ilkovitch, D.: Physics, SVTL, 1958 (in Slovak).

[6] Joseph, R., Schloemann, E.: J. Appl. Phys. 36 (1965), p. 1597.

[7] Brillouin, L., Parodi, M.: Wave Propagation in Periodic Structures, Moscow, 1939 (in Russian).

[8] Stalmachov, V. S., Ignatiev, A. A.: Lectures on Spin Waves, SGU, Sratov, 1983 (in Russian).

[9] Emtage, P. R.: J. Appl. Phys. 49 (1978), p. 4475.

[10] Stalmachov, V. S.: Electronic Waves in High-Frequency Devices with Crossed Fields, SGU, Saratov, 1970.

[11] Damon, R. W., Eshbach, I. R.: J. Phys. Chem. Solids 19 (1961), p. 104.

[12] Stalmachov, V. S., Ignatiev, A. A., Kulikov, M. N.: Radiotechnics and Electronics 26 (1981), p. 2381 (in Russian).

[13] Ganguly, A. K., Webb, D. C.: IEEE Trans. MTT-23 (1075), p. 999.

[14] Landau, L. D., Lifshitz, E. M.: Electrodynamics of Solids, Moscow, 1982 (in Russian).

[15] Vashkovsky, A. V., Gretchushkin, K. V., Stalmachov, A. V.: Radiotechnics and Electronics 30, (1985), p. 2422–2428.

[16] Stalmachov, A. V.: Proc. Conf. "Problems of High-Frequency Electronics", Leningrad (1984), p. 121 (in Russian).

[17] Gretchushkin, K. V., Stalmachov, A. V., Tyulyukin, V. A.: Radiotechnics and Electronics 31 (1986), p. 1487–1494.
[18] Sivukhin, D. V.: Physics, vol.III, Optics, Nauka, Moscow, 1980 (in Russian).
[19] Gurevitch, A. G.: Ferrites at High-Frequencies, Moscow, 1961 (in Russian).
[20] Monosov, I. A.: Non-linear Ferromagnetic Resonance, Moscow, 1971 (in Russian).
[21] Lyusel, U.: Coupled and Paramagnetic Oscillations in Electronics, Moscow, 1963 (in Russian).
[22] Suhl, K.: "Ferrites in Non-linear High-Frequency devices", Moscow 1963 (in Russian).
[23] Shloemann, E., Joseph, R.: "Non-linear Properties of Ferrites in High-Frequency Fields", Moscow, 1963, p. 7 (in Russian).
[24] Vendik, O. G., Kalinikos, B. A., Tchartozhinsky, D. N.: FTT 19, (1977), p. 387 (in Russian).
[25] Adam, J. D., Slitzer, S. N.: Appl. Phys. Lett. 36 (1980), p. 485.
[26] Gurzo, V. V., Prokushkin, V. N., Sharaevsky, Yu. P.: Proc. 5th Soviet Union School on "Spinwave High-Frequency Electronics", Ashkhabad, 1985, p. 23 (in Russian).
[27] Stalmachov, V., Prokushkin, V., Sharaevsky, J.: Intermag Conference Digest, Tokyo, Japan, (1987), p. AG-9.
[28] Gurzo, V., Prokushkin, V., Stalmachov, V., Sharaevsky, J.: Proc. Int. Symp. SWSS, Novosibirsk, USSR, 1986, p. 296.
[29] Benda, O.: Theory of Transmission Lines, EF SVST (University textbook), 1975 (in Slovak).
[30] Vashkovsky, A. V. et al.: Radiotechnics and Electronics 24, (1979), p. 2290 (in Russian).
[31] Vashkovsky, A. V., Gretchushkin, K. V., Stalmachov, A. V., Tyulyukin, V. A.: Letters to ETF 12 (1986), p. 487 (in Russian).
[32] Valiyaevsky, A. B., Vashkovsky, A. V., Stalmachov, A. V., Tyulyukin, V. A.: Radiotechnics and Electronics 33 (1988), p. 1820 (in Russian).
[33] Vashkovsky, A. V., Gretchushkin, K. V., Stalmachov, A. V., Tyulyukin, V. A.: Radiotechnics and Electronics 32 (1987), p. 2295 (in Russian).
[34] Valyaevsky, A. B., Vashkovsky, A. V., Stalmachov, A. V., Tyulyukin, B. A.: ETF 54 (1989), p. 51 (in Russian).
[35] Parekh, J. P., Tuan, H. S.: J. Appl. Phys. 52 (1981), p. 2279.

[36] Marshall, W., Lovesey, S. W.: Theory of thermal neutron scattering, Oxford, Clarendon Press, 1971.
[37] Bacon, G. E.: Neutron Diffraction, 3rd edn., Oxford, Clarendon Press, 1975.
[38] Mazur, P., Mills, D. L.: Phys. Rev. B29 (1984), p. 5081.
[39] Bloor, D., Copland, G. M.: Rep. Progr. Phys. 35 (1972), p. 1173.
[40] Brillouin, L.: Ann. Phys. 17 (1922), p. 88.
[41] Mandelstam, L. I.: Zh. Rad. Fiz. Chem. Obsh. 58 (1926), p. 381 (in Russian).
[42] Fleury, P. A., Porto, S. P. S., Cheesman, L. E., Guggenheeim, H. J.: Phys. Rev. Lett. 17 (1966), p. 84.
[43] Sandercock, J. R., Wettling W.: Solid State Comm. 13 (1973), p. 1729.
[44] Borovik–Romanov, A. S., Kreines, N. M.: Phys. Rep. 81 (1982), p. 351.
[45] Patton, C. E.: Phys. Rep. 103 (1984), p. 251.
[46] Grunberg, P.: Progr. Surf. Sci. 18 (1985), p. 1.
[47] Cottam, M. G., Lockwood, D. J.: Light Scattering in Magnetic Solids, Wiley, New York, 1986.
[48] Hayes, W., London, R.: Scattering of Light by Crystals, Wiley, New York, 1978.
[49] Elliot R.J., London R.: Phys. Lett. 3 (1963), p. 189.
[50] London, R.: J. Phys. C 3 (1970), p. 872.
[51] Cottam, M. G.: J. Phys. C 8 (1975), p. 1933.
[52] Wettling, W., Cottam, M. G., Sandercock, J. R.: J. Phys. C 8 (1975), p. 211.
[53] Sandercock, J. R., Wettling, W.: J. Appl. Phys. 50 (1979), p. 7784.
[54] Sandercock, J. R.: Optics Comm. 2 (1970), p. 73.
[55] Sandercock, J. R.: Proc. 2nd Int. Conf. on Light Scattering in Solids, Flammarion, Paris, 1971, p. 9.
[56] Sandercock. J. R.: Advances in Solid State Physics XV, Pergamon Vieweg, Braunschweig, 1975, p. 183.
[57] Sandercock. J. R. Topics in Appl. Phys. 51 (Ed. Cardona M., Guntherodt G.), Springer Verlag, Berlin, 1982, p. 173.
[58] Camley. R. E., Rahman, T. S., Mills, D. L.: Phys. Rev. B23 (1981), p. 1226.
[59] Griffiths, J. H. E.: Nature 158 (1946), p. 670.
[60] Kittel, C.: Phys. Rev. 73 (1948), p. 155.

[61] Kittel, C.: Phys. Rev. 110 (1958), p. 1295.
[62] Seavey, M. H., Tannenwald, P. E.: Phys. Rev. Lett. 1 (1958), p. 168.
[63] Puszkarski, H.: Progr. Surf. Sci. 8 (1979), p. 191.
[64] Roeschmann, P., Winkler, G.: JMMM 4 (1977), p. 105.
[65] Barak, J.: J. Appl. Phys. 63 (1988), p. 5830.
[66] Barak, J., Ruppin, R., Suss, J.: Phys. Lett. A 108 (1985), p. 423.
[67] Barak, J., Bhagat, S. M., Vittoria, C.: J. Appl. Phys. 59 (1986), p. 2521.
[68] Pankrac, A. I., Smyk, A. F.: Zhurn. Tekh. Fiz. 59 (1989), p. 150 (in Russian).
[69] Kalinikos, B. A., Kovshikov, N. G., Slavin, A. N.: Letters to ZETF 38 (1983), p. 343 (in Russian).
[70] Adam, J. D., Talisa, S. H., Kerestes, J. A.: IEEE Vol. MTT-25 (1989), p. 42.
[71] Sethares, J. C., Weinberg, I. J.: Circ. Syst. Sign. Proc. 4 (1985), p. 42.
[72] Hyben, P., Abraham, A., Cermak, J., Kabos, P.: Acta Phys. Slov. 39 (1989), p. 205.
[73] Beregov, A. S., Kudinov, E. V., Oblamskij, V. G.: IVUZ Radiotechnics 29 (1986), p. 37 (in Russian).
[74] Webb, D. C., Vittoria, C., Lubitz, P., Lessoff, H.: IEEE Trans. Mag-11 (1975), p. 1259.
[75] Gusev, B. N., Civileva, O. A., Gurevic, A. G., Emirjan, L. M., Naronovic, O. B.: Letters ZTF 9 (1983), p. 159 (in Russian).
[76] Stalmachov, V. S., Ignatiev, A. A., Kulikov, M. N.: Radiotech. and Electronics 26 (1981), p. 2381 (in Russian).
[77] Ignatiev, A. A., Lepestkin, A. L.: Proc. Conf. on Dielectric Waveguides and Resonators, Saratov, USSR, 1983.
[78] Mikaeljan: Theory and Applications of Ferrites at Very High Frequencies, Nauka, Moskow, 1963.
[79] Gardiol, F. E.: IEEE Trans. MTT-18 (1970), p. 461.
[80] Montgomery, C. G., Dicke, R. H., Purcell, E. M.: Principles of Microwave Circuits, New York, McGraw-Hill, 1948.
[81] Vlannes, N. P.: J. Appl. Phys. 61 (1987), p. 416.
[82] He, Y. Q., Wigen, P. E.: JMMM 53 (1985), p. 115.
[83] Kabos, P., Wilber, W. D., Patton, C. E., Grunberg, P.: Phys. Rev. B29 (1984), p. 6396.

[84] Wolfram, T., DeWames, R. E.: Progr. Surf. Sci. 2 (1972), p. 233.

[85] Dione, G. F.: J. Appl. Phys. 25 (1974), p. 3621.

[86] Wilber, W. D., Kabos, P., Patton, C. E.: IEEE Trans. Mag-29 (1983), p. 1862.

[87] Grunberg, P., Mayer, C. M., Vach, W., Grimditch, M.: JMMM 28 (1982), p. 319.

[88] Schneider, B.: Phys. Stat. Sol. (b) 51 (1972), p. 325.

[89] Kabos, P., Patton, C. E., Wilber, W. D.: Phys. Rev. Lett. 53 (1984), p. 1962.

[90] Kabos, P., Wilber, W. D., Patton, C. E.: 7th Int. Conf. on Microwave Ferrites, Illmenau, 1984, p. 23–30

[91] Rupp, G., Wettling, W., Smith, R. S., Jantz W.: JMMM 45 (1984), p. 404.

[92] Heinrich, B., Purcell, S. T., Dutcher, J.R., Urguhart, K.B., Cochran, J. F., Arrott, A. S.: Phys. Rev. B 38 (1988), p. 12879.

[93] Sandercock, R.: Solid State Comm. 15 (1974), p. 1715.

[94] Schloemann, J.: J. Phys. Chem. Solids 6 (1958), p. 242.

[95] Roeschmann, P.: Proceedings of Joint MMM-Intermag. Conference, Pittsburgh, 1976, p. 253.

[96] Brower, C. J., Patton, C. E.: J. Appl. Phys. 55 (1982), p. 2104.

[97] Vrehen, Q. H. F.:J. Appl. Phys. 40 (1969), p. 1848.

[98] Inui, T., Ogasawara, N.: IEEE Trans. Mag-13 (1977), p. 1729.

[99] Kabos, P., Kato, t., Mizoguchi, T., Patton, C.E.: IEEE Trans. Mag-20 (1984), p. 1259.

[100] Moosmuller, H., McKinstry, K. D., Patton, C. E.: J. Appl. Phys. 67 (1990), p. 5521.

[101] Benda, O., Kabos, P.: Zbornik prac EF, 1979 (in Slovak).

[102] Benda, O., Kabos, P.: EC 41 (1990), p. 73.

[103] Kabos, P., Patton, C. E.: Advances in Ceramics 16 (1986), p.

[104] Adam, J. D.: Electronics Lett. 6 (1970), p. 718.

[105] Stancil, D. D: J. Appl. Phys. 59 (1986), p. 218.

[106] Kudinov, Beregov,: Research Report, Kiev, 1986 (in Russian).

[107] Stalmachov, V., Ignatiev, A., Lepestkin, A.: Phys. Lett. A 133 (1988), p. 430.

[108] Ignatiev, A. A., Lepestkin, A. N.: Patent USSR AS 1394163, (1988).

[109] Belan, V., Hyben, P., Ceskovic, R., Chikan, S., Kabos, P.: Proc. 10tk ICMF, Szyrk, 1990, p. 461–465.

[110] Patton, C. E., Jantz, W.: J. Appl. Phys. 50 (1979), p. 7082.

[111] Zacharov, V. E., Lvov, V. S., Starobinec, S.s.: Usp. Fiz. Nauk 114 (1974), p. 609.
[112] Nakamura, K., Ohta, S., Kawasaki, K.: J. Phys. C 15 (1982), p. L143.
[113] Gibson, G., Jeffries, C.: Phys. Rev. A 29 (1984), p. 811.
[114] Bertaut, F., Forrat, F.: Compt. Rend. 242 (1956), p. 382.
[115] Geller, S., Gilleo, M/A.: Acta Cryst. 10 (1957), p. 239.
[116] Winkler, G.: Magnetic Garnets, Vieweg & Sohn, Braunschweig/ Wiesbaden, 1981.
[117] Geller, S., Williams, H. J., Espinosa, G. P., Sherwood R.C.: Bell Syst. Tech. J. 43 (1964), p. 565.
[118] Geller, S.: Z. Krystallogr. 125 (1967), p. 1.
[119] Borghese, C.: J. Phys. Chem. Sol. 28 (1967), p. 2225.
[120] Remeika, J. P.: J. Amer. Chem. Soc. 78 (1956), p. 4259.
[121] Nielsen, J. N., Dearborn, E. F.: J. Phys. Chem. Sol. 5 (1958), p. 202.
[122] Nielsen, J. W.: J. Appl. Phys. 29 (1958), p. 390.
[123] Nielsen, J. W.: J. Appl. Phys. 31 Suppl. (1960), p. 51S.
[124] Nielsen, J. W., Lepore, D. A., Leo, D. C.: In Crystal Growth, ed. H. S. Peiser, Pergamon Press, Oxford, 1967, p. 457.
[125] Grodkiewicz, W. H., Dearborn, E. F., Van Uitert, L. G.: In Crystal Growth, ed. H. S. Peiser, Pergamon Press, Oxford, 1967, p. 441.
[126] Van Uitert, L. G., Grodkiewicz, W. H., Dearborn, E. F.: J. Amer. Cer. Soc. 48 (1965), p. 105.
[127] Bennet, G. A.: J. Cryst. Growth 3/4 (1968), p. 458.
[128] Tolksdorf, W.: Cryst. Growth 3/4 (1968), p. 463.
[129] Tolksdorf, W.: Acta Electronica 17 (1974), p. 57.
[130] Schafer, H.: Z. Angew. Chem. 83 (1971), p. 35.
[131] Simsa, Z., Simsova, J., Zemanova, D., Cermak, J., Nevriva, M.: Czech. J. Phys. B 34 (1984), p. 1102.
[132] Parker, R. L.: Solid State Phys. 25 (1970), p. 151.
[133] Cermak, J., Grolmus, J., Smolka, P.: Fine Mech. and Optics 2 (1981), p. 37 (in Slovak).
[134] Mee, J. E., Pullina, G. R., Archer, J. L., Besser, P. J.: IEEE Trans. Mag-5 (1969), p. 717.
[135] Sparks, M.: Ferromagnetic Relaxation Theory, New York, McGraw-Hill, 1964.
[136] Glass, H. L.: J. Cryst. Growth 33 (1976), p. 183.
[137] Hibiya, T.: J. Cryst. Growth 62 (1983), p. 87.

[138] Matthews, J. W., Klockholm E., Plaskett T.S.: in Amer. Inst. Phys. Conf. Proc. No 10 (1972), p. 271.
[139] Nemiroff, M., Hue, H.: IEEE Trans. Mag-13 (1977), p. 1238.
[140] Glass, H. L., Elliot, M. T.: J. Cryst. Growth 27 (1974), p. 253.
[141] Cermak, J., Abraham, A., Fabian, T., Kabos, P., Hyben, P.: JMMM 83 (1990), p. 427.
[142] Glass, H. L., Elliot, M. T.: J. Cryst. Growth 34 (1976), p. 285.
[143] Adkins, L. R., Glass, H. L.: J. Appl. Phys. 53 (1982), p. 8928.
[144] Cermak, J., Kalivoda, L., Nevcriva, M.: Phys. Stat. Sol. (a) 100 (1987), p. 213.
[145] Glass, H. L., Liaw, J. H. W., Elliot, M. T.: Mar. Res. Bull. 12 (1977), p. 735.
[146] Glass, H. L., Stearn, T. S., Adkins, L. R.: Proc. 3th Int. Conf. on Ferrites, 1980, Kyoto, Japan, p. 39.
[147] Dotsch, H., Mateika, P., Roschmann, P., Tolksdorf, W.: Mat. Res. Bull. 18 (1983), p. 1209.
[148] Manasevit, H. M.: US Patent No. 4368098, 1983.
[149] Razeghi, M., Meunier, P. L., Maurel, P.: J. Appl. Phys. 59 (1986), p. 2261.
[150] Abe, M., Tamaura, Y., Goto, Y., Kitamura, N., Gomi, M.: J. Appl. Phys. 61 (1987), p. 3211.
[151] Berry, R. W., Hall, P. M., Harris, M. T.: Thin Film Technology, Toronto, London, Melbourne, Van Nostrand, 1968.
[152] Luder, E.: Bau hybrider Mikroschaltungen, Berlin, Heidelberg, New York, Springer, 1977.
[153] Pfahnl, A.: 1973 Int. Microelectrn. Symp. Dig., 22–23 ct., San Francisco, 1973, p. 5A2-1.
[154] Rikoski, R. A.: Hybrid Microelectronic Circuits — the Thick Film, New York, London, Sydney, Toronto, Wiley, 1973.
[155] Carry, J. R., Rosenberg, R. M., Uhler, R. O.: IEEE Trans. Comp. Hybrids and Man. Tech. CHMT 3 (1980), p. 211.
[156] Vansovskij, V. S.: Magnetism, Nauka, Moscow, 1971.
[157] Chikazumi, S.: Physics of Magnetism, J. Wiley and Sons, New York, 1964.
[158] Narasimba, K. S. V. L.: J. Appl. Phys. 52 (1981), p. 2512.
[159] Erwens, W.: Goldschmidt Inf. 48 (1979), p. 3.
[160] Ray, A. E.: IEEE Trans. Mag–20 (1984), p. 1614.
[161] Sagawa, M., Fujimura, S., Togawa, N., Yamamoto, H., Matsaura, Y.: J. Appl. Phys. 55 (1984), p. 2083.
[162] Cadieu, F. J.: J. Appl. Phys. 61 (1987), p. 4105.

[163] Silvester, P.: Finite Elements for Electrical Engineers, New York, Cambridge University Press, 1983.
[164] Adam, J. D., Daniel, M. R., O'Keeffe, T. W.: Microwave Journal 25 (1982), p. 95.
[165] Adam, J. D., Daniel, M. R.: IEEE Trans. Mag–17 (1981), p. 2951.
[166] Chang, K. W., Owens, J. M., Carter, R. L.: Electr. Lett. 19 (1983), p. 546.
[167] Castera, J. D.: J. Appl. Phys. 55 (1984), p. 2506.
[168] Sethares, J. C.: J. Appl. Phys. 53 (1982), p. 2646.
[169] Brak, J., Lackhish, V.: J. Appl. Phys. 65 (1989), p. 1652.
[170] Kalinikos, B. A., Kovshikov, N. G., Slavin, A. N.: Letters to ZETF 38 (1983), p. 343 (in Russian).
[171] Adkins, L. R.: Circuit Syst. Sign. Proc. 4 (1985), p. 135.
[172] Adkins, R. L., Glass, H. L.: Electronics Lett. 16 (1980), p. 590.
[173] Adkins, R. L., Glass, H. L.: J. Appl. Phys. 53 (1982), p. 8928.
[174] Adkins, R. L., Glass, H. L., Stcarns, F. S., Carter, R. L., Chang, K, W., Owens, J. M.: J. Appl. Phys. 55 (1984), p. 2518.
[175] Adkins, R. L., Glass, H. L.: Proc. IEEE Ultrason. Symp. (1980), p. 526.
[176] Parekh, J. P., Chang, K. W.: Proc. IEEE 71 (1983), p. 685.
[177] Morgenthaler. F. R.: Proc. RADC Microwave Magnetic Workshop 1981, p. 133.
[178] Owens, J. M., Smith, C. V. Jr., Mears, H.: IEEE Trans. MTT-S Dig. (1979), p. 154.
[179] Kallman, H. E.: Proc. IRE (1940), p. 28.
[180] Sethares, J. C., Weinberg, I. J.: Circuit Syst. Signproc. 4 (1985), p. 39.
[181] Kudinov, E. V., Beregov, A. S.: IVUZ Fizika 31 (1988), p. 106 (in Russian).
[182] Wu, H. J., Smith, C. V. Jr., Collins, J. H., Owens, J. M.: Electr. Lett. 13, (1977), p. 610.
[183] Adam, J. D.: IEEE Trans. Mag–23 (1983), p. 3742.
[184] Adam, J. D.: IEEE MTT-S Intern. Symp. Digest (1982), p. 78.
[185] Chang, N. J.: IEEE Trans Mag-18 (1982), p. 1604.
[186] Koike, T.: Ultrasonic Symp. (1980), p. 552.
[187] Sykes, C. G., Adam, J. D., Collins, J. H.: Appl. Phys. Lett. 29 (1976) p. 388.
[188] Collins, J. H., Adam, J. D., Bardai, Z. M.: Proc. IEEE 65, (1977), p. 1090.

[189] Owens, J. M., Smith, C. V., Snapka, E. P., Collins, J. H.: Proc IEEE CH (1978), p. 440.
[190] Castera, J. P., Vollnet, G., Hartemann, P.: Ultrasonic Symp. Proceedings 1 (1980), p. 514.
[191] Carter, R. L., Owens, J. M., Brinlee, W. R., Sam, Y. W., Smith, C. V. Jr.: IEEE MTT-S Int. Microw. Symp. (1981), p. 383.
[192] Tsai, C. S., Young, O.: IEEE Trans. MTT-38 (1990), p. 560.
[193] De, D. K.: J. Appl. Phys. 64 (1988), p. 2144.
[194] Castera, J. P., Hartemann, P.: Circ. Syst. Sign. Proc. 4 (1985), p. 180.
[195] Brinlee, W. R., Owens, J. M., Smith, C. V. Jr., Carter, R. L.: J. Appl. Phys. 52 (1981), p. 2276.
[196] Huijer, E., Ishak, W.: IEEE Trans. Mag-20 (1984), p. 1232.
[197] Chang, K. W., Ishak, W.: Circ. Syst. Sign. Proc. 4 (1985), p. 201.
[198] O'Keeffe, I. W., Paterson, R. W.: J. Appl. Phys. 49 (1978), p. 4886.
[199] Niccoli, G., Chang, K. V.: Electr. Lett. 25 (1989), p. 420.
[200] Balinsky, M. G., Kudinov, E. V., Lubyanov, P. P., Mongolov, B. D.: Proc. 2nd Sem. on functional Magnetoelectronics, Krasnojarsk, USSR, 1986, p. 25.
[201] Poston, T. D., Stancil, D. D.: J. Appl. Phys. 55 (1984), p. 2521.
[202] Balinsky, M. G., Ereshchenko, I. N., Mongolov, B. D.: High Performance MSW elements for High-Frequency Generator Control, Kiev, 1989 (in Russian).
[203] Adam, J. D., Stitzer, S. N.: J. Appl. Phys. Lett. 36 (1980), p. 448.
[204] Stitzer, S. N., Goldie, H., Adam, J. D., Emtage, P. R.: IEEE MTT-S Int. Microw. Symp. Digest (1980), p. 238.
[205] Gurzo, V. V., Prokushkin, V. I. et al.: Radioelektronics 29 (1986), p. 95 (in Russian).
[206] Vashkovsky, A. V., Stalmachov, V. S., Sharaevsky, Yu. P.: Magnetostatic Waves in High-Frequency Electronics, Saratov USSR, 1991 (in Russian).
[207] Panteleev, A. A., Prokushkin, V. I., Sharaevsky, Yu. P.: Proc. Conf. on High-Frequency Technology, USSR, Kuybishev, 1987, p. 109 (in Russian).
[208] Stitzer, S. N., Goldie, H.: IEEE MTT-S Int. Microw. Symp. Digest (1983), p. 326.
[209] Fisher, A. D.: Circ. Syst. Sign. Proc. 4 (1985), p. 263.

[210] Visnovsky, S.: Czech. J. Phys. A 40 (1990), p. 559 (in Czech).
[211] Benda, O.: Ferromagnetism, University textbook, EF SVST.

INDEX